Advanced Structured Materials

Volume 162

Common engineering materials reach in many applications their limits and new developments are required to fulfil increasing demands on engineering materials. The performance of materials can be increased by combining different materials to achieve better properties than a single constituent or by shaping the material or constituents in a specific structure. The interaction between material and structure may arise on different length scales, such as micro-, meso- or macroscale, and offers possible applications in quite diverse fields.

This book series addresses the fundamental relationship between materials and their structure on the overall properties (e.g. mechanical, thermal, chemical or magnetic etc.) and applications.

The topics of *Advanced Structured Materials* include but are not limited to

- classical fibre-reinforced composites (e.g. glass, carbon or Aramid reinforced plastics)
- metal matrix composites (MMCs)
- micro porous composites
- micro channel materials
- multilayered materials
- cellular materials (e.g., metallic or polymer foams, sponges, hollow sphere structures)
- porous materials
- truss structures
- nanocomposite materials
- biomaterials
- nanoporous metals
- concrete
- coated materials
- smart materials

Advanced Structured Materials is indexed in Google Scholar and Scopus.

More information about this series at https://link.springer.com/bookseries/8611

Azman Ismail · Wardiah Mohd Dahalan ·
Andreas Öchsner
Editors

Advanced Materials and Engineering Technologies

Springer

Editors
Azman Ismail
Malaysian Institute of Marine Engineering Technology
Universiti Kuala Lumpur
Lumut, Perak, Malaysia

Wardiah Mohd Dahalan
Malaysian Institute of Marine Engineering Technology
Universiti Kuala Lumpur
Lumut, Perak, Malaysia

Andreas Öchsner
Faculty of Mechanical Engineering
Esslingen University of Applied Sciences
Esslingen am Neckar, Baden-Württemberg, Germany

ISSN 1869-8433 ISSN 1869-8441 (electronic)
Advanced Structured Materials
ISBN 978-3-030-92963-3 ISBN 978-3-030-92964-0 (eBook)
https://doi.org/10.1007/978-3-030-92964-0

This Springer imprint is published by the registered company Springer Nature Switzerland AG
The registered company address is: Gewerbestrasse 11, 6330 Cham, Switzerland

Preface

The chapters in *Advanced Materials and Engineering Technologies* are one of the outputs from the 2nd International Conference on Marine and Advanced Technologies 2021 (ICMAT 2021). Most of the papers reveal the best practice and discuss the theory in relation to multi-disciplinary approaches in materials engineering technology. This book demonstrates various real-world and global engineering problems while touching on evolving design strategies. Among the topics are advanced materials, applied science, marine engineering, and energy application. Throughout the text, the reader can adapt these research findings to the design and analysis of materials in engineering practices. In addition, this book generates interests, ideas, and appropriate teaching support for lifelong the learning process.

Lumut, Malaysia	Azman Ismail
Lumut, Malaysia	Wardiah Mohd Dahalan
Esslingen am Neckar, Germany	Andreas Öchsner

Contents

About the Editors

Azman Ismail is a senior lecturer at the Malaysian Institute of Marine Engineering Technology, Universiti Kuala Lumpur, Malaysia. He received his Ph.D. in Mechanical Engineering from Universiti Teknologi PETRONAS and Master of Engineering in Mechanical-Marine Technology from Universiti Teknologi Malaysia. Prior to that, he was awarded a Bachelor of Engineering (Hons) in Electrical, Electronics, and System from Universiti Kebangsaan Malaysia and Graduate Diploma in Industrial Education and Training from the Royal Melbourne Institute of Technology, Australia. He grooms his technical skill at Victoria University of Technology, Australia, for Advanced Diploma in Construction and Repair Technology (Marine Vessels). He is also active in research and development for welding and joining technologies especially for friction stir welding on tubular sections and flat panels. This also includes green technologies for sustainable marine and coastal development. He is currently leading a research cluster of Advanced Maritime Industries Sustainability at his university. He has published his research findings in indexed journals and chapters and actively competes at international and national-level innovation competitions. In addition to his achievements, he has been a reviewer and an editor for some international journals including Springer. Besides that, he is also active in conservation works as a committee member for the National Eco-Campus Program with the World Wide Fund for Nature of Malaysia (WWF-Malaysia).

Wardiah Mohd Dahalan had received her Ph.D. in Power System from the University of Malaya, Kuala Lumpur, Malaysia, in 2013 and bachelor's degree (Hons) in Electrical and Electronics Engineering from the University of Dundee, Scotland, UK, in 1996. She is currently appointed as a senior lecturer at the Department of Marine Electrical and Electronics Engineering of University, Kuala Lumpur (UniKL-MIMET). As the head of Research and Innovation department since 2015, she actively administers all research activities in UniKL-MIMET. Supervising of internal and external research grants and organization of all activities related to innovation such as exhibitions, competitions, and conferences are her core activities at UniKL-MIMET. She actively secures many research grants, nationally and internatonally. At the same time, she also shares her expertise by becoming either the author or co-author of the publications of local as well as international journals, books, and proceedings especially in the area of power system and energy. Her deep research interest includes network reconfiguration, optimization techniques, and renewable energy. She is also a member of IEEE, Rina-IMARest, Malaysian Society for Engineering and Technology (MySET), and Malaysia Board of Technologist (MBOT).

Andreas Öchsner is a full professor for lightweight design and structural simulation at the Esslingen University of Applied Sciences, Germany. Having obtained a Dipl.-Ing. degree in Aeronautical Engineering at the University of Stuttgart (1997), Germany, he served as a research and teaching assistant at the University of Erlangen-Nuremberg from 1997 to 2003 while working to complete his Doctor of Engineering Sciences (Dr.-Ing.) degree. From 2003 to 2006, he was an assistant professor at the Department of Mechanical Engineering and the head of the Cellular Metals Group affiliated with the University of Aveiro, Portugal. He spent seven years (2007–2013) as a full professor at the Department of Applied Mechanics, Technical University of Malaysia, where he was also the head of the Advanced Materials and Structure Lab. From 2014–2017, he was a full professor at the School of Engineering, Griffith University, Australia, and the leader of the Mechanical Engineering Program (the head of Discipline and the program director).

Chapter 1
The Analysis of Mechanical Properties for Coconut Composites Using the Taguchi Method

Mohamad Sazali Said, Mohd Shahrizan Yusoff, Nurrasyidah Izzati Rohaizat, Nurul Nai'my Wan, Siti Sara Mohd Ali, and Muhammad Iqbal Adnan

Abstract In recent years, industrial pollution is a currently big issue. Research toward biodegradable materials such as coconut shell reinforced composites has increased due to environmental awareness and global warming concerns that grow day by day. The researcher is triggered to implement material that is eco-friendly. The research has been conducted to determine the mechanical properties of coconut composites by using the Taguchi method. The mechanical properties of tensile strength, modulus elasticity, and elongation were investigated in studying whether they are affected by the weight percentage and speed. The mold is the redesign and fabricated based on ASTM D638-14. According to the composition of the coconut composite 30, 40, and 50%, the tensile specimen have been fabricated. The tensile test has been conducted to determine the mechanical properties of the coconut composite using a universal testing machine (UTM), and the result was analyzed using the Taguchi Method L9. According to the Taguchi analysis, the optimum parameters are 30% weight percentage while speed is 7 mm/min. The weight percentage 30% has

M. S. Said (✉) · M. S. Yusoff · S. S. M. Ali · M. I. Adnan
Manufacturing Section, Universiti Kuala Lumpur Malaysian Spanish Institute, Kulim Hi-Tech Park, 09000 Kulim Kedah, Malaysia
e-mail: msazali@unikl.edu.my

M. S. Yusoff
e-mail: mshahrizan@unikl.edu.my

S. S. M. Ali
e-mail: sara.ali@s.unikl.edu.my

M. I. Adnan
e-mail: iqbal.adnan@s.unikl.edu.my

N. I. Rohaizat · N. N. Wan
Mechanical Section, Universiti Kuala Lumpur Malaysian Spanish Institute, Kulim Hi-Tech Park, 09000 Kulim Kedah, Malaysia
e-mail: nurrasyidah.izzati@unikl.edu.my

N. N. Wan
e-mail: nurulnaimy@unikl.edu.my

A. Ismail et al. (eds.), *Advanced Materials and Engineering Technologies*, Advanced Structured Materials 162,
https://doi.org/10.1007/978-3-030-92964-0_1

better tensile test (6.24 N/mm^2), modulus of elasticity (2097.07 N/m^2), and elongation at failure (0.19%) compared to weight percentage (40 and 50%) and speed (5 mm/min and 6 mm/min), respectively. Overall, the weight percentage and speed can affect the mechanical properties of coconut composites.

Keywords Coconut composite · Tensile test · Tensile strength · Taguchi method · Mechanical properties

1.1 Introduction

Industrial pollution is a current big issue because some industries usually cause the pollution. The effects of industrial pollution are vast, causing water contamination, releasing toxins into the soil and the air, and causing environmental disasters. In recent years, environmental awareness and global warming are also growing day by day and become a reason why worldwide researchers are triggered to implement and apply eco-friendly materials. Natural fibers are the best alternative due to the fact that they are cheaper, more economical and environmentally friendly [1].

The researchers have shown interest in materials that are eco-friendly and materials that reduce carbon dioxide emissions toward the environment. It makes the researchers more interested in biodegradable plant fibers such as coconut shell reinforced of composites for the automotive and industrial sectors, even for the construction sector [2]. It gains more considerable interest due to renewability, reduction in weight and cost, and less abrasiveness to equipment. The coconut shell is an agricultural waste that is very useful as an alternative material in manufacturing applications.

Besides, coconut shell is an essential natural filler that had been originally provided from tropical countries like Malaysia, Indonesia, Thailand, Sri Lanka, India, etc. Coconut particles have gained interest due to non-toxic, not easily combustible, and high hardness. This is supported by study of Said which revealed that composites are made up of two or more readily distinct materials with different physical or mechanical properties [3]. It has become the popular reinforcement of composite; it also has high strength, even light in weight, less density, and low-cost composite [4]. The coconut shell also can be found without cost, and the revolution goods use up to 99.7% recycled materials. Natural fibers like coconut shell also have proven to be better reinforcement in polymer matrices because they reduce the density and cost of the resultant composite [5]. So, when high forces are applied to coconut composites, it produces high fracture loads, and failure measurements of composites would also produce lower durability.

Several literature reviews have been made to define the composite's composition to fabricate more substantial and high-strength composites. According to Somashekhar, three different weight percentages and epoxy resin have been used to form coconut composites such as 30, 40, and 50% specimens based on ASTM [6]. The speeds that should be used depending on the shape of the specimen can be in the range of 5–500 mm/min based on ASTM. ASTM D638-14 is used for fabrication of the coconut composite specimens [7].

This research will highlight the use of coconut-activated carbon as raw material with epoxy resin and hardener to observe the mechanical properties before implementation for manufacturing application [8]. The tensile strength, modulus elasticity, and elongation are determined while doing a tensile test using a universal testing machine [9]. The result is analyzed using the Taguchi method. This research also reveals whether the weight percentage and speed have significant effects on the mechanical properties of coconut composites and achieve the objective of this research.

1.2 Methodology

The experiment started with the procurement of the coconut-activated carbon, epoxy resin, hardener, and a silicon rubber mold (SRM). The SRM used is a rectangular shape according to standard ASTM D638-14 with dumbbell shape samples for tensile testing [10]. Figure 1.1 shows the silicone rubber mold, hardener, and epoxy resin. Thus, Fig. 1.2 shows the schematic drawing of tensile specimen according to ASTM D638-14.

Fig. 1.1 Silicone rubber mold (SRM), hardener, and epoxy resin

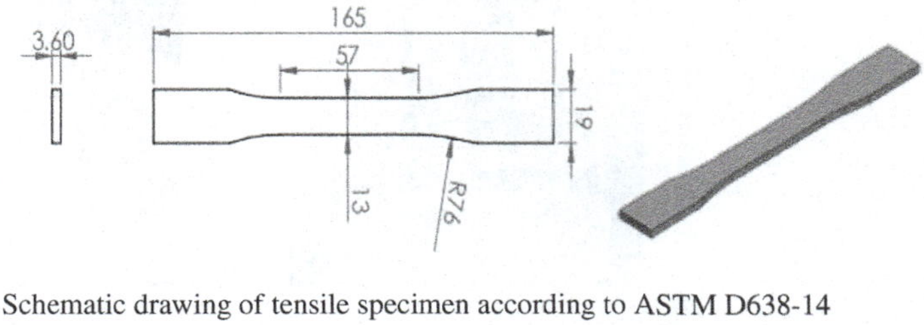

Fig. 1.2 Schematic drawing of tensile specimen according to ASTM D638-14

Table 1.1 Coconut composite composition

Specimen	Coconut-activated carbon (%)	Epoxy resin and hardener (%)
1	30	70
2	30	70
3	30	70
4	40	60
5	40	60
6	40	60
7	50	50
8	50	50
9	50	50

Firstly, the coconut-activated carbon was weighted using a digital weight scale. The total number of sample specimens that need to be prepared for testing is nine samples with different weight percentages. For prepared specimens, 30, 40, and 50% of coconut-activated carbon, and 70, 60, and 50% of epoxy resin are uniformly mixed and poured into silicone rubber mold (SRM). Finally, the mixture of the coconut composite is left at room temperature for 24 h until it hardened. After 24 h, the coconut composite specimens were taken out and were ready for testing. Table 1.1 provides the coconut composite composition.

The tensile test has been done to determine the tensile strength that will occur under tension load. It is the measurement of the material up to which extent it can bear the loads and resist the load without failure. The Quasar 100 universal testing machine (UTM) was used with different speeds (5 mm/min, 6 mm/min, 7 mm/min) for tensile testing as shown in Fig. 1.3. Dumbbell shape specimens are prepared according to ASTM D638-14, visible in Fig. 1.4, and were used during the tensile test [10]. Nine specimens have been tested to determine the mechanical properties.

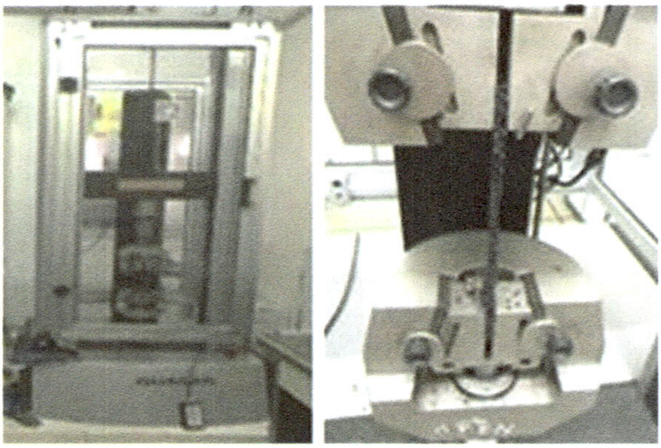

Fig. 1.3 Universal testing machine (UTM) for tensile testing

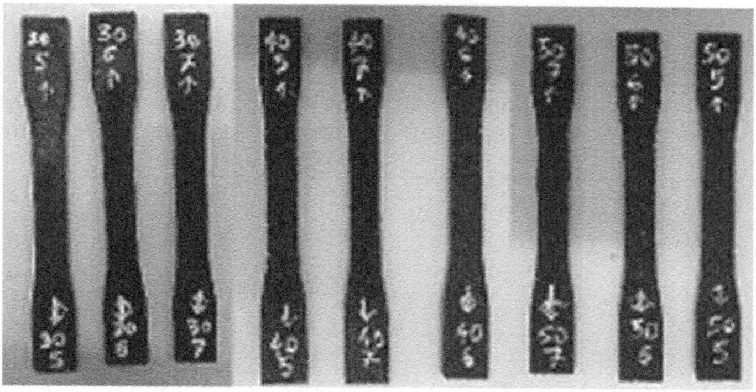

Fig. 1.4 Dumbbell tensile test specimen ASTM D638-14

The contribution of this study is evident as the resulting outcomes can be capitalized as guidelines to elongate and increase the resistance of coconut composite based on ASTM D683-14. The load (*P*) is noted until the rupture of the work piece takes place. The computerized UTM will calculate the tensile strength, modulus elasticity, and elongation at failure of the coconut composite specimen at the breakpoint.

1.3 Results and Discussion

This investigation of the coconut composite specimens is fabricated according to the L9 orthogonal array and according to ASTM D638-14 by using the Minitab software to evaluate the mechanical properties such as the tensile strength, modulus elasticity, and elongation at failure with different weight percentages and speeds. Table 1.2 exhibits the experimental results.

Table 1.2 and Fig. 1.5 present the effect of weight percentage and speed on the tensile strength of coconut composite specimen. The tensile strength result shows that the coconut composite specimens with a weight percentage of 30% have the maximum tensile strength, compared to 40% and 50%. However, the best speed for tensile testing of the coconut composite is 7 mm/min.

The tensile strength of the coconut composite decreases with increase in the weight percentage of the coconut-activated carbon with the matrix of composites [11]. Agreed with NE Naveen earlier work, the tensile strength decreases as the fiber volume fraction increases, and this concept also agrees with the coconut-activated carbon weight percentage [12]. In addition, the bonding between the coconut-activated carbon and matrix forms a strong strength of coconut composite.

Figure 1.6 shows the maximum elongation that was analyzed using the Taguchi method. The weight percentage is 50% and a speed of 5 mm/min is the optimum parameter for elongation. Therefore, the maximum elongation was determined as 11.4% is the utmost. Among the plausible explanations for these findings is that

Table 1.2 Experimental results

Experimental tensile testing results					
Specimen	Weight percentage (%)	Speed (mm/min)	Tensile strength (N/mm^2)	Modulus of elasticity (N/mm^2)	Elongation at failure (%)
1	30	5	2077.33	2077.33	0.36
2	30	6	1404.03	1404.03	0.13
3	30	7	2097.07	2097.07	0.19
4	40	5	1132.06	1132.06	2.82
5	40	6	886.90	886.90	4.12
6	40	7	1339.39	1339.39	6.39
7	50	5	1077.79	1077.79	11.41
8	50	6	880.16	880.16	7.52
9	50	7	1106.09	1106.09	4.50

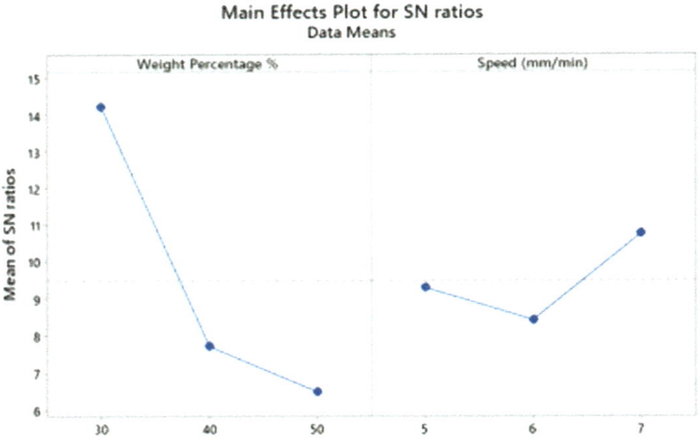

Fig. 1.5 Tensile strength graph result

the result reversed if compared to tensile strength and modulus elasticity due to the speed that applies. The slow testing allows the coconut composites relaxation and higher elongation values. The elongation happened at a slow speed before it starts to rupture. Secondly, the composition of composites is being also the factor that the result is reversed. According to Prabhu, as the composition of the coconut composites increases, the composites' elongation will decrease because when the percentage of coconut-activated carbon increases, it will become more brittle [13].

At the end of the result and discussion, the optimum parameters are obtained when the composition of the weight percentage is 30%, while speed is 7 mm/min, tensile strength is 6.24 N/mm^2, modulus elasticity is 2097.07 N/m^2, and elongation at

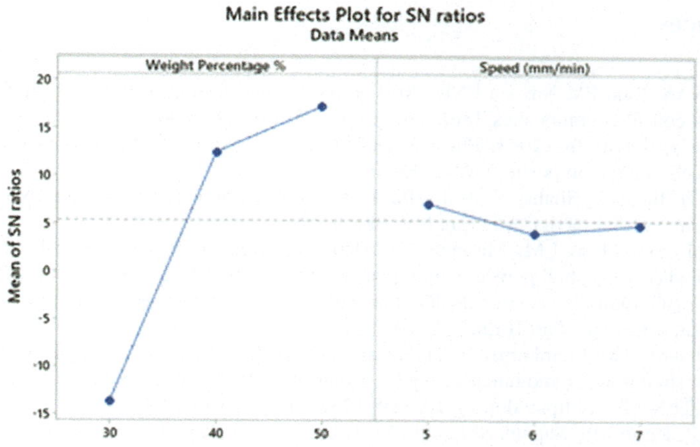

Fig. 1.6 Elongation graph result

failure is 0.19%. The most significant parameter is the weight percentage and speed that affect the mechanical properties of the coconut composite.

1.4 Conclusion

This research aimed to investigate the mechanical properties of coconut composite. The analysis of the mechanical properties of coconut composite using the Taguchi method was studied experimentally through tensile testing and Taguchi analysis. These findings enhance the understanding of coconut composite on mechanical properties. As a consequence, this research successfully redesigned and fabricated the mold based on ASTM D638-14 and fabricated the coconut composite specimens with different weight percentages. The mechanical properties have been studied and observed from the tensile test results collected. The result from the tensile test is analyzed to determine the optimum parameters of coconut composite by using the Taguchi method. This study set out to determine that the weight percentage and speed significantly affect the mechanical properties of coconut composite using Taguchi analysis. Finally, this research can conclude that the coconut composites with a 30% composite composition and 7 mm/min speed are the optimum parameters for manufacturing applications due to eco-friendly materials that reduce carbon dioxide emissions.

Acknowledgements This work is ostensibly supported by the Universiti Kuala Lumpur (UniKL) under UniKL Excellent Research Grant Scheme (UERGS), UER20015 project.

References

1. Priya NAS, Raju PV, Naveen PNE (2014) Experimental testing of polymer reinforced with coconut coir fiber composites. Int J Emerg Technol Adv 4:435–460
2. Bhaskar J, Singh VK (2013) Physical and mechanical properties of coconut shell particle reinforced-epoxy composite. J Mater Environ 4:227–232
3. Said MS, Iqbal A, Shabri N et al (2021) Valuation of industrial water quality on coconut composite as filter media. Adv Eng Res 200:52–56
4. Andezai AM, Masu LM, Maringa M (2020) Investigating the mechanical properties of reinforced coconut shell powder/epoxy resin composite. Int J Eng Res 13:2742–2751
5. Adeniyi AG, Onifade DV, Lghalo JO et al (2019) A Review of coir fiber reinforced polymer composites. Compos Part B Eng 176:107–305
6. Somashekhar TM, Premkumar N, Vighnesha N et al (2018) Study of mechanical properties of coconut shell powder and tamarind shell powder reinforced with epoxy composite. IOP Conf Ser Mater Sci Eng. https://doi.org/10.1088/1757-899X/376/1/012105
7. Salleh Z, Rosdi MS, Sapuan SM et al (2013) Mechanical properties of coconut carbon fibre/ epoxy composite material. Int J Mech Eng 2:55–62
8. Salleh Z, Mun'aim MA, Idrus M et al (2014) Mechanical properties of activated carbon (AC) coconut shell reinforced polypropylene composites encapsulated with epoxy resin. APCBEE Procedia 9:92–96
9. Sindhu P, Chouhan SS (2018) Mechanical properties of reinforced epoxy composite using waste coconut shell charcoal. Int Res J Eng Tech 5:2373–2377
10. Kumar S, Deka K, Suresh P (2016) Mechanical properties of coconut fiber reinforced epoxy polymer composites. Int RJ Eng Tech 3:1334–1336
11. Salleh Z, Yusop MYM, Rosdi MS (2013) Mechanical properties of activated carbon coir fibers reinforced with epoxy resin. J Mech Eng Sci 5:631–638
12. Naveen PNE, Yasaswi M (2013) Experimental analysis of coir fiber reinforced polymer composite materials. Int J Mech Eng Rob 2:10–18
13. Prabhu R, Amim AK, Dhynchandra (2015) Development and characterization of low cost polymer composites from coconut coir. AM J Mater Sci 5:62–68

Chapter 2
Subsurface Characterization of Heat Treated Subtractive and Additive Manufactured (SLM) Ti6Al4V Specimens

Sangeeth Suresh, Ruganethan Sandarsekran, Mohd Zaid Abu Yazid, and Darulihsan Abdul Hamid

Abstract The widespread use of additive manufacturing (AM) for Ti6Al4V components have demanded an equally extensive study on their comparison with the subtractive manufactured counterparts. Subtractive manufacturing (SM) of Ti6Al4V has its own limitations due to the peculiar properties of the material in the rolled condition. AM process too has its own set of limitations that include porosity, brittleness, and high residual stresses leading to premature failure. Heat treatment (HT) is an option to counter these drawbacks to a certain extent. This study analyzes the effects of HT on the mechanical properties, surface and subsurface characteristics of both AM and SM specimens. A stress relieving process followed by HT above β transus temperature and an aging process enhances the hardness in both counterparts. However, surface levels of SM specimens prove better than AM specimens for wear-resistant applications. AM specimens require the removal of surface to depths between 100 and 200 μm to act as a wear-resistant surface. AM specimens attain a β phase prior to HT and hence a stress relieving operation would suffice in case its being used for applications that require a tough core.

Keywords Ti6Al4V · SLM printing · HT · Subsurface · Microstructure

S. Suresh (✉) · R. Sandarsekran · M. Z. A. Yazid · D. A. Hamid
Univeristi Kuala Lumpur-Malaysia Italy Design Institute, Cheras, Kuala Lumpur, Malaysia
e-mail: sangeeth@unikl.edu.my

R. Sandarsekran
e-mail: ruganethan.sandarsekran@s.unikl.edu.my

M. Z. A. Yazid
e-mail: mzaiday@unikl.edu.my

D. A. Hamid
e-mail: darulihsan@unikl.edu.my

A. Ismail et al. (eds.), *Advanced Materials and Engineering Technologies*,
Advanced Structured Materials 162,
https://doi.org/10.1007/978-3-030-92964-0_2

2.1 Introduction

The use of titanium grade 5 alloy (Ti6Al4V) as a replacement for heavier metals in various automotive, aerospace, marine, medical, and other applications is widely attributed to its significant mechanical properties, including its high strength to weight ratio, resistance to wear and corrosion, high hardness, and excellent workability in high temperatures [1]. The characteristics that make Ti6Al4V a difficult-to-machine material are, its low thermal conductivity that accumulates heat at the machined surface causing variations in the surface properties, its high hardness that reduces tool life and causes strain hardening, its chemical reactivity causing rapid deterioration of the coatings on the cutting tools [2, 3]. Products built through an AM process could vary in volume from its SM counterpart slightly due to unavoidable porosities and inhomogeneities that could eventually affect their mechanical properties [4, 5]. A comparison between the SM and AM methods is shown in Fig. 2.1. This study focusses on the selective laser melting AM process as depicted in Fig. 2.2, in which the laser beam heats up the powder particles till they melt and bond with each other [6].

Major drawbacks with the SM process are the enormous wastage of material and limitations in terms of intricate and hidden geometries that cannot be machined [7–9]. Residual stress is yet another factor to be considered both in the SM and the AM methods, because of the influence of thermal energy in both the processes. In AM, the temperature gradients are produced by the laser beam. In such cases a martensitic alteration is necessary in the form of a stress relieving operation [10]. Further HT and aging processes have been recommended for both SM and AM components to enhance the hardness and wear resistance of the components. The β transus temperature (996 °C) is defined as the temperature at which the α phase of Ti6Al4V converts into a β phase with a proportional crystalline structural transformation from a hexagonal close packed (hcp) to a body centered cubic (bcc) structure causing increase in hardness [11]. Surface distortions as shown in Fig. 2.3, and stress accumulation in the machined specimens, both the AM and the SM counterparts, warrant further HT processes to enhance the lost surface and bulk properties [12–14]. This study delves into the optimization of HT parameters to measure the microhardness and

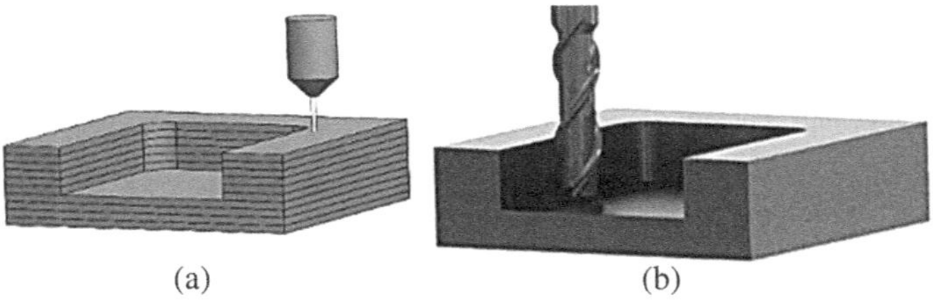

Fig. 2.1 **a** SLM printing in layers (CAD); **b** post-printing-layers fused together; **c** raw material for SM; **d** SM—milling process

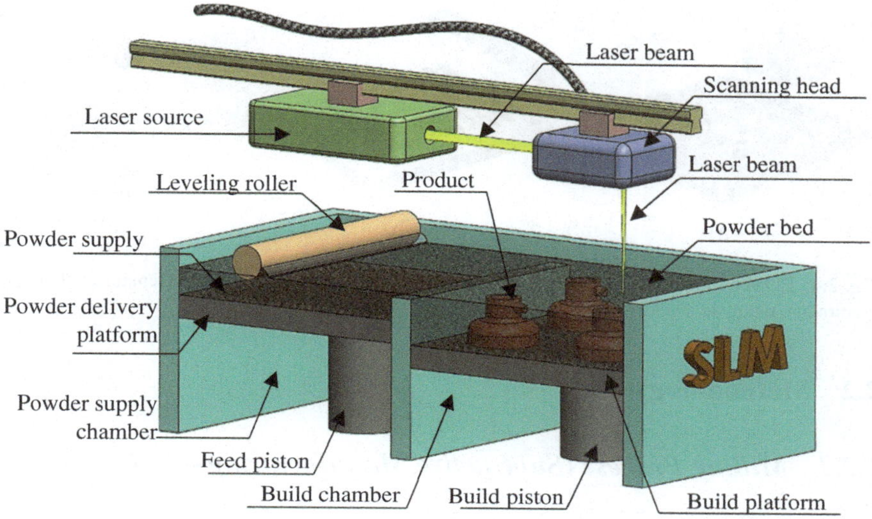

Fig. 2.2 SLM AM process with powder feedstock

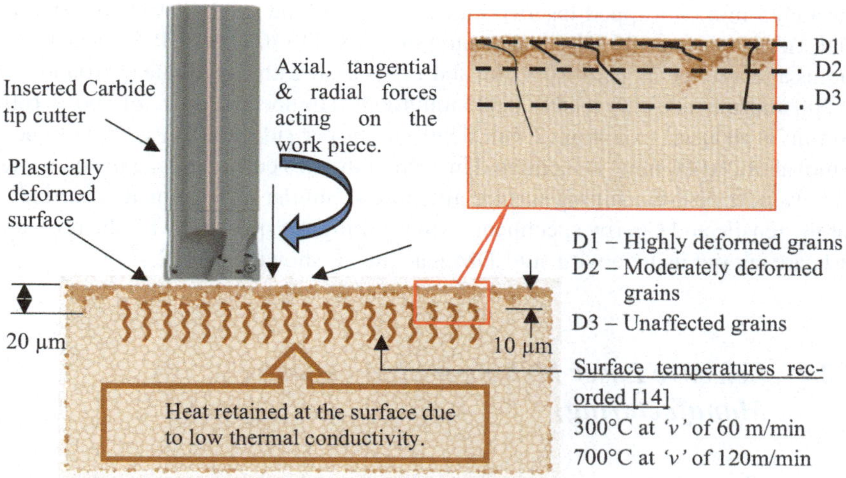

Fig. 2.3 Surface deformations observed in machining of Ti6Al4V

microstructure of both the SM and AM variants, at surface and subsurface levels paving groundwork for further tribological analysis.

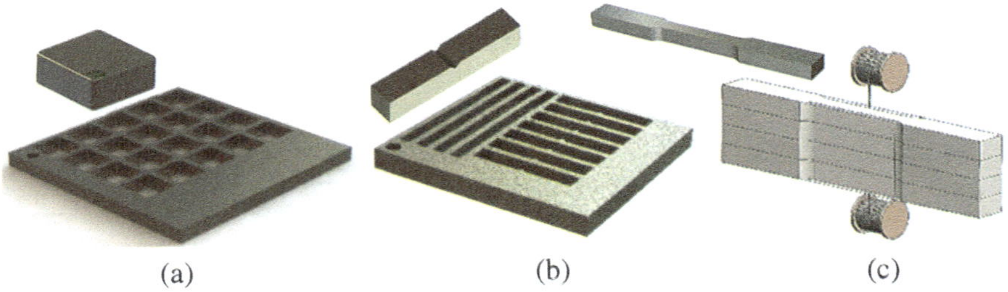

Fig. 2.4 EDM wire-cut of **a** surface characterization and tribo-test specimens, **b** tensile and Charpy specimens, **c** tensile stack wire-cut

2.2 Methodology

2.2.1 Milling Process (Subtractive Manufacturing—SM)

The milling process is carried out on a DMU 50 series three-axis machine. The cutting speed (v) is kept at a magnitude of 60 m/min (chosen to avoid overheating and adhesions at higher cutting speeds [15]), resulting in a spindle speed of 765 RPM for a Ø25 mm two-tipped inserted cutter, using carbide inserts PVD coated with a TiN-TiAlN coating (85 HRC). A Sumitomo–AXMT170508PEER-E tip with a tip radius and a relief angle of R0.8 mm and 11° is used with a feed rate (f) rounded off to 200 mm/min at a chip load of 0.125 mm/tooth. The specimen size of 100 × 100 × 10 mm is reduced to 6 mm (2 mm depth of cut per side) thickness. A volumetric removal of 20,000 mm^3 is facilitated in each of the two cuts to induce residual stress and evaluate post-machining surface alterations. Similar arrangement is carried out for the tensile and Charpy specimens. After milling the specimens for the respective tests are cut out using a wire cut EDM machine as shown in Fig. 2.4.

2.2.2 Selective Laser Melting—SLM (Additive Manufacturing)

In the SLM process the interactivity between the powder stock material in the exposed layer and the laser beam transpires in microseconds wherein the powder stock melts and fuses with the immediate product layer below it. The SLM process is carried out on an SLM® 280^{HL} which has a build chamber of 280 × 280 × 350 mm with a dual laser system. With a build rate up to 55 cm^3/h, it can build a feature as thin as 0.15 mm at a scanning speed of 10 m/s. The Ti6Al4V powder particle size ranges from 20 to 63 μm. Inert argon gas inside the build chamber prevents oxidation on the specimens. The laser beam scans with a power of 275 W producing a heat of 2710 °K exceeding the melt temperatures of most metals and composites. This study uses the

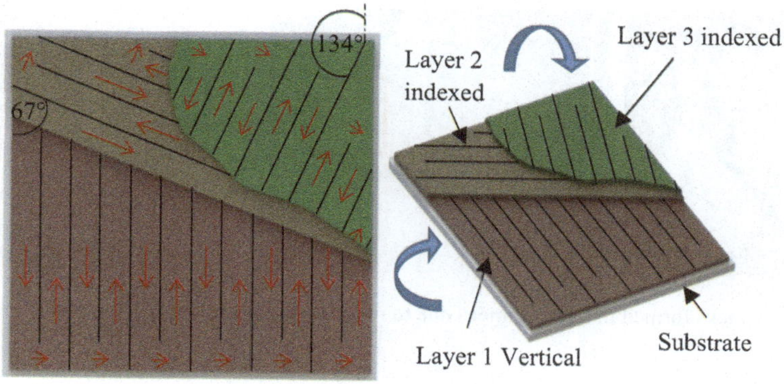

Fig. 2.5 Meander-path layer build up

meander scan path due to the small specimen size helping in a faster build rate and better homogeneity. The meander scan path starts with a vertical or horizontal path followed by a pre-programmed indexing to an angle of 67° in consecutive layers (67°,134°, 201°.... till the specimen height is achieved) as shown in Fig. 2.5 [16]. The process is carried out at a build speed of 15 cm^3/h and a scanning speed of 700 mm/s. The thickness of the layer is 40 μm and the spacing between the hatches is 60 μm. The laser power is maintained at 150 W. The laser beam diameter at the powder surface is Ø0.20 mm.

2.2.3 Stress Relieving and HT—(SM and AM)

In milling (SM), the depth of cut is intentionally set to 2 mm per side to induce residual stresses and cause plastic deformation on the surface as shown in Fig. 2.3. Tool geometry and tool life are also major determinants of residual stresses in machined specimens [17]. At a constant depth of cut (ap) of 5 mm, an increase in cutting speed (v) and feed rate (f) have shown a transformation from compressive to tensile stresses mainly noticed at subsurface levels of 10–20 μm. At lower subsurface levels negligible stresses have been observed. At lower cutting speeds (v) the tensile residual stress magnitudes of 520 MPa at the surface and compressive tensile stress of 300 MPa are recorded below the surface. However, as the cutting speed increases, only compressive stress magnitudes of 300–750 MPa are observed, both, at the surface and below the surface, with no tensile stresses detected along the feed direction. A similar trend is also noticed in the cutting axis perpendicular to the feed direction. The residual stress values have been recorded at depths up to 0.08 mm [18]. Residual stresses within the specimens in SLM could result in deteriorated mechanical properties, cracks, and distortions as shown in Fig. 2.6. AM specimens have also recorded tensile stress of 100–300 MPa at their surfaces and compressive stresses

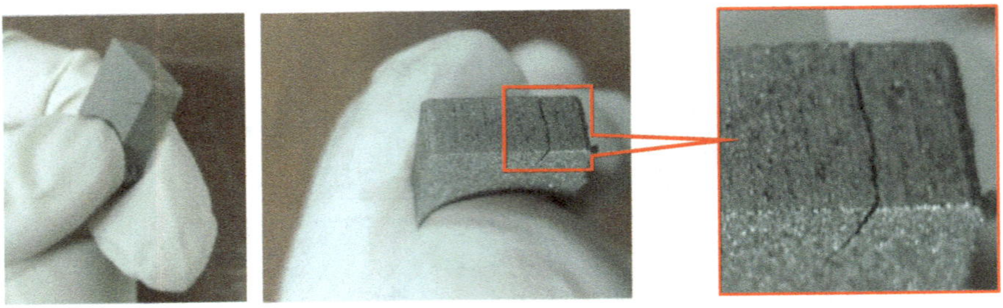

Fig. 2.6 Cracks formed in the specimens due to residual stresses in SLM

Table 2.1 HT experimental set up

Process	Temperature (°C)	Cooling method	Duration (h)
Hardening	950 1050 1150	Water Quenching Air Cooling (220 m^3/min approx.) Furnace Cooling	1
Aging	500 550 600	Air Cooling (220 m^3/min approx.)	4

of 38–80 MPa at the subsurface [19–22]. Since both processing methods exhibit almost similar residual stresses, the specimens from both processes are subjected to a constant stress relieving process in a HT furnace with optimized parameters, also to enhance its ductility [23]. The specimens are stress relieved through a cycle of heating processes in the furnace starting with a ramp up to 350 °C for a duration of one hour followed by a dwell period of 30 min at the same process. The second part of the cycle consists of a ramp up to 850 °C for a duration of one hour followed by furnace cooling. The stress relieving process is followed by a HT process to not only enhance the mechanical properties of the specimens but also regain their lost surface properties. An experimental design is adapted for HT, to validate the effects of heating above and below the β transus temperature of 995 °C. β transus temperature is the temperature above which phase changes are observed in the crystalline structure of Ti6Al4V specimens [24]. Optimization of the HT parameters is carried out in a Nabertherm N-11 h furnace. The preliminary experiment design is shown in Table 2.1. The hardness of the specimens is measured using a Highwood Digital Microhardness Tester.

2.2.4 *Finishing Process*

Surface characterization of specimens demands intensive surface finishing, starting with a surface grinding process. The depth of cut in grinding is maintained at 10 μm per cut (to avoid further buildup of residual stresses) at a cross feed of 0.5 m/min at a spindle speed of 3000 RPM under a water-soluble synthetic oil as cooling fluid. The grinding process is followed by polishing using emery sheets from a grit size of 600 for roughing to 2500 for a finish polishing operation, at a disk speed of 800 RPM. Polishing process is succeeded by an acetone immersion and ultra-sonic cleaning in distilled water to remove off any surface impurities. The specimens are chemical etched in Kroll's reagent (solution of water 93%, nitric acid 6%, and hydrofluoric acid 1%) to expose the surface for surface characterization.

2.2.5 *Surface Characterization*

Surface and subsurface microstructure is analyzed using optical microscope (Optika B600 MET) and scanning electron microscope (Hitachi SU1510 SEM). Surface composition and concentration of the alloying elements are inspected through an energy dispersive X-ray (Oxford Xplore 30 EDX) test.

2.3 Results and Discussion

The raw material hardness of SM specimens was 330 HV and the AM specimens at 360 HV. After the milling operation, no considerable variations in hardness have been observed. Stress relieving and further HT have shown the microhardness results to be favoring the combination of a hardening temperature of 1050 °C as represented in Table 2.2. The 10 μm depth is considered as the 'surface' here after the removal of HT scales. 100 and 200 μm are the subsurfaces. Post-heat treatment and the hardness of the AM specimens were observed to be higher than the SM counterparts in all the three temperature settings. The tensile strength of the AM specimens is almost the same regardless of the HT temperatures. The toughness of both variants at 1050 °C

Table 2.2 Microhardness values comparison between SM and AM specimens

Process	Hardening (60 min) + Water Quenching (°C)	Aging	HV at surface depth $Z = -10\ \mu m$	HV at surface depth $Z = -100\ \mu m$	HV at surface depth $Z = -200\ \mu m$
SM	950	500 °C (4 h) + Furnace cooling	392	455	404
	1050		553	497	469
	1150		410	461	427
AM	950		410	412	408
	1050		420	418	424
	1150		460	448	438

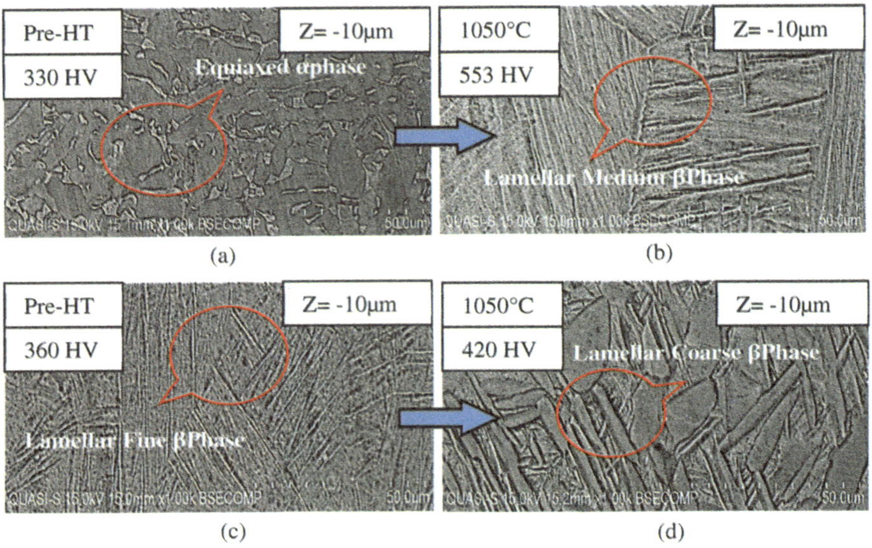

Fig. 2.7 Microstructural comparison 10 μm depth **a** SM–Pre-HT, **b** SM–Post-HT, **c** AM–Pre-HT, **d** AM–Post-HT

is better than at 1150 °C. In the SM specimens, a transformation from the α phase to the β phase is observed, while in the AM specimens, both pre-HT and post-HT specimens exhibit a β phase as shown in Fig. 2.7. The obvious reason for this observation is the processing temperature. SM process has shown surface temperatures of 300–600 °C during the machining. These temperatures are still under the β transus temperature of 995 °C and hence cannot trigger a phase change. However, the process temperature of the AM process, at a laser power of 275 W peaks to as high as 2500 °C which is 2½ times greater than the β transus temperature. The AM (SLM) specimens can be seen dominated by the lamellar or columnar structures which represent a martensitic state with a BCC crystalline structure. Martensitic structures also exhibit higher tensile strength due to a higher hardness, but a lower ductility making it brittle [25]. A lamellar structure is also observed in the post-heat-treated SM specimens, which exhibit an obvious phase change from the initial equiaxed structure depicting a ferrite state with a BCC structure as seen in Fig. 2.7. The lamellar structure of the AM specimens transforms from fine to a coarse structure when heat treated. A coarser lamella indicates a more brittle surface prone to premature failure in applications characterized by frictional wear. In SM specimens medium grained lamellae are observed in the post-heat-treated SM specimens which indicates a less brittle surface. At the subsurface levels, finer lamellar grains are observed in both specimens indicating a less brittle surface which could be considered for applications requiring interaction between surfaces leading to a slower wear rate.

2.4 Conclusion

Applications of Ti6Al4V, as discussed earlier, ranges from wear-resistant functions like turbine blade disks, piston rings, etc., to functions that require high toughness like fasteners, connecting rods, springs, retainers, etc. Both SM and AM specimens exhibit different properties at different surface and subsurface levels. A higher hardness and tensile strength with a finer lamellar β phase surface would be appropriate for wear-resistant functions. For SM components, a fine lamellar surface is achieved through a HT above the β transus temperature (1050 °C) followed by a water quench and aging process to a temperature of 500 °C and furnace cooled. However, for AM specimens, the same set of HT processes followed by a surface removal to a depth between 100 and 200 μm either through a surface grinding, bead blasting or etching operation is recommended for wear-resistant surfaces. SM components do not have wear resistance requirements but need a tougher core could use an α phase HCP crystalline structure, which is obtained by HT below the β transus temperature (950 °C) followed by the same set of quenching and aging processes. This helps in retaining a tougher core. However, AM components develop a β phase BCC structure during the 3D printing stage itself due to the high temperature laser beam and hence retain the same crystalline structure. A stress relieving process on these components would be sufficient in applications that require a tougher core. Further study to analyze the wear resistance of the surface and the subsurface levels, their correlation to the surface microstructure, composition, and concentration of the alloying elements, will be studied through tribology experiments.

References

1. Liu C, Goel S, Llavori I, Stolf P, Giusca CL, Zabala A, Rabinovich GSF (2019) Benchmarking of several material constitutive models for tribology, wear, and other mechanical deformation simulations of Ti6Al4V. J Mech Behav Biomed Mater 97:126–137
2. Meilinger A, Torok I (2013) The importance of friction stir welding tool. Prod Process Syst 6(1):25–34
3. Liu S, Shin YC (2019) Additive manufacturing of Ti6Al4V alloy: a review, Mater Des 164:107552
4. Fan Z, Feng H (2018) Study on selective laser melting and HT of Ti-6Al-4V alloy. Results Phys 10:660–664
5. Huang R, Riddle M, Graziano D, Warren J, Das S, Nimbalkar S, Masanet E (2016) Energy and emissions saving potential of additive manufacturing: the case of lightweight aircraft components. J Clean Prod 135:1559–1570
6. Beretta S, Romano S (2017) A comparison of fatigue strength sensitivity to defects for materials manufactured by AM or traditional processes. Int J Fatigue 94:178–191
7. Patterson AE, Messimer SL, Farrington PA (2017) Overhanging features and the SLM/DMLS residual stresses problem: review and future research need. Technologies 5(2):15
8. Braian M, Jönsson D, Kevci M, Wennerberg A (2018) Geometrical accuracy of metallic objects produced with additive or subtractive manufacturing: a comparative in vitro study. Dent Mater 34(7):978–993

9. Suresh S, Abdul Hamid D, Yazid MZA, Che Daud NR, Mohd DNNL (2017) An investigation into the impact of flexural elastic modulus on the dimensional instability in CNC turning of titanium alloy grade 5-Ti6Al4V. Mech. Eng. J. 2:73–85
10. Suresh S, Hamid DA, Yazid MZA, Nasuha N, Ain SN (2017) Tool geometry and damage mechanisms influencing CNC turning efficiency of Ti6Al4V. AIP Conf Proc 1901(1):040004
11. Bejjani R, Bamford E, Cedergren S, Archenti A, Rashid A (2020) Variations in the surface integrity of Ti-6Al-4V by combinations of additive and subtractive manufacturing processes. Materials 13(8):1825
12. Khorasani AM, Gibson I, Ghaderi A, Mohammed MI (2019) Investigation on the effect of HT and process parameters on the tensile behaviour of SLM Ti-6Al-4V parts. Int J Adv Manuf Technol 101(9):3183–3197
13. Patil S, Jadhav S, Kekade S, Supare A, Powar A, Singh RKP (2016) The influence of cutting heat on the surface integrity during machining of titanium alloy Ti6Al4V. Procedia Manuf. 5:857–869
14. Sutter G, List G (2013) Very high-speed cutting of Ti–6Al–4V titanium alloy–change in morphology and mechanism of chip formation. Int J Mach Tools Manuf 66:37–43
15. Abdalla AO, Amrin A, Muhammad S, Hanim MA (2017) Effect of HT parameters on the microstructure and microhardness of Ti-6Al-4V alloy. AIP Conf Proc 1865(1):030001
16. Iqbal A, Suhaimi H, Zhao W, Jamil M, Nauman MM, He N, Zaini J (2020) Sustainable milling of Ti-6Al-4V: investigating the effects of milling orientation, cutter' s helix angle, and type of cryogenic coolant. Metals 10(2):258
17. Pitassi D, Savoia E, Fontanari V, Molinari A, Luchin V, Zappini G, Benedetti M (2018) Finite element thermal analysis of metal parts additively manufactured via selective laser melting. Intech123–156
18. Shan C, Zhang M, Zhang S, Dang J (2020) Prediction of machining-induced residual stress in orthogonal cutting of Ti6Al4V. Int J Adv Manuf Technol 1–11
19. Yang D, Liu Z, Ren X, Zhuang P (2016) Hybrid modeling with finite element and statistical methods for residual stress prediction in peripheral milling of titanium alloy Ti-6Al-4V. Int J Mech Sci 108:29–38
20. Ahmad B, Van DVSO, Fitzpatrick ME, Guo H (2018) Residual stress evaluation in selective-laser-melting additively manufactured titanium (Ti-6Al-4V) and inconel 718 using the contour method and numerical simulation. Addit Manuf 22:571–582
21. Ali H, Ghadbeigi H, Mumtaz K (2018) Effect of scanning strategies on residual stress and mechanical properties of selective laser melted Ti6Al4V. Mater Sci Eng A 712:175–187
22. Ali H, Ghadbeigi H, Mumtaz K (2018) Processing parameter effects on residual stress and mechanical properties of selective laser melted Ti6Al4V. J Mater Eng Perform 27(8):4059–4068
23. Yang T, Liu T, Liao W, Zhang C, Duan S, Chen X (2018) Study on deformation behavior of Ti6Al4V Parts fabricated by selective laser melting. MATEC Web Conf. 237:01008
24. Khorasani AM, Ian G, Moshe G, Guy L (2018) A comprehensive study on surface quality in 5-axis milling of SLM Ti-6Al-4V spherical components. Int J Adv Manuf Technol 94(9):3765–3784
25. Pilchak AL, Sargent GA, Semiatin SL (2018) Early stages of microstructure and texture evolution during beta annealing of Ti-6Al-4V. Metall Mater Trans A Phys Metall Mater Sci 49(3):908–919

Chapter 3
Determining the Alternative Type of Wood for Class a Fire Crib Test According to UL 2166 Standards

Jum'azulhisham Abdul Shukor, Muhammad Al-Hapis Abdul Razak, Azmi Hassan, Rahim Jamian, and Ahmad Majdi Abdul Rani

Abstract Wood is one of the primary combustible fuels used for Class A fire tests for clean agent fire suppression systems. The species classified in the standards, on the other hand, did not originate locally and are hardly found. It is overpriced to obtain in small quantities for the sake of the experimental test. Thorough investigations were suggested to validate the use of the proposed material, including verifying the species of the specimen, burning test, and conducting wood moisture content verification using oven-dry method to ensure the quality of combustible fuel for the final test as per standard requirement. The aim of this study is to justify the use of unspecified wood to meet the appropriate requirements for performing a fire crib test. As a result, the pinewood specifications were nearly and within the minimum criterion for performing the fire crib test.

J. A. Shukor (✉)
Manufacturing Engineering Technology Section, Univerisiti Kuala Lumpur Malaysia Italy Design Institute, Wilayah Persekutuan Kuala Lumpur, 119, Jalan 7/91, Taman Shamelin Perkasa, 56100 Kuala Lumpur, Malaysia
e-mail: jumazulhisham@s.unikl.edu.my

J. A. Shukor · M. A.-H. A. Razak
Manufacturing Section, Univerisiti Kuala Lumpur Malaysian Spanish Institute, Kulim Hi-Tech Park, 09000 Kulim, Kedah, Malaysia
e-mail: alhapis@unikl.edu.my

A. Hassan
Electrical, Electronics and Automation Section, Univerisiti Kuala Lumpur Malaysian Spanish Institute, Kulim Hi-Tech Park, 09000 Kulim, Kedah, Malaysia
e-mail: azmi.hassan@unikl.edu.my

R. Jamian
Department of Mechanical Engineering Technology, Faculty of Engineering Technology, Universiti Tun Hussein Onn Malaysia, Hab Pendidikan Tinggi Pagoh, Kampus Pagoh, KM 1, Jalan Panchor, 84600 Panchor, Johor, Malaysia
e-mail: rahimj@uthm.edu.my

A. M. A. Rani
Mechanical Engineering Department, Universiti Teknology PETRONAS, Persiaran UTP, 32610 Seri Iskandar, Perak, Malaysia
e-mail: majdi@utp.edu.my

A. Ismail et al. (eds.), *Advanced Materials and Engineering Technologies*, Advanced Structured Materials 162,
https://doi.org/10.1007/978-3-030-92964-0_3

Keywords Fire crib test · Oven-dry method · Pine · Pinaceae · Moisture content

3.1 Introduction

A total flooding fire extinguishment test is performed to determine the effectiveness of a novel extinguishment agent [1]. The purpose of the fire crib test, two types of test fuel, was highlighted either kiln-dried wood for Class A or heptane for Class B fires. For Class A fire, the combustible fuel must be either fir or spruce with standard moisture content as required by standards. However, the wood to be used is difficult to obtain locally. These imported softwoods have seldom been used by local traders and manufacturers. Importing a small quantity is not economically viable. However, an option is suggested to use based on one of its family. Softwood, specifically pine has widely been used as wooden pallets and is available locally.

In some cases, this family species may contain similar characteristics. But to differentiate the semi-finished cutting kiln wood, it requires time, some technical expertise, references, or some high-technology equipment to verify it. It takes the identification of the wood species to become challenging. The objective of this research is to investigate, verify the type of wood, and determine its moisture content for the purpose of fire crib test.

The criteria to be investigated of the selected wood are benchmarked from valid references. The one has been proposed to be the combustion fuel, response for the fire crib test are according to its experimental characteristics including charred layer, mass loss rate, maximum burning rate, and the ratio between relative burning rate and the time. In addition, the percentage of moisture of these woods is being investigated which has significant influence on the time of burning.

The moisture content of the wood plays an important role in the size of the fire and burning rates. The less moisture content means the wood takes less time and easy to burn out while the higher moisture content, the wood takes time to burn out. This also affects the charring condition during the burning process [2]. In this manner, the moisture content of the pine in this test must be taken using the oven-dry method and control within the standards requirement. The moisture can be determined either by experimenting using a high-technology equipment [3], portable moisture meter or through a traditional experimental method [4, 5]. The use of vibration technique is unpreferable for the semi-process or processed woods [6]. In this study, the experimental model using the oven-dry method is accurate [3] and preferred with actual data compared with the estimation results from the devices.

The weight of moisture contained in a piece of wood is expressed as a percentage of the dry weight of its oven-dry weight. The water content has been removed from this procedure, and the weight of wood has been reported.

Fig. 3.1 Softwood specimen identified for the study

3.2 Methodology

The final goal of this research is to conduct a total flooding fire suppression test to determine the effectiveness of a new gaseous fire suppression agent on Class A fires. In this case, the main combustion fluid for this test has been chosen as kiln-dried wood. The interaction of burning material and extinguishing agent will be evaluated.

Because of the difficult situation of obtaining the specified type of wood, an imported species of wood, pine is recommended. Its properties must be verified within the specifications needed in the requirements beforehand. The previously unknown softwood but easily find locally has been identified as shown in Fig. 3.1.

Before the fire crib test may be performed, a sample must be taken and be classified in terms of its reaction to the fire. Its moisture content must be tested to the standard requirements.

3.2.1 Identification of Wood Sample

A preliminary study of the spruce and fir families was carried out. Spruce (*Picea spp.*) and fir (*Abies spp.*) are all medium to high, needle-bearing, ever- green, cone-producing trees with pyramidal configurations. Because of their cone development, they are known as conifers. Later, it is easy to identify all family members belong to this family. These woods are members of the Pinaceae family, along with a few other varieties of woods.

Expert assistance to ensure the validity of the softwood specimen collected, as shown in Fig. 3.2, is convenient, fast, and cost effective. For clarification, samples

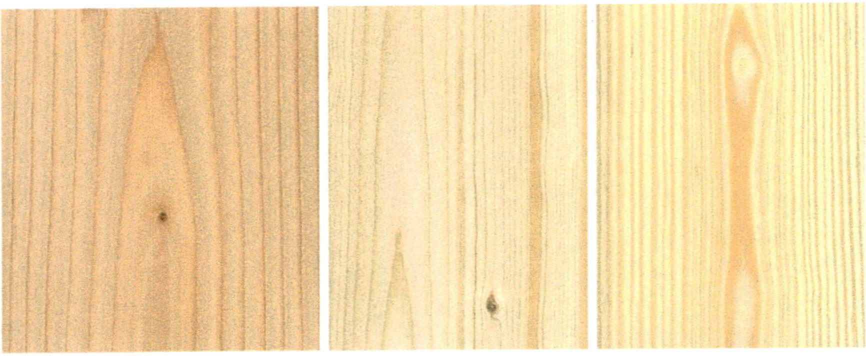

Fig. 3.2 From left: Fir, Spruce, Pine

Table 3.1 Wood Family of Pinaceae and its scientific name

Wood name	Scientific name
Spruce	*Picea spp.*
Fir	*Abies spp.*
Pine	*Pinus spp.*
Cedar	*Cedrus*
Hemlock	*Conium maculatum*
Larch	*Larix*

have been distributed to the Forest Research Institute Malaysia (FRIM) and Jabatan Perhutanan Negeri Kedah. The samples were compared using the Forestry Databook and The Global Wood Density Database, as well as their independent judgments and official confirmation.

3.2.2 *Study of Spruce, Fir, and Its Family*

Two approaches were used to analyze the specimen for this study, which included evaluating the data on the moisture content of green softwoods and the possible reaction of woods to fire (Table 3.1).

3.2.2.1 Reaction of Wood to the Fire

The reaction of wood to the fire was assessed from a previous study. These wood samples were taken from the trunk of the tree to examine the changes of properties including mass loss, relative burning rate, charred layer, and (a)/(b) ratio [7]. The distance between the source of heat and the test specimen was set at 40 mm such

Table 3.2 Comparison of average moisture content of selected Wisconsin conifers softwoods [8]

Softwood (Wisconsin species)	Average Moisture Content (%)
Cedar, eastern red	33
Fir, balsam	88
Hemlock, eastern	97
Pine, red	32
Spruce, black	52
Tamarack	49

that the temperature mat manifested itself and the accuracy of the measurement was assured. The specimens were exposed to radiant heat for 3 min after 15 min pre-heating from 70 to 1100 °C.

3.2.2.2 Moisture Content

An assessment of the moisture content of these woods was also performed to confirm that the moisture content of pine is fairly within the range [8, 9]. Table 3.2 compares the average moisture content of various green softwoods collected from the handbook.

3.2.3 Pine Moisture Content Test

The oven-drying method is selected for determining the moisture content of wood and is widely used in the forestry industry [9, 10]. Malaysian Standards MS837:2006 Solid Timber–Determination of Moisture Content (Clauses 4 and 5) Oven-Dry Method was used by the researcher to perform a wood moisture content test. To conduct the fire crib test the moist content of the combustion fluid must be controlled according to the standard and must be between 9 and 13% to pass the Class A fire crib test later.

The test specimen was weighted before drying process in a ZDR-5030 forced air -drying oven with 160 mm × 250 mm × 120 mm chamber, with a constant temperature of 103 °C ± 2 °C or holding between 100 and 105 °C for 24 h and the dried specimens were re-weighed [11].

Species	Pine (*Pinus sp.*)
Sample dimension ($w \times l \times d$):	95 × 130 × 20 mm
Initial weight (wi):	128.84 g
Expected moisture content:	9.0–13.0%

The weight of moisture content in pine is calculated as the percentage of the moisture differential weight before and after oven-dry treatment divided by the oven-dry weight [5, 10]. The moisture content equation is projected in Eq. (3.1).

$$MC = \frac{(w_i - w_o)}{w_o} \times 100\% \tag{3.1}$$

where MC is the percentage of moisture content, w_i is the weight before oven-dry process in grams, and w_o is the burning weight after oven-dry process in grams.

The characteristics of the fire and its moisture content were determined and compared with spruce and fir to verify that pine is appropriate for the fire crib test. The outcomes of the sufficiently generated data were carefully analyzed.

3.3 Results and Discussion

3.3.1 Type of Wood Identification

Based on the findings of the published study mentioned by FRIM and Jabatan Perhutanan Negeri Kedah, it was confirmed that the specimen used for the analysis was pine, or Pinus sp. in the botanical name. It has been discovered that pine, spruce, and fir in common are softwoods that originate from the same family tree and have identical appearances, including their needle and cone.

3.3.2 Reaction of Wood to the Fire

Since pine is in the same family tree with spruce and fir, it still cannot be declared suitably for conducting the fire crib test. As mentioned, the same species may naturally have similar or different characteristics.

The reaction of identified wood (specifically pine, fir, spruce) to fire was extensively conducted. In this study, properties in relation to fire including mass loss, charred layer, burning rate, maximum relative burning rate, the ratio between relative burning rate and the time necessary to achieve its peak (a/b ratio), and density have been acknowledged.

The outcomes of the experimental burning of three reviewed and suggested *Pinaceae* family woods are shown in Table 3.3 (Fig. 3.3).

Based on Table 3.3 and Fig. 3.4, it can be concluded that pine is highly recommended to be used as alternative for the purpose of fire crib test. This is due to the

Table 3.3 Results of the experimental burning of three Pinaceae family woods

Type of Wood	Weight/mass loss [%]	Charred layer [%]	Relative burning rate [% s − 1]	Max. relative burning rate [s]	(a)/(b) ratio	Density [kg m − 3]
Fir	11.35	7.37	0.000913	145	6.3	500.71
Pine	13.71	9.12	0.001244	157.5	7.9	400.93
Spruce	17.77	22.94	0.001692	165	10.25	345.98

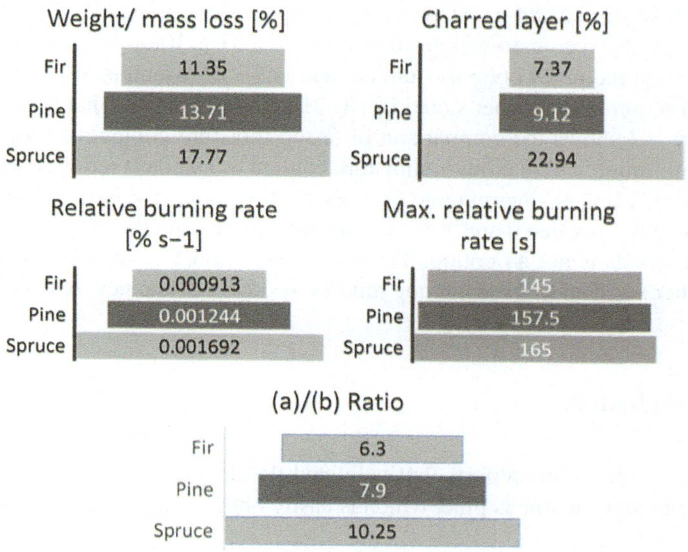

Fig. 3.3 Results of the burning test of three *Pinaceae* woods

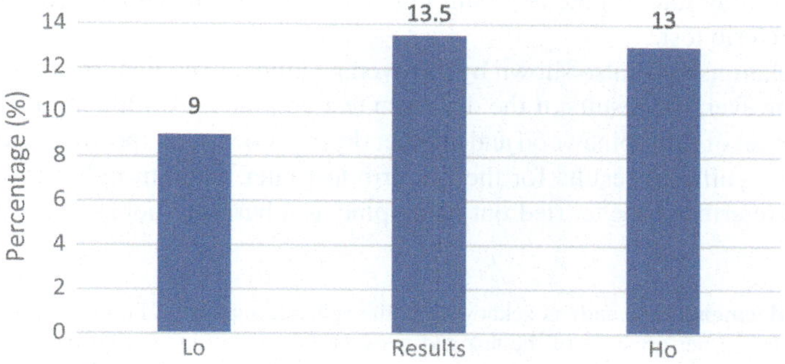

Fig. 3.4 Pine moisture content using oven-dry method and comparison made for the controlled upper and lower limits (9–13%)

limitation obtaining the specified type of combustible fuel either spruce or fir for the purpose of the fire crib test.

3.3.3 *Results of Pine Moisture Rate for Fire Crib Test*

Based on the wood moisture test using oven-dry method, the initial weight of the specimen taken is 128.84 g. After given a constant temperature of 103 °C and 24 drying hours, the specimen weight reduced to 113.57 g. It has been confirmed and concluded that the pinewood moisture content is 13.5% which is 5% more than the maximum requirement as per stated in UL 2166. Which means that this lot is over dry. Figure 3.3 shows the comparison of tested moisture content of pine compare with the minimum and maximum moisture content as standard requirement.

This set of wood is unacceptable for the total flooding fire crib test. The moisture content must be between 9 and 13%. In this case, increasing the moisture content of this set of woods is not an option. This set of wood must be rejected and replaced with another set. The same technique must be used to test the new batch of wood.

3.4 Conclusion

According to the expert reports that validated the database, the wood found from the wooden pallet sample is pine, which is easily obtained locally for the making of pallet.

It has been revealed through burning experimental results that the fire test on pine has significant results in which the characteristics of mass loss, charred layer, and relative burning rate of pine are within the regulated parameters which can be used for the fire crib test.

The relationship is also shown by the average moisture content from the previous study. However, the results of the dry-oven test on pine are a little high, suggesting that a new set of fresh pinewood and another drying process, as specified, are required to ensure significant results for the fire crib test later. It is strongly recommended that the fire crib test be carried out using pine as a burning fuel as alternative to fir or spruce.

Acknowledgements The authors acknowledge the valuable support of Forest Research Institute Malaysia for the performance of the test and assistance from Jabatan Perhutanan Negeri Kedah on the research mentioned above. Also, a special gratitude to HORI UniKL MSI for funding the publishing of this article.

References

1. Diab MT, Haelssig JB, Pegg MJ (2020) The behavior of wood crib fires under free burning and fire whirl conditions. Fire Saf J 112:102941
2. Bartlett AI, Hadden RM, Bisby LA (2019) A review of factors affecting the burning behavior of wood for application to tall timber construction. Fire Technol 55:1–49
3. Han Y, Fang J (2020) Spectral detection of wood moisture. IOP Conf Ser Earth Environ Sci 598:1–7
4. Samuelsson R, Burvall J, Jirjis R (2006) Comparison of different methods for the determination of moisture. Biomass Bioenergy 30:929–934
5. Reeb J, Milota M (1999) Moisture content by the oven-dry method for industrial testing weight of water weight of wood MC. WDKA, pp 66–74
6. Taniwaki M, Akimoto H, Hanada T, Tohro M et al (2007) Improved methodology of measuring moisture content of wood by a vibrational technique. Wood Mater Sci Eng 2:77–82
7. Osvaldova LM (2018) Wood of coniferous trees: Reaction to fire. Int J Wildland Fire 131–151
8. Ross RJ (2010) Wood handbook: wood as an engineering material. USDA forest service, forest products laboratory. Gen Tech Rep FPL-GTR 190:1–509
9. Samuelsson R, Burvall J, Jirjis R (2006) Comparison of different methods for the determination of moisture content in biomass. Biomass Bioenergy 30:929–934
10. Mahmud SZ et al. (2017) Physical and mechanical properties of juvenile wood from Neolamarckia cadamba planted in West Malaysia. Maderas Cienc. y Tecnol 19:225–238
11. Carneiro JS et al (2018) The oven-drying method for determination of water content in Brazil nut. Biosci J 34:595–602

Chapter 4
Prediction of Malaysian Talus Bone Morphology Using Artificial Intelligence

Rosdi Daud, Nurazlina Sulaeman, Mas Ayu Hassan, and Arman Shah Abdullah

Abstract Talus fractures keep on presenting to a difficult and generally experienced gathering of injuries. This published report shows that not all the current bone implants are the ideal counterpart for the specific population. Along these lines, this investigation received a three-dimensional (3D) estimation way to deal with given exact information to the anatomical morphology of talus bone. Seventy-four Malaysian healthy subjects experienced computerized tomography (CT) arthrography. 3D computerized talar models were generated and three morphological boundaries predicted through Mimics and Solidworks software. Sagittal Talar radius (STRa), Throchlea Tali length (TTL), Talar Anterior width (TaAW) are the most part chosen. Information investigation was directed by determination of information test through Matlab programming. In this way, the information was obtained dependent on the artificial intelligence (AI) forecast of the talus bone morphometric. While, the AI strategy demonstrated a more noteworthy limit of forecast in regards to the low level of mistake and high correlative qualities since the average percentage errors of the predicted talus bone morphology parameters are around 10% which 11.3% for STRa, 12.95% for TaAW, and 9.45% for TTL. AI is an exceptionally exact prescient

R. Daud (✉) · M. A. Hassan
Automotive Engineering Center, Universiti Malaysia Pahang, 26600 Pekan, Pahang, Malaysia
e-mail: rosdidaud@ump.edu.my

M. A. Hassan
e-mail: masszee@ump.edu.my

R. Daud · N. Sulaeman
Faculty of Mechanical and Automotive Engineering Technology, Universiti Malaysia Pahang, 26600 Pekan, Pahang, Malaysia

M. A. Hassan
Faculty of Manufacturing and Mechatronic Engineering Technology, Universiti Malaysia Pahang, 26600 Pekan, Pahang, Malaysia

A. S. Abdullah
Faculty of Technical and Vocational, Universiti Pendidikan Sultan Idris, 35900 Kuala Kubu Bharu, Perak, Malaysia
e-mail: armanshah@fptv.upsi.edu.my

A. Ismail et al. (eds.), *Advanced Materials and Engineering Technologies*,
Advanced Structured Materials 162,
https://doi.org/10.1007/978-3-030-92964-0_4

technique and can be utilized as helping instruments in developing bone implant specifically for Malaysian patient and for Asian patient in general.

Keywords Talus fracture · Bone morphometric · Artificial intelligent

4.1 Introduction

Nowadays, healthcare innovation is considered one of the most significant aspects in improving and saving countless lives around the world. Medical technology is one of an enormous field where advancement plays an important role in sustaining health. As technology in the world of healthcare keeps on developing, the technological innovations keep on to be furnished with better approaches to improve the quality of care conveyed to patients and improve the state of global healthcare. An artificial implant is generally used to replace a damaged ankle joint (tibiotalar joint), where the shinbone (tibia) lays on top of a bone of the foot, the talus [1, 2]. In any case, the smooth cartilage on the surface of the bones erodes by time which is inevitably results in pain, inflammation, and swelling of the joint [3]. So that, ankle replacement surgery is a way that can be chosen to supplant damaged joint to take out those pain and swelling [4]. Nowadays, cadaver specimens, radiographs, and CT images have for the foremost part been chosen to obtain detailed morphological data of talus but it is restricted to the availability of bones [1]. Cadaver specimens are difficult to obtain, making it difficult to oversee huge scope studies. In spite of the fact that, radiographs can be easily obtained from clinical cases, contrasts in the angles of posture and tube projection will cause fundamentally variability in structural variables [5]. CT images, on the other hand, are extremely consistent because they can be reconstructed to a 3D model through computer software. To date, 2-dimensional (2D) plain radiography is the most well-known technique used to access ankle morphometric [6]. However, this method has a significant limitation since it cannot give 3-dimensional (3D) details of the anatomy. One investigation detailed an unsatisfactory intrinsic error when utilizing plain radiographic estimations for lower leg situating [7]. Estimation through cadaveric investigation should be possible yet is restricted to the accessibility of bones. Technologic propels in high-goal processed tomography (CT) presently allows 3D assessment of bone morphometric.

In present study, CT scanning was performed by a TOSHIBA/Aquilion ONE scanner [5, 7]. The CT images were sent out to Mimics (v19.0, Materialize, Belgium) in advance imaging and correspondences in medicine (DICOM) format [7]. Through the "threshold, region grow, edit masks, calculate 3D and distance" capacities in Mimics software, the images were recreated to 3D models. In total, 13 parameters could be determined for measurements. Subsequently, intended to introduce the morphometric of the superior articular surface of the trochlea bone utilizing 3D models created from CT scans. Morphometric measures comparing to those of the X-ray investigation were performed on the focuses lying on the two anatomical planes, which expected as the highest quality level for the estimations on the X-ray pictures [6]. A near factual investigation between sexual orientations and between the left- and right-hand talus bones was performed using the t-test [7]. Linear regression analysis was performed to decide the relationship between the deliberate morphometric boundaries. Morphometric information were contrasted and the elements of the talar segments from 4 distinctive total ankle arthroplasty (TAR) inserts to decide the similarity [7].

Although artificial replicas have been widely invented today, most of them have only met with European population size and population criteria [1, 5]. If an Asian person needs the artificial implant, then the physician will have some difficulty in identifying the right size and fit the existing printed bone for the Asian patient in need due to human genetic variation. Therefore, sometimes they have to remake them to make them fit to the Asian patients especially Malaysian. Thus, the aim of this study is to investigate the alternative method to obtain talus bone morphology using an artificial intelligence method.

4.2 Methodology

4.2.1 Procedure for Collecting Data/Measurement of Talus Bone Morphology

Seventy-four Malaysian healthy subjects experienced computerized tomography (CT) arthrography of the lower limb. The CT images were exported to Mimics software in digital imaging and medical communication (DICOM) format. The software created a 3D model, which included both soft tissue and bone material. In order to separate the bone from soft tissue, we used the "threshold, region grow and modify masks" function in Mimics software, and the images were built to 3D models. The threshold tool was used to separate the lower limb bones from the bones of the foot. The trim tool was then used to create a solid surface model of the talus bone, and these models were saved as.IGS files. The.IGS of the talus bone models were then imported into the SolidWorks software. It was capable of performing 2D and 3D measurements of the bone's model, and these capabilities were used in order to make morphological measurements of individual bones as well as combination of bones. The which the morphological measurements of talus bone are shown in Fig. 4.1.

4.2.2 Prediction of Talus Bone Morphology via Artificial Intelligence

MATLAB software is a multi-worldview numerical processing condition and inclusive a programming language created by MathWorks. MATLAB permits grid controls, plotting of capacities and information, usage of calculations, making of

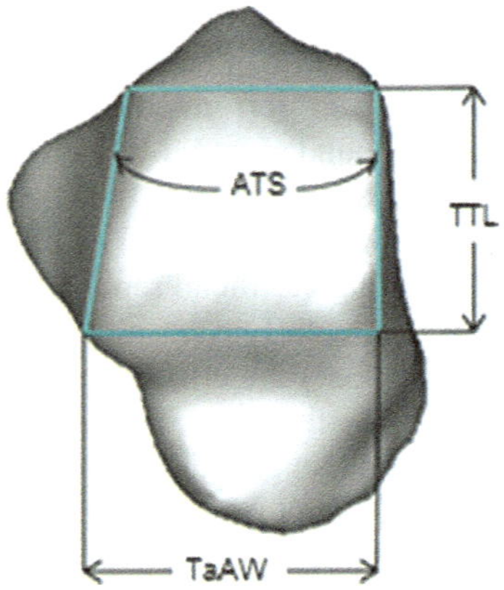

Fig. 4.1 Indications of the ankle morphological parameters

UIs, and interfacing with programs written in different languages. The MATLAB software was introduced to establish the reliability of the measurements. All of the data pertaining that gathered were included in the reliability of the measurement method. For the morphometric measurement, the "nntool" toolbox was utilized to train the neural network. The "nntool" command window was chosen to create some networks as shown in Fig. 4.2 once the inputs and targets data were imported. Selection of network type, training function, number of layer, and others was chosen here before experiencing the network training. What can clearly be seen in Fig. 4.3 is the variability of graphs, which shows the train info of neural networks and their performance. Training process of each networks would be continued and repeated until the regression graph approaches to 1.0.

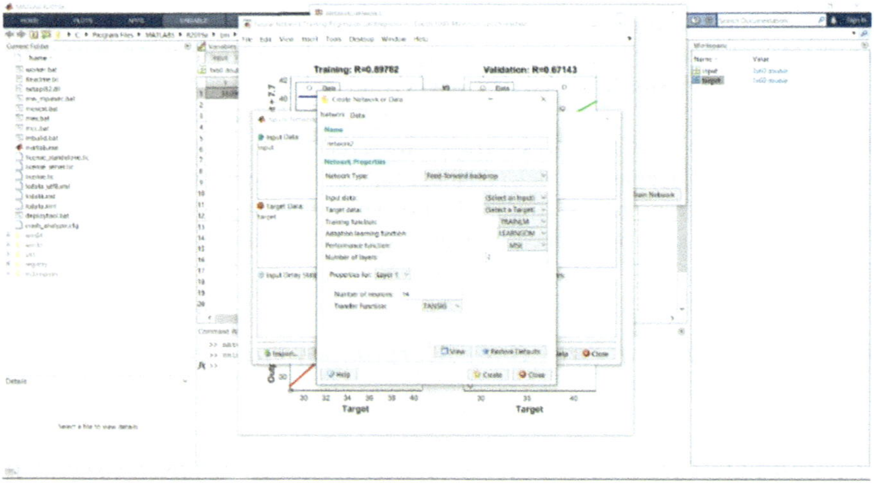

Fig. 4.2 Network creation

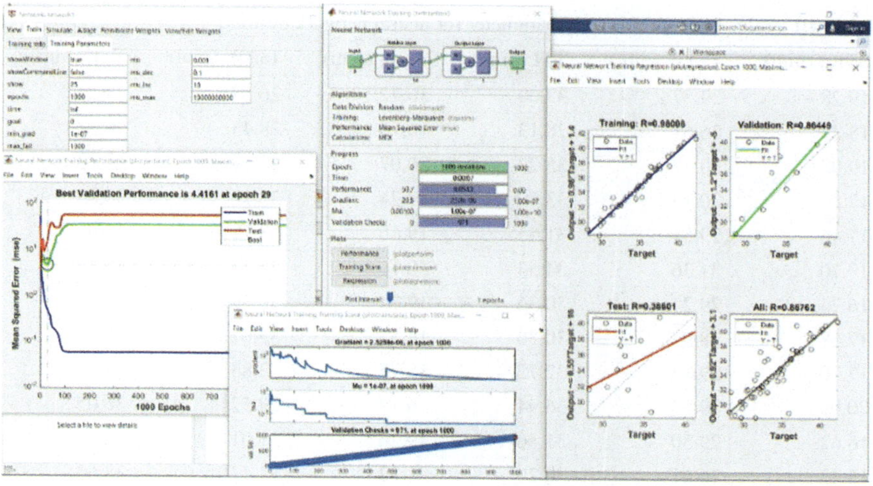

Fig. 4.3 Full data of trained network

This study utilized the CT scan data of 74 participants consisting of male and female adults with a ranges age of 19 to 30 years old. The ranges of body weight and height were 39 kg to 118 kg and 141 cm to 182 cm, respectively. Table 4.1 shows a total of 60 out of 74 subjects' talus bone morphology parameters, which is STRa, TaAW, and TTL were utilized for the neural network training. From the trained neural network, they were then used for talus bone morphology parameter prediction testing.

4.3 Result and Discussion

After the neural network successful trained, we used that network to predict the sagittal talar radius (STRa), talar anterior width (TaAW), and trochlea tali length (TTL). Average error of predicted sagittal talar radius (STRa) is 11.3%, average error of talar anterior width (TaAW) is 12.95%, and average error of trochlea tali length (TTL) is 9.45% as shown in Tables 4.2, 4.3, and 4.4, respectively.

Although the prediction average of the talus morphology parameters looks acceptable since the error percentage is around 10% but a few prediction percentage errors seem up to 30%. Thus definitely this prediction result can still be debated and of course, one possible explanation for these may be due to the lack of adequate details data of the volunteered subjects. The information of their age, height, and weight seem not to fully satisfy the significant criteria as the input data for network development in the neural network system (artificial intelligence). Further data collecting regarding on subjects anthropometric measurements such as measurement of the foot

Table 4.1 Talus bone morphology parameter for neural network training

STRa (mm)	TaAW (mm)	TTL (mm)	STRa (mm)	TaAW (mm)	TTL (mm)
19.28	29.57	33.09	16.72	26.18	28.84
15.27	26.61	28.13	16.10	28.43	29.57
19.08	30.4	35.23	20.07	29.71	35.19
17.80	30.12	33.70	19.24	33.16	36.93
17.46	29.54	31.55	19.66	33.78	35.66
19.70	31.36	34.53	20.16	31.23	40.32
16.74	29.21	30.36	19.74	36.59	37.13
17.58	28.97	30.36	18.57	33.67	36.25
17.70	29.13	32.72	18.75	33.83	34.63
20.99	29.13	36.94	19.19	29.57	36.05
16.62	29.30	31.46	19.26	31.01	36.22
16.43	26.61	30.20	21.22	32.42	38.63
17.99	29.70	33.29	19.84	30.39	36.43
19.70	27.88	31.36	20.00	29.41	38.85
17.48	28.33	32.18	19.42	31.35	37.52
19.28	30.31	34.69	18.94	34.04	34.18
18.37	27.15	33.18	19.55	32.62	37.57
19.48	30.31	34.20	21.26	32.82	40.75
17.33	26.45	31.78	22.78	31.99	42.39
17.89	30.45	31.37	20.19	37.14	39.23
16.74	27.62	31.03	17.01	35.85	32.35
16.84	26.30	29.78	19.21	26.53	34.85
18.55	26.55	32.97	18.51	30.59	36.26
20.34	31.10	34.46	18.05	33.98	34.99
18.98	28.96	32.06	19.52	32.10	36.98
16.65	26.65	30.57	18.32	32.95	32.39
18.08	28.27	32.50	19.33	30.15	36.23
16.62	26.70	29.64	20.23	30.2	35.92
15.96	27.05	28.33	21.3	31.62	40.06
17.73	25.65	29.83	21.66	36.35	38.70

length and width high probably will enhance the prediction performance using the artificial intelligence method.

Morphometry of the lower leg complex is regularly done utilizing 2D plain radiographs. In any case, this 2D method has an intrinsic constraint while dissecting unsymmetrical 3D objects, for example, the talus bone. For instance, the wedged state of the talar in the cross over the plane, which is more extensive anteriorly than posteriorly on the prevalent surface, cannot be seen utilizing plain radiographs [1].

Table 4.2 Predicted sagittal talar radius, STRa

Target (mm)	Prediction (mm)	Error (%)
19.47	21.95	12.74
21.26	21.79	2.71
19.56	20.34	3.98
20.61	15.27	25.9
19.31	15.81	18.12
19.43	17.23	11.30
17.35	21.27	22.60
18.61	22.57	21.26
18.80	15.32	18.00
18.20	17.91	1.58
18.44	17.63	4.42
19.67	20.39	3.64
23.11	20.77	10.10

Table 4.3 Predicted talar anterior width, TaAW

Target (mm)	Prediction (mm)	Error (%)
37.28	40.75	9.32
38.59	33.00	14.48
36.88	39.20	6.30
38.66	34.70	10.20
36.06	34.39	4.62
36.49	35.46	2.80
33.45	39.29	17.46
35.40	41.81	18.1
35.55	29.98	15.67
32.09	31.83	0.80
33.57	34.05	1.43
35.38	36.86	4.17
36.07	28.72	20.37
41.97	38.99	7.09

At any rate, nine estimations are required for satisfactory advancement of inserts for lower leg substitution, and three of these must be obtained from a 3D view [1]. In any case, the measure of information that can be obtained constrained to the accessibility of bones. This study digitized their bodies since certain estimations must be made through 3D models. Subsequently, this study completely utilized CT required

Table 4.4 Predicted trochlea tali length, TTL

Target (mm)	Prediction (mm)	Error (%)
32.04	33.39	4.30
37.70	37.13	1.51
33.68	34.26	1.71
32.92	26.05	20.87
34.91	25.72	26.32
33.82	25.71	23.99
33.28	26.59	20.12
29.10	33.08	13.66
33.11	25.69	22.43
30.59	29.28	4.26
29.74	28.72	3.42
32.28	37.08	14.88
30.65	37.14	21.17
36.90	35.93	2.64

estimations to embed advancement through 3D perception. Linear regression analysis was performed to decide if the segment factors, for example, age, stature, and weight file impact the size of the trochlear bone. The data reported here appear to support the assumption that demonstrated no relationship between these factors and the deliberate morphologic boundaries, aside from height, which possibly indicated a positive connection with the STRa, TaAW, and TTL.

4.4 Conclusion

It is can be concluded that prediction of talus bone morphology via artificial intelligence is possible to be done since the average percentage errors of the talus bone morphology parameters are around 10% which 11.3% for STRa, 12.95% for TaAW, and 9.45% for TTL. Thus, this artificial intelligence is exceptionally exact prescient technique that can be utilized as helping instruments in designing bone implant specifically for Malaysian patient and for Asian patient in general.

Acknowledgements The authors would like to thank Ministry of Higher Education for providing financial support under Fundamental Research Grant Scheme (FRGS) No. FRGS/1/2018/TK03/UMP/02/5 (University reference RDU190114) and Universiti Malaysia Pahang for laboratory facilities as well as additional financial support under Internal Research Grant RDU190327.

References

1. Daud R, Abdul Kadir MR et al (2013) Three-dimensional morphometric study of the trapezium shape of the trochlea tali. J Foot Ankle Surg 52(4):426–431
2. Cenni F, Leardini A, Cheli A (2012) Position of the prosthesis components in total ankle replacement and the effect on motion at the replaced joint. Int Orthop 36(3):571–578
3. Damani Y (2011) Talus morphology and its functional implications on the ankle joint. Thesis M.S., Mechanical Engineering and Mechanics, Drexel University
4. Schwartz AM, Runge WO et al (2020) Fractures of the talus: current concepts. Foot & Ankle Orthop. https://doi.org/10.1177/2473011419900766
5. Zhang K et al (2019) Measurement of proximal tibial morphology in northeast Chinese population based on three-dimensional reconstruction computer tomography. Medicine. https://doi.org/10.1097/MD.0000000000017508
6. Stagni R, Leardini A, Catani F, Cappello A (2004) A new semi-automated measurement technique based on X-ray pictures for ankle morphometry. J Biomech 37(7):1113–1118
7. Reis HC, Bayram B, Seker DZ (2016) A semiautomatic segmentation approach to biometric measurement of the talus bone of sedentary women and ballerinas using CT images. Asian Biomed 10(5):455–459

Chapter 5
Finite Element Method Study: The Effect of Insertion Torque and Angle on Pedicle Screw Loosening

Rosdi Daud, Muhammad Syakir Isa, Mas Ayu Hassan, and Arman Shah Abdullah

Abstract Pedicle screws are widely used for the treatment of spinal instability by spine fusion. This treatment can be performed at any level in the spine (cervical, thoracic, or lumbar) and prevents any movement between the fused vertebrae. However, screw loosening is a major problem of spine fusion, contributing to delayed patient recovery. This appears to be a minor problem for fixation and fusion of healthy, non-osteoporotic bone. Screw loosening happens when insertion torque and angle are not suitable for the pedicle screw which affects the screw pullout strength. Based on the optimum torque and angle in finite element analysis, loosening of screw in spine fusion can be minimized by increasing the screw pullout strength. The highest pullout strength can minor the loosening of the screw. Four insertion angles of the pedicle screw were used which are 0°, 10°, 20°, and 30°. Besides that, the effect of insertion torque on pullout strength can be seen on the equivalent von Mises stress value when applied three values of moment to pedicle screw-bone with constant insertion angle. The lowest stress for pedicle screw can give a better fixation with bone and thus can increase the pullout strength. We found that the insertion angle of 10° gives higher pullout strength of pedicle screw-synthetic bone which indirectly will minimizing the effect of the screw loosening from bone. Besides that, insertion

R. Daud (✉) · M. A. Hassan
Automotive Engineering Centre, Universiti Malaysia Pahang, 26600 Pekan, Pahang, Malaysia
e-mail: rosdidaud@ump.edu.my

M. A. Hassan
e-mail: masszee@ump.edu.my

R. Daud · M. S. Isa
Faculty of Mechanical and Automotive Engineering Technology, Universiti Malaysia Pahang, 26600 Pekan, Pahang, Malaysia

M. A. Hassan
Faculty of Manufacturing and Mechatronic Engineering Technology, Universiti Malaysia Pahang, 26600 Pekan, Pahang, Malaysia

A. S. Abdullah
Faculty of Technical and Vocational, Universiti Pendidikan Sultan Idris, 35900 Kuala Kubu Bharu, Perak, Malaysia
e-mail: armanshah@fptv.upsi.edu.my

A. Ismail et al. (eds.), *Advanced Materials and Engineering Technologies*,
Advanced Structured Materials 162,
https://doi.org/10.1007/978-3-030-92964-0_5

torque with 1200 Nm gives a lower equivalent von Mises stress which causes a lower effect on fixation between screw and bone and thus has potential in minimizing the loosening to pedicle screw.

Keywords Pedicle screw · Finite element analysis · Screw loosening · Insertion torque and angle · Optimization

5.1 Introduction

Pedicle screw-rod constructs are commonly used for the treatment of fractures, tumors, and degenerative disease in the spine. When osteoporosis is present, gaining adequate screw purchase during these procedures can be challenging [1]. The weakened screw-bone interface can be explained by a direct correlation between bone mineral density and pedicle screw fixation [1]. Fusion is therefore more difficult to achieve than in healthy bone [2] causing a longer duration of loading on the instrumentation, [3] which increases the risk of construct failure. Position and numbers of radial holes, cement injection technique were also investigated to increase the pullout performance especially for osteoporotic vertebrae [4, 5]. Screw pullout is the standard method of testing screw fixation within bone. In the spine it gives information on the stability of a pedicle screw-rod construct via screw purchase. Pedicle screw failure, however, typically does not occur with simple pullout. Instead, loosening, due to fatigue, weakens the screw-bone interface leading to pullout. More attention is being paid to understanding the role of fatigue in pedicle screw failure [6]. Achieving optimal fixation of the screw-bone interface is vital to clinical outcome.

To determine the optimal fixation technique for pedicle screws, numerous studies have been undertaken to evaluate how specific factors affect loosening and pullout strength. Among others, researchers and physicians have evaluated insertion techniques, screw characteristics, augmentation, bone quality, and morphometry. Generally, a correlation between pullout strength and bone mineral density (BMD) has been noted by many researchers [6, 7]. This correlation separates bone quality into healthy and osteoporotic bone, and each type has unique needs that should be addressed in screw design and insertion technique. Trajectory, insertion technique, and screw design have a large effect on screw purchase and may provide a good alternative technique when a surgical limitation is present [1, 4, 6]. The extrapedicular technique, for example, may be a viable option for patients with morphometry that makes adequate screw purchase a challenge.

There is some debate whether an extrapedicular screw trajectory decreases fixation strength [8] or whether it remains the same [9]. The differences may be due to sample size, in which a small sample size increases the likelihood of a type II error. Insertion technique can be altered to optimize fixation strength. Lateral misdirection may increase the likelihood of fracturing the wall of the pedicle and therefore should be avoided. Single screw and pullout strength tests were carried out according to American Society for Testing and Materials F 543-07 on foam models to test the effect of insertion angle using polyaxial pedicle screws. Rigid polyurethane foams are widely used as a substitute for cadaver spinal bone because of their consistent and homogeneous structural properties. Grade 10 (160 kg/m^3) represents osteoporotic a block dimensions of 120 mm × 60 mm × 40 mm. For good results in osteoporotic bone, there is a benefit of using insertion angle in 10° to minimize screw loosening [6].

Many surgeons prefer a pedicle screw with high insertion torque because it gives good tactile feedback of bony purchase. By contrast, although there is a strong correlation between insertion torque and BMD, screw loosening is not objectively predicted by insertion torque in the clinical setting. But some journals reported that loosened screws are associated with decreased pullout strength and extraction or high torque [1, 4, 10, 11]. Therefore, the aim of this paper is to investigate the effect of insertion torque and angle on pedicle screw loosening by using the finite element method.

5.2 Methodology

Developing a screw with accurate thread style is crucial in achieving best results among the shape because the most popular size, shape, and pitch can vary with the supported specific anatomy. The pedicle screw was constructed using SolidWorks software with a diameter of 6.0 mm and a length of 45 mm (Fig. 5.1). This software was capable of performing 2D and 3D measurements of the bone's model, and these capabilities were used in order to make morphological measurements of individual talus bones as well as combination of bones as shown in Fig. 5.1.

The synthetic bone is also constructed using the SolidWorks software. The material of the pedicle screw model was considered elastic and isotropic which requires two parameters to describe the properties which are E (elastic modulus) and ν (Poisson's ratio). There were two types of materials used in this project which are titanium alloy (grade 23) for the pedicle screw and polyurethane for the synthetic bone. Table 5.1 shows the properties of the titanium alloy and polyurethane.

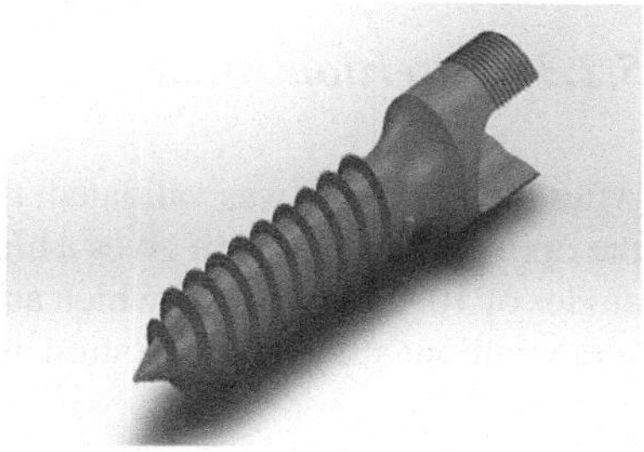

Fig. 5.1 3D model of cannulated pedicle screw

Table 5.1 Properties of titanium alloy (Grade 23) and polyurethane

Properties	Titanium alloy	Polyurethane
Young's modulus, E (GPa)	104.5	0.78
Tensile strength, σ_x (MPa)	860.0	33.0
Yield stress, σ_y (MPa)	820.0	133.0
Poisson's ratios, ν	0.3	0.3

5.2.1 Finite Element Method

A static structural load analysis performed using the ANSYS software. A static structural analysis determines the displacements and stresses in structures or components caused by loads that do not induce significant inertia and damping effects. The finite element analysis began with the 3D-model of implant that was prepared in the SolidWorks software and exported to the ANSYS software. Then the model was meshed using tetrahedral elements for the simulation. The element size has been taken 1 mm on the pedicle screw. In many regions tetrahedral meshes were generated and provided a bonded interface conditions.

5.2.2 Finite Element Model Validation

Validation method is a technique for assessing either the results of a structural analysis (model) are correct or not. It is mainly used in settings where the goal is prediction, and one wants to estimate how accurately a predictive model will perform in practice. FE models must be validated and updated to assess the quality of FE models and increase confidence in the results. The result of FEA (pullout strength) was compared to the experimental result done by previous researchers [6]. They did pullout test on the foam with the inserted screw using a BiSS Nano-25 universal testing machine. A tensile load of 5 mm/min was applied to the test specimen. The FE model is considered valid if the gap of compared results were less than 15%.

5.2.3 Insertion Angle

After the FE model was validated, the new insertion angles were assigned in getting the optimum angle which gives a higher pullout strength. Figure 5.2 shows insertion angles of 10°, 20°, and 30° which are based on the schematic representative of single screw pullout study configuration as reported by Venkatesh et al. [6].

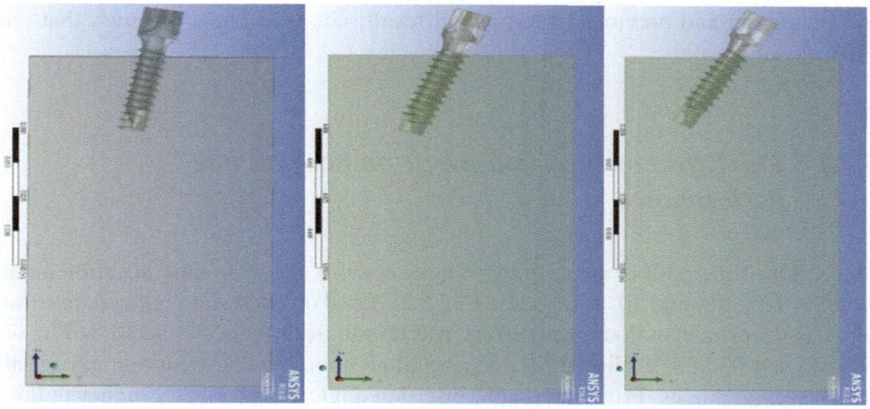

(a) (b) (c)

Fig. 5.2 Insertion angles **a** 10°, **b** 20°, **c** 30°

5.2.4 Insertion Torque

Some of the published journals prove that loosened screw was affected by decreased pullout strength and extraction or high torque. This means that low torque gives less loosening of the screw. The torque was represented by the moment of 1200, 1400, and 1600 Nm and applied in the FE to get the equivalent von Mises stress (EVM) value on the screw-bone interface. The EVM value can represent the pullout strength value since both are significantly correlated.

5.3 Result and Discussion

FEA revealed that the reaction force obtained to pull the pedicle screw for the insertion angle 0° was a similar value with the result obtained by previous researcher [6]. Table 5.2 shows the summary of the validation result with the percentage error between

Table 5.2 FE Model validation

	Experimental results by [6]	Result by FEM
Tensile load (mm/min)	5	5
Reaction force (N)	690	733.96
Present of error (%)	6.0%	

our prediction and previous experimental result, i.e., 6%. Thus, it proves that our FEA model is valid.

5.3.1 The Effect of Insertion Angle on Screw-Bone Performance

The result of the pullout strength of pedicle screw-bone using new insertion angle which is 10°, 20°, and 30° is shown in Fig. 5.3. Based on the result obtained, reaction force (can represent pullout strength) for insertion angle of 10°, 20°, and 30° is 776.87, 609.55, and 335.60 N. Figure 5.4 shows pullout strength distribution graph for all four insertion angles. From the graph, insertion angle of 10° gives a higher pullout strength than other angles.

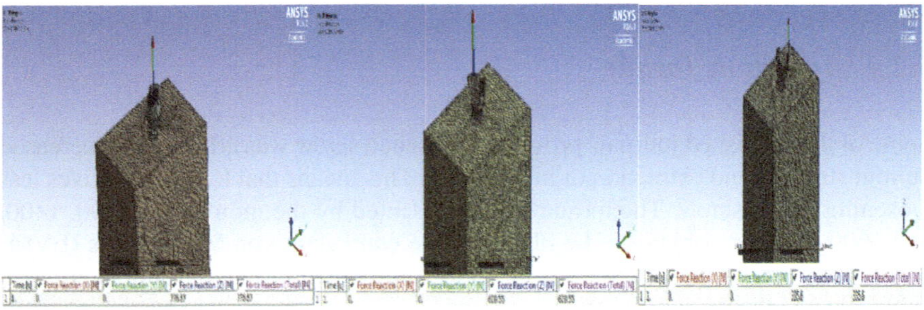

(a) (b) (c)

Fig. 5.3 **a** Reaction force for 10° insertion angle, **b** 20°, and **c** 30°

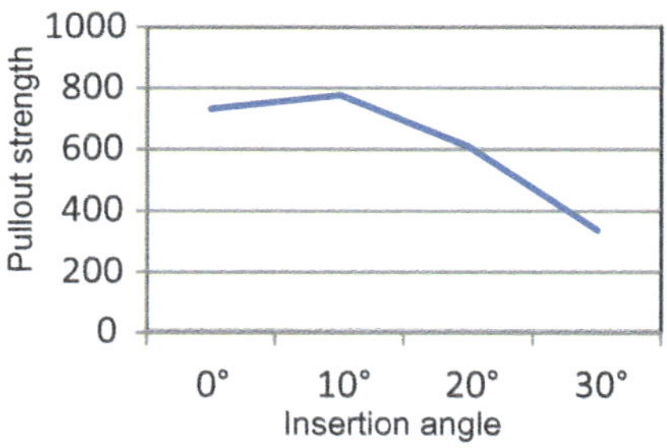

Fig. 5.4 Graph of reaction force (pullout strength) versus insertion angle

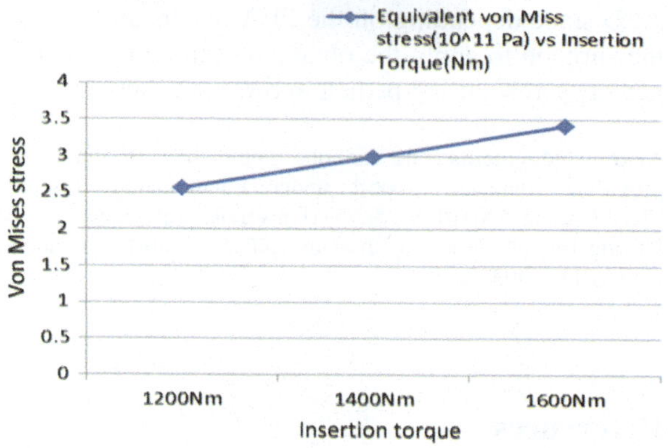

Fig. 5.5 Graph of von Mises stress versus insertion torque

5.3.2 The Effect of Insertion Torque on Screw-Bone Performance

The FEA results (von mises stress) distributed on pedicle screw due to three values of torque which are 1200, 1400, and 1600 Nm are shown in Fig. 5.5. The effect of insertion angle and torque for screw loosening can be predicted using the finite element method due to the fact that the validation error is only 6%. Thus, Fig. 5.4 shows that an insertion angle of 10° gives a higher pullout strength of pedicle screw-synthetic bone which indirectly will minimize the effect of the screw loosening from the bone. The effect of torque on stress value of pedicle screw are shown in Fig. 5.5 which the stress values of pedicle screw seem proportional with the values of torque. When a stress for pedicle screw is lower, the fixation of screw to bone is really fixed and gives minor loosening to pedicle screw.

5.4 Conclusion

In this paper, the investigation on the effect of insertion angle and torque on pedicle screw is done by doing finite element analysis. The FEA was done for four insertion angles of pedicle screw which are 0°, 10°, 20°, and 30° and also by applying three different moment of 1200, 1400, and 1600 Nm in the numerical simulation. Based on the FE results obtained, it is proven that the investigation on the effect of insertion angle and torque on pedicle screw can be done using the finite element method which also can be considered valid due to the gap of compared results between our FEA result and experimental result by previous researcher were less than 15%. Thus the optimum insertion torque and angle obtained also can be considered valid in minimizing pedicle screw loosening. The optimum value for insertion torque is 1200 Nm while 100 is the optimum insertion angle value. However, fabrication and further testing still need to be conducted in order to evaluate this optimum design of

AFO since depending on the FEA results alone it cannot give high confidence for the surgeon to apply the obtained optimum value of insertion angle and torque on pedicle screw during pedicle screw-spine surgery.

Acknowledgements The authors would like to thank Ministry of Higher Education for providing financial support under Fundamental Research Grant Scheme (FRGS) No. FRGS/1/2018/TK03/UMP/02/5 (University reference RDU190114) and Universiti Malaysia Pahang for laboratory facilities as well as additional financial support under Internal Research Grant RDU190327.

References

1. Balikci T, Kiyak G, Heydar AM et al (2018) Mid-length pedicle screws in posterior instrumentation of scoliosis. Asian Spine J 12(1):3–11
2. White KK, Oka R, Mahar AT, Lowry A et al (2008) Biomechanical comparison of different anchors (foundations) for the pediatric dual growing rod technique. Spine J 8(6):933–939
3. Furderer S, Scholten N, Coenen O et al (2011) In-vitro comparison of the pullout strength of 3 different thoracic screw fixation techniques. J Spinal Disord Tech 24(1):E6-10
4. Zakaria FA, Daud R, Mas-Ayu H et al (2017) The Effect of position and different size of radial hole on performance of cannulated pedicle screw. MATEC Web Conf. https://doi.org/10.1051/matecconf/201710813001
5. Varghese V, Krishnan V, Saravana KG (2018) Testing pullout strength of pedicle screw using synthetic bone models: is a bilayer foam model a better representation of vertebra? Asian Spine J 12(3):398–406
6. Krishnan V, Varghese V, Saravana KG (2016) A finite element analysis based sensitivity studies on pull out strength of pedicle screw in synthetic osteoporotic bone models. IEEE EMBS Conference on Biomedical Engineering and Science (IECBES 2016), pp 382–387
7. Cho W, Cho SK, Wu C (2010) The biomechanics of pedicle screw-based instrumentation. J Bone Joint Surg Br 92(8):1061–1065
8. Wang H, Wang H, Sribastav SS et al (2015) Comparison of pullout strength of the thoracic pedicle screw between intrapedicular and extrapedicular technique: a meta-analysis and literature review. Int J Clin Exp Med 8(12):22237–22245
9. Laura B E (2013) Optimization of pedicle screw depth in the lumbar spine: biomechanical characterization of screw stability and pullout strength. Thesis M.S., Bioengineering, The University of Toledo
10. Christodoulou E, Chinthakunta S, Reddy D et al (2015) Axial pullout strength comparison of different screw designs: fenestrated screw, dual outer diameter screw and standard pedicle screw. Scoliosis. https://doi.org/10.1186/s13013-015-0039-6
11. Kang SH, Kim KT et al (2011) A case of pedicle screw loosening treated by modified transpedicular screw augmentation with polymethylmethacrylate. J Korean Neurosurg Soc 49(1):75–78

Chapter 6
The Effect of PLA/HA Coating Thickness on Crack Formation and Corrosion Performance

Mas Ayu Hassan, Zubaidah Zamri, Rosdi Daud, Norizah Redzuan, and Izman Sudin

Abstract Surface modification of metallic implants is often required to facilitate positive interaction between the implant and the surrounding hard tissue. In this study, a polymer-ceramic composite coating of polylactic acid/hydroxyapatite (PLA/HA) was successfully deposited on a Co–Cr–Mo alloy by the dip coating method in chloroform suspension at room temperature. The effect of various PLA/HA dipping layers was studied and the dip coating process parameters were optimized in order to obtain a homogeneous, crack free, densely packed and adhesive coating. It is found that PLA/HA-coated substrate with 3 dipping layers were denser and less crack sensitive compared to 6 dipping layers. Although it is hypothesized that a coarser coated surface helps to facilitate ingrowth of osseous tissue in human body, but current findings show opposite manners due to the fact that a higher corrosion rate was obtained. The coated substrate with 6 dipping layers also were found more profound to micro-cracks and delamination with a lower microhardness value compared to coated substrate with 3 dipping layers.

Keywords Co–Cr–Mo · PLA/HA coating · Biomaterial · Surface morphology · Dip coating

M. A. Hassan (✉) · Z. Zamri
Faculty of Manufacturing and Mechatronic Engineering Technology, Universiti Malaysia Pahang, 26600 Pekan, Pahang, Malaysia
e-mail: masszee@ump.edu.my

R. Daud
Faculty of Mechanical and Automotive Engineering Technology, Universiti Malaysia Pahang, 26600 Pekan, Pahang, Malaysia
e-mail: rosdidaud@ump.edu.my

N. Redzuan · I. Sudin
Department of Materials, Manufacturing and Industrial Engineering, School of Mechanical Engineering, Universiti Teknologi Malaysia, 81310 Skudai, Johor, Malaysia
e-mail: norizah@utm.my

I. Sudin
e-mail: izman@utm.my

A. Ismail et al. (eds.), *Advanced Materials and Engineering Technologies*, Advanced Structured Materials 162,
https://doi.org/10.1007/978-3-030-92964-0_6

6.1 Introduction

Co–Cr–Mo alloys are widely used for dental and orthopaedic prostheses owing to their favourable mechanical properties, low cytotoxicity, good corrosion resistance and biocompatibility [1, 2]. One of the main requirements for implant longevity is high corrosion resistance. The formation of an oxide layer on Co–Cr–Mo surface occurs rapidly upon contact with atmosphere. This oxide layer is considered instrumental in decreasing the dissolution rate of toxic metal ions in the biological environment [3, 4]. Moreover, the presence of an oxide layer also enhances the bioactivity of Co–Cr–Mo in body fluid as it provides sites for the deposition of calcium and phosphate compounds via an ionic exchange process with apatite from the bone tissue [5]. However, the oxide layer does not promote the formation of a hydroxyapatite layer, which is integral for the rapid fixation of an implant.

Several studies have reported excessive levels of Co, Cr and Mo ions in the vicinity of the implanted Co–Cr–Mo implants. The released of these ions will later combine with biomolecules and result in adverse biological reaction [2]. The coating of bioactive ceramic on metal implants is an effective approach to solve corrosion related problems and improve the biocompatibility of metallic implants. Hydroxyapatite (HA) is a synthetic material, which promotes osteointegration and thus accelerates tissue fixation at the implant surface during the early stages of implantation [6]. Dip coating is a promising technique which consists of three stages; dipping, withdrawing and drying (Fig. 6.1) [7]. This technique offers numerous advantages such as inexpensive setup, process simplicity, uniformity of deposition, low processing temperature and the ability to coat complex shapes and patterns. Additionally, the coating amount and thickness can be controlled by adjusting the concentration of suspension, the number of dips and varying the withdrawal speed. Heat treatment of

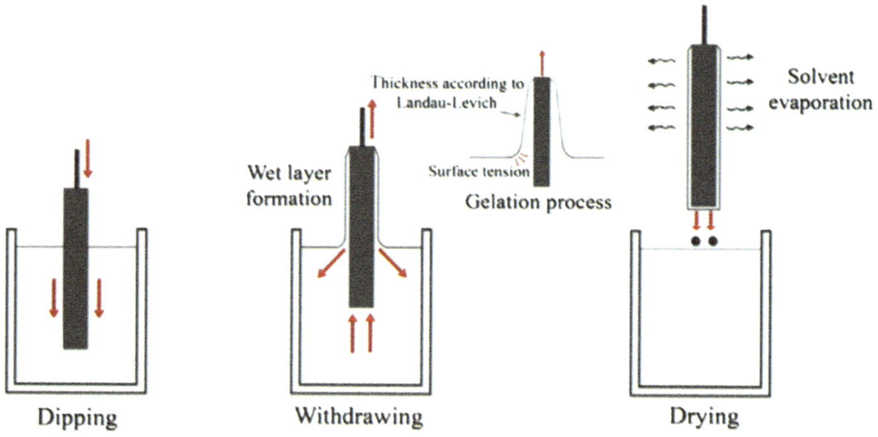

Fig. 6.1 Schematic diagram of dip coating process

coated substrate is often required to densify the coating layer, to increase the coating-implant bonding and eliminate porosity. However, too high a sintering temperature can lead in degradation of the metallic substrate and also phase transformation of HA into a non-crystalline phase which increases the dissolution rate in body fluid [8, 9]. Furthermore, thermal stresses originating from the difference in thermal coefficient between the ceramic coating and metallic implant result in the formation of micro-cracks and delamination of the coating from the substrate. Polylactic acid (PLA) is a semi-crystalline aliphatic polymer with good biocompatibility, sustained biodegradability and remarkable mechanical properties [1, 10]. With regards to aforementioned evidences, therefore this study was presented to report the deposition of highly compact polymer-ceramic composite coating of PLA/HA onto Co–Cr–Mo alloy, which does not require post-depositional heat treatment for densification. The effect of the PLA/HA coating thickness on the corrosion rate was also studied using the potentiodynamic polarization test in circulated simulated body fluid.

6.2 Methodology

In the present study, a cobalt-chromium-molybdenum alloy (Co–Cr–Mo) bar with ASTM standard of F1537 and chemical compositions of (in wt%): C: 0.24; Cr: 29.6; Mo: 6.5; Si: 0.7; Ni: 0.1; Fe: 0.12; Mn: 0.7; N: 0.16 and Co: balance was used. The alloy bar was cut using a precision cutter into disc shape with dimension of 14 mm diameter and 2 mm thickness (Buehler, Isomet 4000). The variation in surface roughness was reduced using #500 grit SiC paper in wet grinding method. The surface roughness of the alloy discs was then measured using a surface profilometer (Mitutoyo SJ-301) and 0.1 ± 0.02 μm surface roughness was achieved. Then PLA/HA coating deposition process was carried out using a HTWL-01 Desktop Dip Coater (MT1 Cooperation, USA) at room temperature and under controlled humidity. The substrates were assigned for dipping and withdrawal steps in the prepared PLA/HA slurry with fixed speed of 200 mm/min. This step was repeated for 3 and 6 dipping layers with an interval of 1 min for drying step in every cycle. This step is mainly used to obtain a desired coating thickness of PLA/HA on the substrates. The substrates were then left to dry in the vacuum furnace at a temperature of 50 °C for 1 h to evaporate the residual solvent and densify the coating layer. This step is crucial to improve the bonding strength of PLA/HA coating onto the surface of the Co–Cr–Mo substrate. Substrates that were coated with 3 layers and 6 layers of PLA/HA coating will be recognized as PLA3 and PLA6, respectively. Single Vickers hardness indentation method was performed at five different locations on the coated PLA3 and PLA6 substrate to obtain the average hardness of the coated layer. A load of 5 N with dwell of 10 s was applied in this test. Meanwhile, for the corrosion testing, potentiodynamic polarization diagrams were determined using a classical three-electrode cell with specimen as a working electrode with an exposed area of about 0.402 cm^2, a saturated calomel electrode (SCE) as a reference electrode and graphite as a counter electrode by employing Princeton Applied Research, US potentiostat (VersaSTAT

3-300). The coated PLA3 and PLA6 substrates were exposed to the simulated body fluid (Hank's solution) at circulated condition for 30 min prior to generation polarization. Megastudio Software measured the impedance in the frequency range of 5 MHz to 100 kHz, using a perturbation amplitude of 20 mV around the corrosion potential. The corrosion potential (E_{corr}) and corrosion current density (i_{corr}) were calculated from the intersection of the cathodic and anodic Tafel curves using the Tafel extrapolation method. The test was repeated for three times for each of sample condition.

6.3 Results and Discussion

Representation of the FESEM images of the PLA/HA coating on Co–Cr–Mo alloy is shown in Fig. 6.1. The surface morphology and cross-section of PLA3 substrate clearly shows that there were no micro-cracks or undulation noticed in the substrate (Fig. 6.1a) and the coating surface looks more uniform as compared to PLA6 substrate (Fig. 6.1d). Further observations on the cross-section of PLA3 substrate and PLA6 substrate showed that the thickness of PLA/HA coating formed at 3 dipping layers and 6 dipping layers was 49.14 μm and 76.39 μm, respectively. Based on Fig. 6.1b, the PLA/HA coating with 3 layers is denser which is indicating the formation of a compact composite material on the substrate. This compact structure is proven to be highly essential to provide better barrier to prevent harmful ions such as Co and Cr, migrating from implant to the host tissues [11, 12]. In Fig. 6.1c, the micro-cracks were evident on 6 dipping layers of PLA/HA coating and widely spread on the surface. This phenomenon occurs due to cohesive strength between PLA and HA which becomes weaker as thicker coatings were applied. It is believed that failures happened at the PLA-HA interfaces due to weaker chemical bonding. Similar findings were also reported by other researchers [1]. Closed-up image on micro-cracks in PLA6 substrate was profound towards the metal substrate (Fig. 6.2).

Table 6.1 shows the hardness measured on the coated substrates surface after the dip coating process. Based on the data obtained, the hardness of PLA3 substrates is about 2 times higher compared to PLA6 substrate which is 8.02 N and 4.18 N, respectively. This happened due to PLA6 substrate has massive surface cracks and pores on the coated surface. The polarization test was employed to study the corrosion resistance performance of two different PLA/HA dipping layer on Co–Cr–Mo alloy surface. The related polarization curves are plotted in Fig. 6.4. While the values of the corrosion current density (i_{corr}) and corrosion potential (E_{corr}) obtained from the curves analyses were summarized in Table 6.2. It is observed that the E_{corr} for PLA6 substrate (−221.404 mV) is much higher than PLA3 substrate (−110.169 mV). It is also observed that Co-Cr-Mo substrate with 3 dipping layers of PLA/HA has lower corrosion rate compared to substrate with 6 dipping layers (0.015743 mmpy vs. 0.049434 mmpy). This phenomenon happened due to the denser and uniform PLA/HA coating on the PLA3 surface as mentioned in earlier discussions. The dense

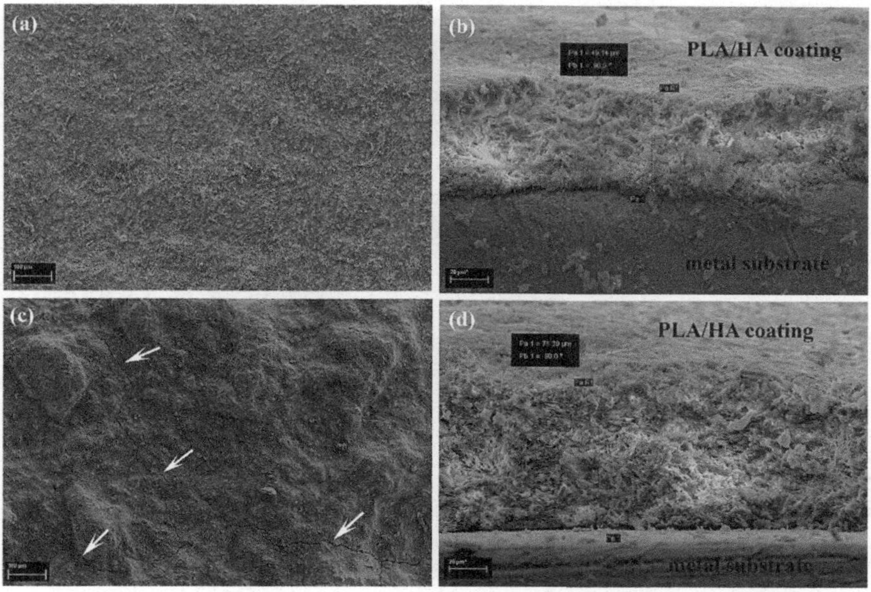

Fig. 6.2 Surface morphology and cross-section views of PLA/HA coating on Co–Cr–Mo alloy, **a** and **b** for PLA3 substrate, **c** and **d** for PLA6 substrate

Table 6.1 Vickers hardness measurement on PLA/HA-coated surface of Co–Cr–Mo substrate

Vickers hardness (HV) measurement			
Substrate	Applied load (N)	Average HV values	PLA/HA coating thickness (μm)
PLA3	5	8.02	49.14
PLA6	5	4.18	76.39

coating is believed to be able to act as a barrier to prevent chemical attack and thus protect the alloy from corrosion (Fig. 6.3).

Table 6.2 Tafel results for PLA3 and PLA6 substrates in Hank's solution

Dipping sample	E_{corr} (mV)	i_{corr} (nA)	Corrosion rate (mm/year)
3 dipping layers	−110.169	8.897	0.015743
6 dipping layers	−221.404	92.918	0.049434

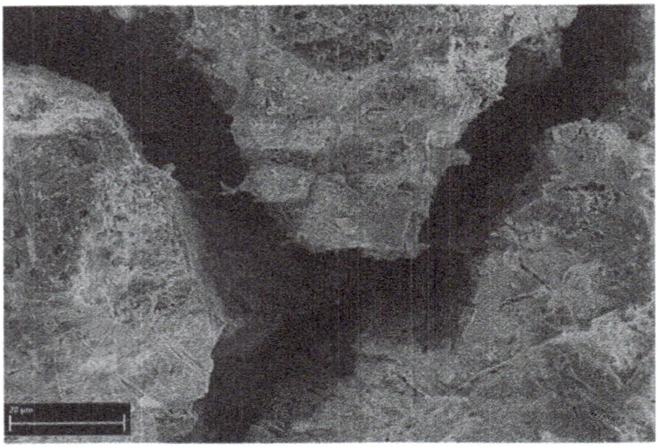

Fig. 6.3 Closed-up images of micro-cracks in PLA6 substrate

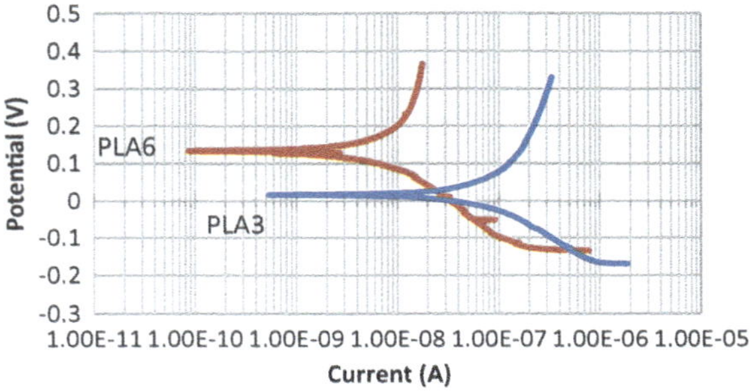

Fig. 6.4 Corrosion rate for PLA3 and PLA6 substrates in Hank's solution

6.4 Conclusion

The thickness of PLA/HA coating has a direct influence on the formation of micro-cracks as well as on the corrosion performance on the Co–Cr–Mo alloy. Thinner and denser PLA/HA coating on PLA3 substrate without micro-cracks is obtained on Co–Cr–Mo alloy which is able to reduce the corrosion rate (0.015743 mmpy vs. 0.049434 mmpy). Thicker and weaker cohesive strength on PLA6 substrate that consists with massive micro-cracks promotes higher corrosion rate. This phenomenon happened due to the simulated body fluid managed to penetrate into the bulk material and caused excessive release of metal ions.

Acknowledgements The authors would like to thank the Ministry of Higher Education for providing financial support under Fundamental Research Grant Scheme (FRGS) No. FRGS/1/2018/TK03/UMP/03/1 (University reference RDU190130) and Universiti Malaysia Pahang for laboratory facilities as well as additional financial support under Internal Research Grant RDU192309. Special thanks to Universiti Teknologi Malaysia (UTM) Skudai for providing additional financial assistance under CRG 26.0 grant programme.

References

1. Liu S et al (2020) Current applications of poly (lactic acid) composites in tissue engineering and drug delivery. Compos B Eng 199:108238
2. Mas-Ayu H et al (2019) Improving biocompatibility of cobalt based alloy using chemical etching and mechanical treatment. Materwiss Werksttech 50:254–259
3. Suci DAW et al (2020) Implant surface modification strategies through antibacterial and bioactive components. In: Biopolymer-based formulations. Elsevier, pp 647–673
4. Mas-Ayu H et al (2018) In-vitro biocompatibility study of hydroxyapatite coated on Co-Cr-Mo with oxide interlayer. J Teknol 80(1):35–42
5. Zuchuat J et al (2020) CoCrMo alloy as biomaterial for bone reconstruction in oral and maxillofacial surgery: a scoping review. J Oral Res 9:212–219
6. Harun WSW et al (2018) A comprehensive review of hydroxyapatite-based coatings adhesion on metallic biomaterials. Ceram Int 44(2):1250–1268
7. Mohd YMF et al (2014) Dipcoating of poly (ε-caprolactone)/hydroxyapatite composite coating on Ti_6Al_4V for enhanced corrosion protection. Surf Coat Tech 245:102–107
8. Yang C, Guo YK, Zhang ML (2010) Thermal decomposition and mechanical properties of hydroxyapatite ceramic. Trans Nonferr Metal Soc 20(2):254–258
9. Bandyopadhyay A et al (2019) Additively manufactured calcium phosphate reinforced CoCrMo alloy: bio-tribological and biocompatibility evaluation for load-bearing implants. Addit Manuf 28:312–324
10. Salerno A et al (2014) Macroporous and nanometre scale fibrous PLA and PLA–HA composite scaffolds fabricated by a bio safe strategy. RSC Adv 4(106):61491–61502
11. Mas-Ayu H et al (2014) Influence of carbon concentrations in reducing Co and Cr ions release in cobalt-based implant: a preliminary report. Adv Mat Res 62–466
12. Wang Q, Zhang L, Dong J (2010) Effects of plasma nitriding on microstructure and tribological properties of CoCrMo alloy implant materials. J Bionic Eng 7(4):337–344

Chapter 7
Rudder Stock Vibration in Different Materials on Boat Hulls

Md Redzuan Zoolfakar, Nur Ashikin Ayub, and Shareen Adlina Shamsuddin

Abstract Vibration is known as one of the energy losses experienced by a ship propulsion system. Excessive vibration might occur resulting in the negative effect to the system leading responsible bodies such as the International Maritime Organization (IMO) have set a certain guidelines and requirements in reducing vibration. The rudder stock is one of the mechanisms that experiences vibration due to the energy loss in the propulsion system. This paper will discuss on the vibration level generated by the rudder stock in different materials on the boat hull. There are some parameters that have been taken into account in this research such as speed of motor, size of propeller, rudder stock position from the center of the propeller, and the 30° angle of turn for both port and starboard side. Different parameters should generate different vibration effects that will be transmitted to the rudder stock.

Keywords Vibration · Rudder stock · Hull vibration · Propeller excitation · Energy loss

7.1 Introduction

A ship propulsion system can be defined as the mechanism or system used to drive a ship or boat across the open water [1]. The ship propulsion system may be electric or mechanical. A mechanical propulsion system involves the use of a diesel engine to drive the propelling shaft of the ship, whereas an electrical propulsion is a system consisting of a prime mover such as a steam turbine, diesel engine, etc., and a generator, electrical motor, and related equipment such as measuring instruments, converters that are used to drive the system [2]. The entire energy provided by the

M. R. Zoolfakar (✉) · N. A. Ayub · S. A. Shamsuddin
Marine and Electrical Engineering Technology, Universiti Kuala Lumpur Malaysian Institute of Marine Engineering Technology, Lumut, Malaysia
e-mail: redzuan@unikl.edu.my

S. A. Shamsuddin
e-mail: shareen@unikl.edu.my

A. Ismail et al. (eds.), *Advanced Materials and Engineering Technologies*, Advanced Structured Materials 162,
https://doi.org/10.1007/978-3-030-92964-0_7

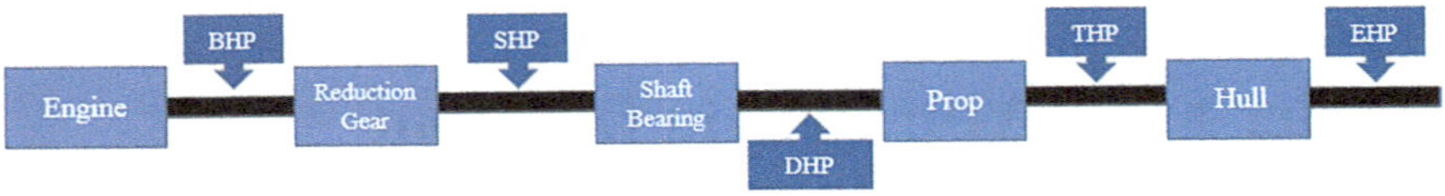

Fig. 7.1 Simplified diagram of a ship's drive train

main engine is essentially transferred to a simple rotary movement of the mechanical mechanism of the propeller shafting in an ideal propulsion system [3].

According to [4], a drive train of a ship propulsion system consists of five main components: (1) BHP-brake horsepower, (2) SHP- shaft horsepower, (3) DHP- delivered horsepower, (4) THP- thrust horsepower, and (5) EHP-effective horsepower. Figure 7.1 shows a simplified diagram of a ship's drive train. The biggest energy losses in the ship propulsion system are the thermodynamic and mechanical engine losses. These cause the loss of approximately 60 percent of the fuel energy before it becomes rotational power at the engine output or the brake horsepower [4]. As a result, additional motions, such as longitudinal, transverse, and/or torsional vibrations, are produced in response to the propeller shafting simple rotational motion. These extra motions are inefficient from the perspective of energy transfer and delivery in the whole propulsion system. These motions are responsible for the dissipation of kinetic energy in rotational motion as well as the deposition of internal energy in structural materials. Fatigue loss occurs as these energies reach critical limits, and the mechanism is characterized by residual energy processes of thermal and vibroacoustic nature [3].

Naval architects and marine engineers have long been fascinated by vibration because of its negative impact on passenger and crew comfort as well as the ship's structure [5]. Hull vibration, transmitted in the form of structural noise in all ship spaces, is induced by the action of the machinery and equipment, but the propeller is usually the main source of vibration [6, 7]. The rotation of the propeller induces differential pressure in the surrounding water, which are conveyed as hydroacoustic waves to the hull plating above the propeller and excites it [7]. Excessive vibrations can contribute to fatigue damage to the vessel's structure and mechanism sections or failure of the onboard machinery [8].

The concern about the complication between the hull and the propulsion system and the transmission of vibration via couplings and bearings are seen as something serious [9]. However, the propeller operation transmitted the vibration to the hull structure of the vessel in two different ways (1) the vibration is transferred or distributed to the hull through the shaft line and the bearing (thrust and journal bearing) and (2) the vibration oscillation is transferred directly to the hull by the water or through the vibration of the rudder stock [10].

The vibration level generates by the different rudder stock materials is observed, and the results will be measured and compare based on the parameters used in the experiment.

7.2 Methodology

Generally, the study required the researcher to develop an experimental rig to run the experiment. This rig is important to help the researcher to collect data from the testing before analyzing it. A sensitive analysis method is used in the tabulation of the data in this project. The methodology used for this study is arranged accordingly so that it can produce a result within the time given. It also defines the project's series flow and the design proposed to achieve the desired outcome. It contains a flow chart that shows the data collection procedure from project parameters as well as data analysis processes. The flowchart in Fig. 7.2 indicates all the stages that need to be done by the researcher to obtain the results that are parallel with the aim and objectives of the project.

The experiment is conducted by following all the procedures starting from the experiment planning, project implementation, and the data collection process. The data will be recorded by adhering to all the variables and translated into a graph form for analysis purposes.

7.2.1 Experimental Planning

Experimental planning in this project shows all the parameters that need to be measured leading to the development of the experiment. This includes all the details of the parameters and the materials and equipment required in the project. With this experiment planning, the researcher will get a clear picture of how to conduct the experiment.

There are mainly five parameters which have been considered to conduct this experiment. First, the experiment in this project will be conducted by using three different types of rudder stock. The materials selected for the rudder stock are stainless steel, aluminum, and carbon composite.

Second, different diameters of the 3-blade propeller will be used in this experiment. An even increase of diameter from 40, 44, and 48 mm will be used. The aim of using the different diameters of propeller is to observe the different vibrations induced from the trust produced by the propeller toward the rudder stock.

Third, the experiment will be carried out by running the motor at three different revolutions which are 900, 1200, and 1500 rpm. The next variable is regarding to the position of the rudder stock. The vibration of the rudder stock will be measured at different locations. The location is calculated by the percentage of the propeller diameter (%Dp). The location will be set at 59% Dp, 65% Dp, and 71% Dp.

Lastly, the vibration of the rudder stock will be measured at different turning angles. The angle will be measured from 0°, 10°, 20°, and 30° for both port and starboard side.

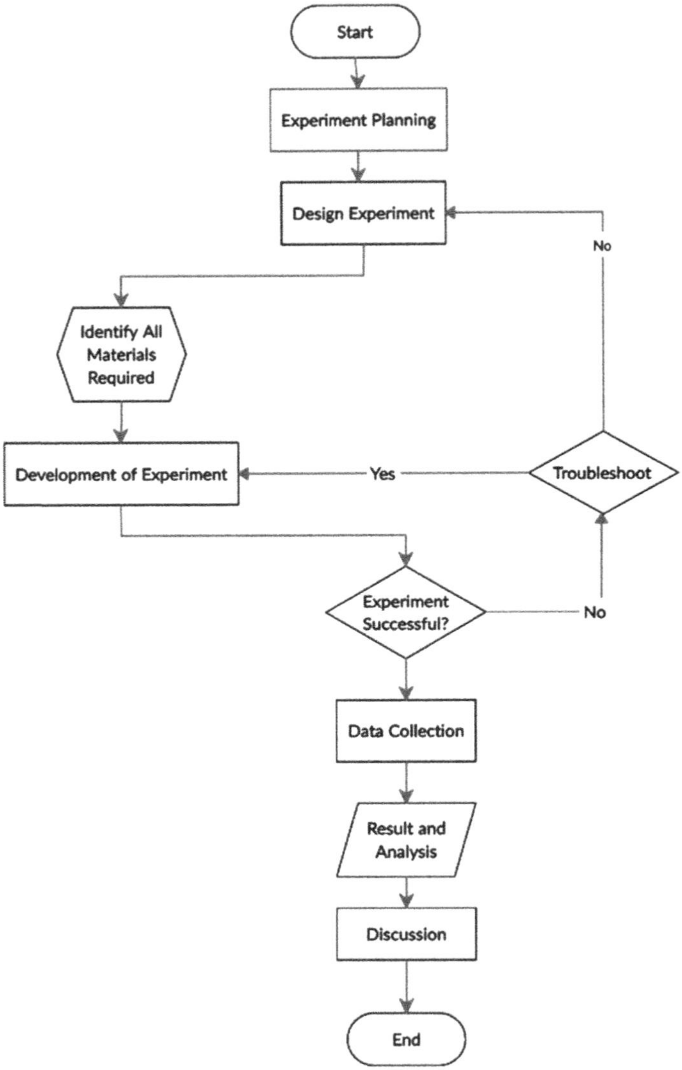

Fig. 7.2 Flowchart of experiment

7.2.2 Development of the Experiment

The flowchart in Fig. 7.3 is important to ensure that the flow of the research is correct, and the experiment can be carried out according to the schedule. The development of the experiment is carried out after all the materials and the equipment required is identified.

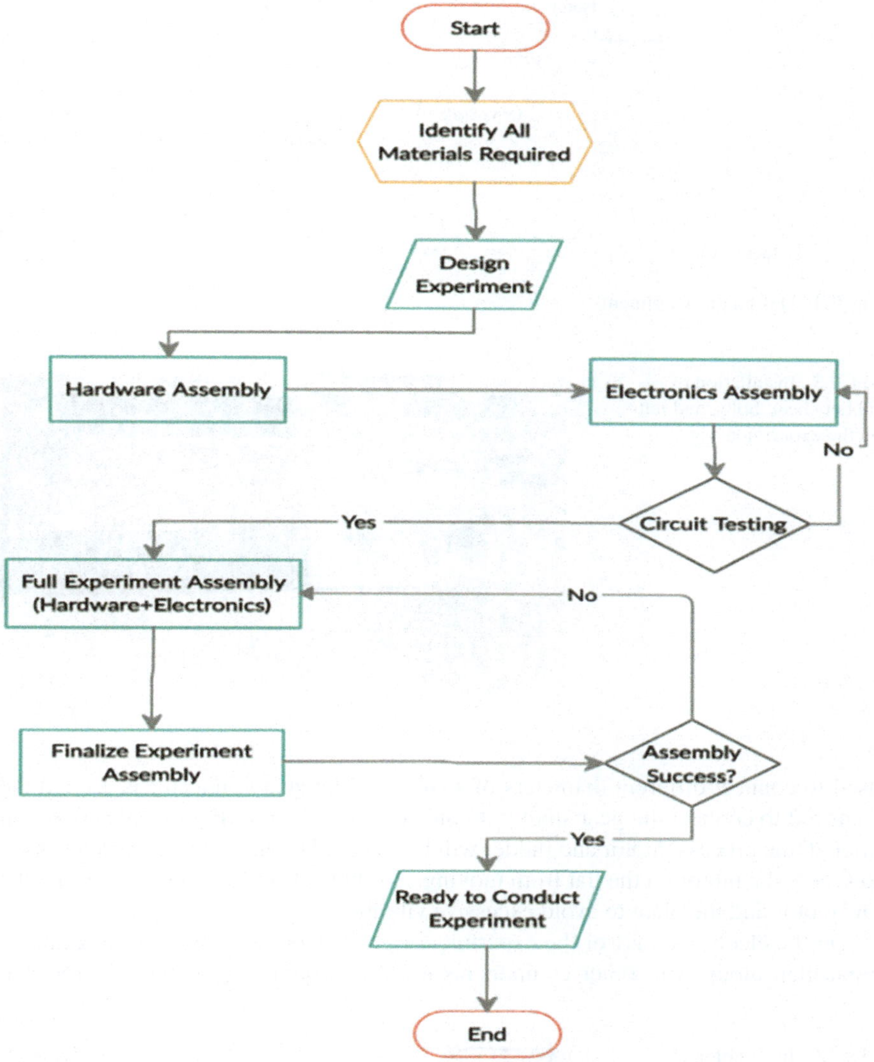

Fig. 7.3 Development of experiment flowchart

Hardware assembly is the first part in the development of the experiment. Figure 7.4 shows the design of the experiment. During construction of the rig, a rubber sheet is used as the shock or vibration absorber between the plate and the tank to absorb excessive vibration when running the experiment. Bolts and nuts are also used to secure and tightened the plate with the tank top.

The shaft is mounted on a wooden stand in horizontal position as shown in Fig. 7.5. The end of the shaft is connected with a shaft connectors where the connector are

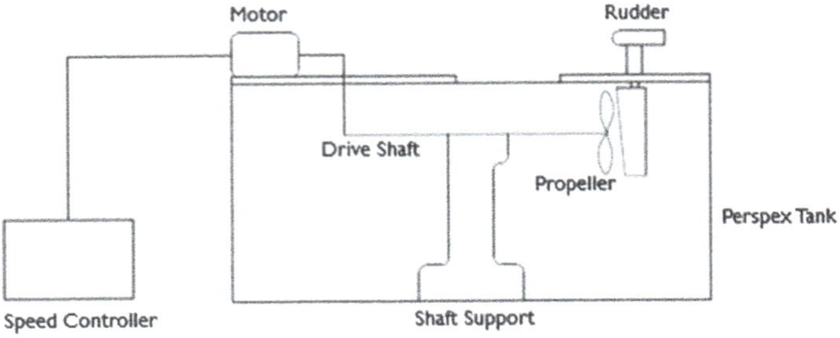

Fig. 7.4 Design of experiment

Fig. 7.5 Installation of rubber sheet, bolts, and nuts on the experiment rig

used to connect different diameters of shaft rod. The gear connector is used at the shaft rod to connect the gear shaft with the vertical shaft rod. Figure 7.6 shows that after all the process, motor and rudder will be installed to the plate. A bracket is used to secure the motor on the flat from moving. A rubber sheet also placed between the DC motor and the plate to avoid excessive vibration.

On the electronic part of the experiment rig, an Arduino board acts as a microcontroller, along with other components such as a diode, capacitor, resistor, and

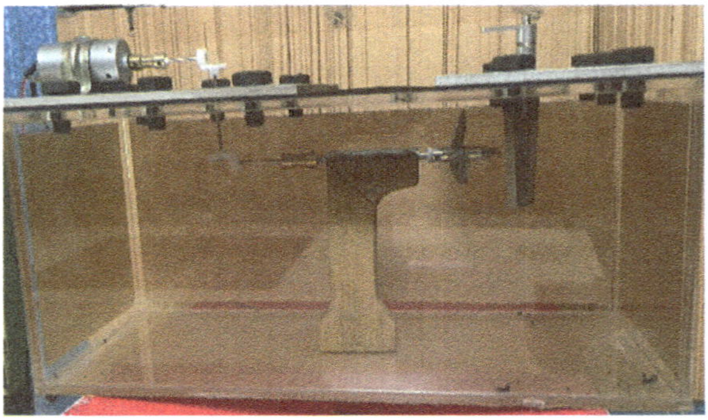

Fig. 7.6 Installation of motor, propeller, and rudder

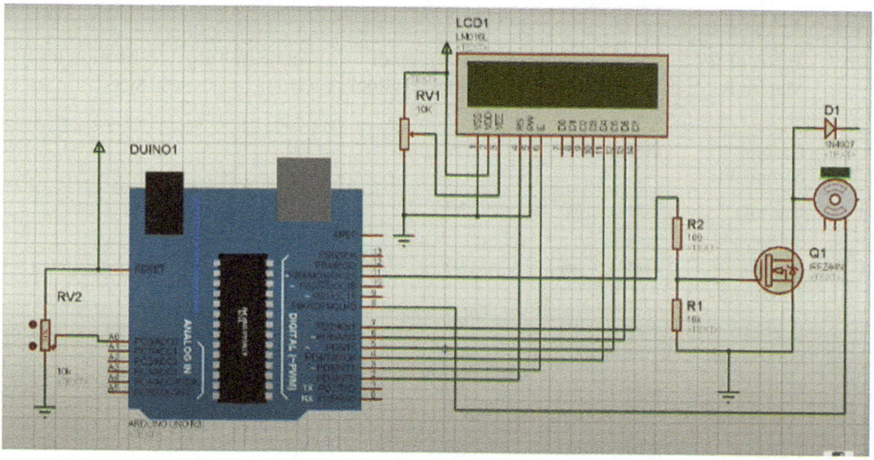

Fig. 7.7 Schematic diagram for speed controller

connecting cable, which are connected on a solderless bread board. This experiment employs a 12 V DC motor and a 9 V rechargeable battery to power the electronic module. The LCD panel is used to display the desired rpm value set on the microcontroller. The electronic connection is carried out by following the schematic diagram in Fig. 7.7. Finally, testing on the speed controller is conducted to ensure that the controller is correctly connected and functioning to run the experiment.

7.2.3 Method to Collect Data

A sensitive analysis method will be used for the data collection process of the experiment. The data recorded throughout the experiment will be tabulated by referring to Table 7.1. The data will be recorded by following to all the variables and transformed into a graph form for analysis purpose.

7.3 Results and Discussion

The results obtained will be discussed according to the parameters that have been set from the beginning of the research. The data will all be recorded by using the sensitive analysis method in Table 7.2 and transformed into the graph form. The aim of using this method is to get clearer information related to the results that have been obtained. The graphs in Figs. 7.8, 7.9, 7.10, 7.11 and 7.12 show the results of the experiment.

Table 7.1 Tabulation of data

No.	Types of RS Materials	RPM	Diameter (mm)	Position of Rudder Stock (% Dp)	Angle of Turn (°)	Vibration level (m/s^2)
1	Carbon composite	900	40	59	P,0	
2	Carbon composite	900	40	59	P,10	
3	Carbon composite	900	40	59	P,20	
4	Carbon composite	900	40	59	P,30	
5	Carbon composite	900	40	59	S,0	
6	Carbon composite	900	40	59	S,10	
7	Carbon composite	900	40	59	S,20	
8	Carbon composite	900	40	59	S,30	
9	Carbon composite	900	40	65	P,0	
10	Carbon composite	900	40	65	P,10	
11	Carbon composite	900	40	65	P,20	
12	Carbon composite	900	40	65	P,30	
13	Carbon composite	900	40	65	S,0	
14	Carbon composite	900	40	65	S,10	
15	Carbon composite	900	40	65	S,20	
16	Carbon composite	900	40	65	S,30	
17	Carbon composite	900	40	71	P,0	
18	Carbon composite	900	40	71	P,10	
19	Carbon composite	900	40	71	P,20	
20	Carbon composite	900	40	71	P,30	

(continued)

Table 7.1 (continued)

No.	Types of RS Materials	RPM	Diameter (mm)	Position of Rudder Stock (% Dp)	Angle of Turn (°)	Vibration level (m/s^2)
21	Carbon composite	900	40	71	S,0	
22	Carbon composite	900	40	71	S,10	
23	Carbon composite	900	40	71	S,20	
24	Carbon composite	900	40	71	S,30	
25	Carbon composite	900	44	59	P,0	
26	Carbon composite	900	44	59	P,10	
648	Stainless steel	1500	48	71	S,30	

The vibration level is measured in units of acceleration (m/s^2) because acceleration is commonly used as a method of treating objects rotating at high speeds and is considered a useful parameter for measuring the probability of dynamic fracturing. As for the negative value on the graph, it is used as the indication of the angle of turn for the port side of the propeller.

Based on Fig. 7.8, the graph shows that the vibration level generated by the rudder stock increased because as the rpm of the motor increased, the faster the rotation of the propeller. The rotation of the propeller induced differential pressure in the surrounding water which contributed to the vibration of the rudder stock. The faster rotation of the propeller creates higher differential pressure in the surrounding water. The graph also shows that the vibration level of the rudder stock increased gradually when the angle of turn of the rudder blade increased. When the rudder blade is turned, one side of the rudder blade has higher tendency to be exposed to the water pressure that is created by the rotation of the propeller than the other side.

Figure 7.9 shows that the vibration level of the carbon composite rudder stock is higher when the rudder stock is placed to the nearest position from the propeller. There is no vibration detected when the distance of the rudder stock to the propeller is 71% of the propeller diameter while the vibration starting to present when the distance is 59% of the diameter. This is due to the direct influence of the water pressure created by the rotation of the propeller that reaches the rudder plate increased when the rudder location is approaching the propeller. As it had been stated earlier in this chapter, the graph shows that the vibration level is slightly different where the port side is higher than the starboard side. The difference appears because the type of propeller used in this experiment is a positive displacement propeller where the rotation takes place in a clockwise direction. Thus, this rotation causes higher water pressure created on the port side of the propeller.

Table 7.2 Tabulation of result

No.	Types of RS Materials	RPM	Diameter (mm)	Position of Rudder Stock (% Dp)	Angle of Turn (°)	Vibration level (m/s^2)
1	Carbon composite	900	40	59	P,0	0.3
2	Carbon composite	900	40	59	P,10	0.5
3	Carbon composite	900	40	59	P,20	0.7
4	Carbon composite	900	40	59	P,30	0.9
5	Carbon composite	900	40	59	S,0	0.3
6	Carbon composite	900	40	59	S,10	0.4
7	Carbon composite	900	40	59	S,20	0.5
8	Carbon composite	900	40	59	S,30	0.7
9	Carbon composite	900	40	65	P,0	0.2
10	Carbon composite	900	40	65	P,10	0.4
11	Carbon composite	900	40	65	P,20	0.5
12	Carbon composite	900	40	65	P,30	0.7
13	Carbon composite	900	40	65	S,0	0.2
14	Carbon composite	900	40	65	S,10	0.3
15	Carbon composite	900	40	65	S,20	0.4
16	Carbon composite	900	40	65	S,30	0.6
17	Carbon composite	900	40	71	P,0	0
18	Carbon composite	900	40	71	P,10	0.2
19	Carbon composite	900	40	71	P,20	0.3
20	Carbon composite	900	40	71	P,30	0.5

(continued)

Table 7.2 (continued)

No.	Types of RS Materials	RPM	Diameter (mm)	Position of Rudder Stock (% Dp)	Angle of Turn (°)	Vibration level (m/s^2)
21	Carbon composite	900	40	71	S,0	0
22	Carbon composite	900	40	71	S,10	0.1
23	Carbon composite	900	40	71	S,20	0.2
24	Carbon composite	900	40	71	S,30	0.4
25	Carbon composite	900	44	59	P,0	0.5
26	Carbon composite	900	44	59	P,10	0.7
648	Stainless steel	1500	48	71	S,30	0.5

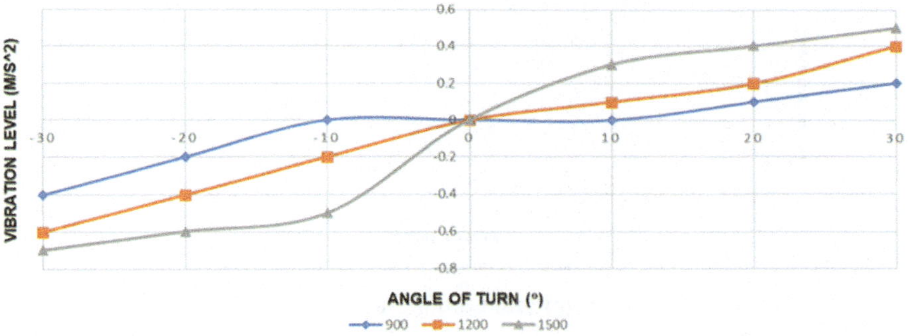

Fig. 7.8 Rpm of motor vs the angle of turn of the rudder blade for stainless steel rudder stock with 48 mm of propeller diameter at 71% of rudder stock position of the propeller diameter

Referring to Fig. 7.10, the graph shows that when the rpm of the motor increases, the vibration level produced by the rudder stock increases. As mentioned in Fig. 7.8, the vibration level generated by the rudder stock increased because as the rpm of the motor increases, the faster the rotation of the propeller. The faster rotation of the propeller creates the higher differential pressure in the surrounding water. As for the angle of turn, the vibration level produced by the rudder stock increases when the angle of turn increases. This is because the higher the angle of turn causes a bigger surface area of the rudder blade to be exposed to the water pressure produced by the rotation of the propeller. However, as mentioned in Fig. 7.9, the vibration level generates on the port side is slightly higher than the starboard side because the propeller used in this experiment is a positive displacement propeller where the rotation takes place in a clockwise direction.

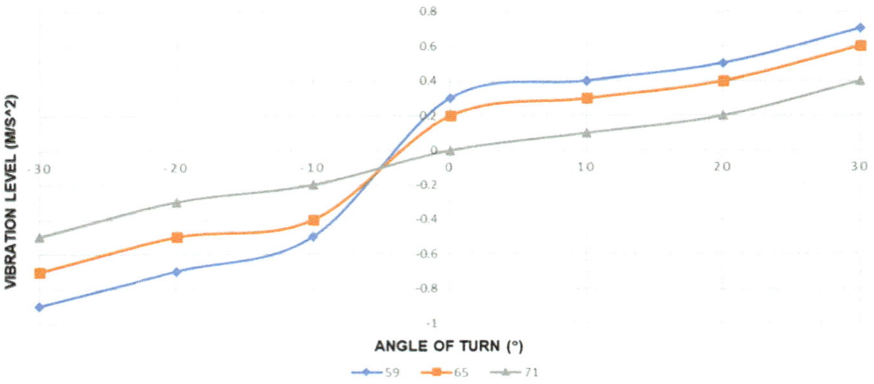

Fig. 7.9 Rudder stock position at different percentage of propeller diameter versus the angle of turn of the rudder blade for carbon composite rudder stock at 900 rpm with 40 mm of propeller diameter

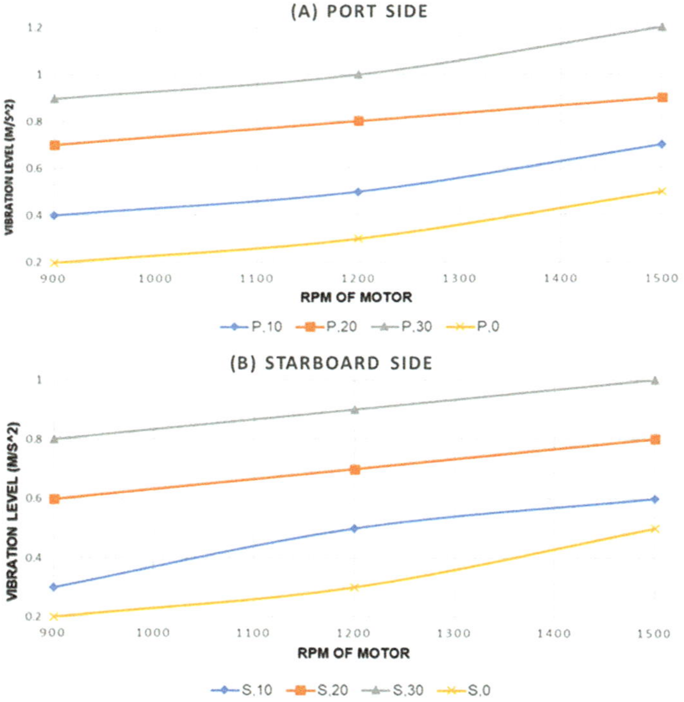

Fig. 7.10 Angle of turn versus rpm of motor for aluminum rudder stock positioned at 59% of propeller diameter with 44 mm propeller diameter at both **a** Port side and **b** Starboard side of propeller

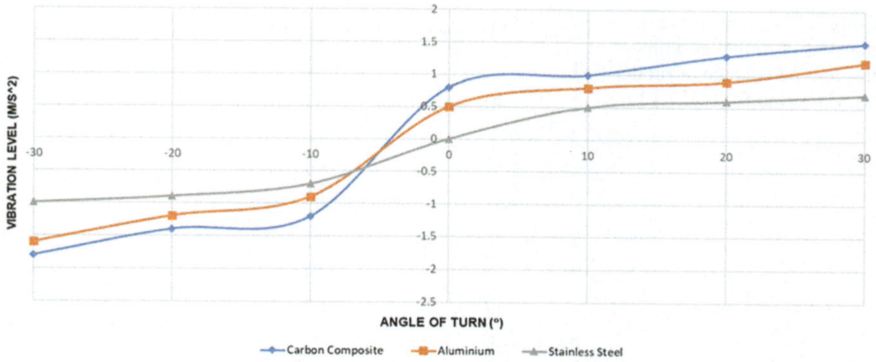

Fig. 7.11 Types of rudder stock materials versus the angle of turn of the rudder blade for rudder stock positioned at 65% of 48 mm propeller diameter at 1500 rpm of motor

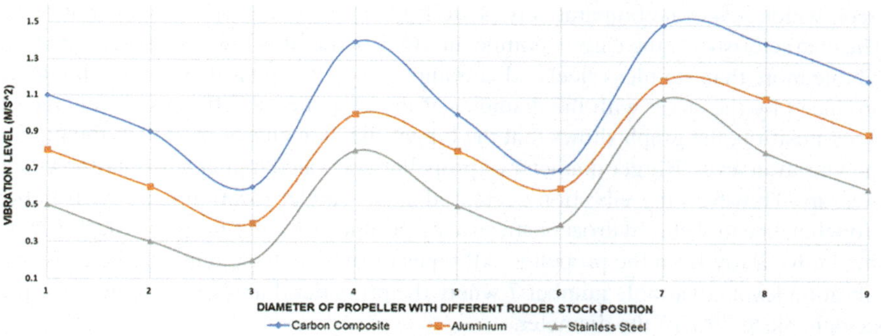

Fig. 7.12 Types of rudder stock material vs diameter of propeller with different rudder stock position at 1200 rpm measured at 30° angle of turn on port side of propeller

Based on Fig. 7.11, the graph shows that stainless steel rudder stock generates the least vibration level compared to carbon composite and aluminum rudder stock. This is because the properties of the stainless steel make it comparatively powerful than the other rudder stock materials. Comparing the vibration level generated by the rudder stocks at different angles of turn, the stainless steel rudder stock generates no vibration when the rudder is at 0° while carbon composite generates the highest. However, the vibration started to increase as the angle of turn increases. As mentioned in Fig. 7.10, this is because when the rudder blade is turned, a higher angle of turn causes a bigger the surface area of the rudder blade to be exposed to the water pressure produced by the rotation of the propeller.

Figure 7.12 shows that an increase in the diameter of the propeller causes the increase in the vibration level generated by the rudder stock. The details representing the diameter of the propeller with different percentage of rudder stock position on the graph are listed in Table 7.3. Here, the best rudder stock materials with the least

Table 7.3 Details of the diameter of propeller with different percentage of rudder stock position for Fig. 7.12

No.	Diameter (mm)	Position of Rudder Stock (%Dp)
1		59
2	40	65
3		71
4		59
5	44	65
6		71
7		59
8	48	65
9		71

vibration effect are the stainless steel rudder stock, while the highest is the carbon composite stock. As mentioned in Fig. 7.11, this is due the properties of the stainless steel, which makes it comparatively powerful than the other rudder stock materials. The characteristic of the carbon composite stock where it is lower in weight tends to vibrate more than stainless steel and aluminum stock. Comparing the vibration level produced by the stock with the diameter of the propeller at different percentage of stock position, the graph shows that the bigger the diameter of propeller, the higher the vibration level. Bigger diameter of propeller generated higher surrounding water pressure. However, the vibration decreased as the rudder location is away from the propeller because the hydroacoustic waves produced by the propeller decrease as they move away from the propeller. All types of rudder stock generate their highest vibration level at variable number 7 where the propeller diameter is 48 mm, and the stock is place 59% of the diameter.

7.4 Conclusion

This paper presented an experiment of rudder stock vibration in different materials on boat hull. Three different types of rudder stock material were used in the experiment. The rudder stock had the same diameter and blade area. Particularly, the vibration level generated by different rudder stock materials at different parameters has been studied.

During the experiment, the rudder stock was placed at three different locations in the distance from the center of the propeller equal to 59, 65, and 71% of the diameter of the propeller. The vibration level for all types of rudder stock material increases as the distance of the rudder stock is approaching the propeller.

As for rpm of the motor, the result shows that the increase in rpm of motor causes the increase in the vibration level generated by the rudder stock. The faster rotation of the propeller creates the higher differential pressure in the surrounding water.

A big difference of the rudder stock vibration can be observed when the diameter of propeller used in the experiment increases. Bigger diameter of propeller creates bigger water pressure that is transmitted to the rudder blade that generated vibration on the stock.

The experiment is conducted at different angles of turn for both port and starboard side of the propeller. As the angle of turn is gradually increased, the increase of vibration level is observed. This is because the higher angle of turn causes a bigger surface area of the rudder blade to be exposed to the water pressure produce by the rotation of the propeller.

Based on all the parameters, all types of rudder stock vibration experience different vibration levels as the parameters are changed. However, during the experiment, stainless steel stock generated the least vibration level compared to aluminum and carbon composite stock while carbon composite stock has the highest tendency to experience vibration. Here, it can be concluded that the stainless steel stock is the best rudder stock material with the least vibration effect.

Acknowledgements This research is conducted as one of the requirements that must be completed by the researcher that supported by UniKL MIMET with guidance from experts and for this, we would like to express our gratitude to anyone who was deliberately or inadvertently involved in the completion of this study. We also thank the readers for their positive feedback and helpful ideas in writing this paper.

References

1. Wikipedia, the free encyclopedia (2021) Marine propulsion. https://en.wikipedia.org/wiki/Marine_propulsion. Accessed 20 April 2021
2. Krcum M (1971) Ship propulsion system, maritime faculty-split. Zrinskofrankopanska 38, HR-21000 Split, Croatia. https://www.bib.irb.hr/10968/download/10968.krcum971.doc. Accessed on 13 April 2021
3. Korczewski Z (2018) A method to assess transverse vibration energy of ship propeller shaft for diagnostic purposes. Polish Marit Res 24(4):102–107
4. Adji S (2002) Resistance and propulsion,course objectives. In: Resistance and powering of ship. https://www.usna.edu/NAOE/_files/documents/Courses/EN400/02.07%20Chapter%207.pdf. Accessed on 18 April 2021
5. Guides AS (2009). The fundamentals of ship vibration. https://doi.org/10.1111/j.1559-3584.1948.tb02755.x
6. Burnside OH et al (1979) A design procedure for minimizing propeller induced vibration in hull structural elements. https://apps.dtic.mil/sti/pdfs/ADA079443.pdf. Accessed on 18 April 2021
7. Delft TU (2011) Prediction of propeller-induced hull-pressure fluctuations proefschrift. https://ris.utwente.nl/ws/portalfiles/portal/15293908/2011_PhD_Van_Wijngaarden.pdf. Accessed on 19 April 2021
8. Pirovsky C et al (2012) Numerical modeling of ship's vibrations and model validation by full-scale dynamic test. In: International Conference on Noise and Vibration Engineering 2012 (ISMA 2012). International Conference on Uncertainity Structural Dynamics, vol 4, pp 3223–3237

9. Zhang C et al (2019) Experimental research on the vibration of ship propulsion shaft under hull deformation excitations on bearings. Shock Vibr. https://doi.org/10.1155/2019/4367061
10. Sukhanov S O et al (2000) Reduction of propeller induced hull vibration. In: 29th International Congress and Exhibition on Noise Control Engineeing, August 1–4

Chapter 8
Natural Superhydrophobic Material as Alternative Substance in Antifouling Paint

Md Redzuan Zoolfakar, Muhammad Zaihasrul Zahari, and Fatin Zawani Zainal Azaim

Abstract Biofouling is the accumulation of several non-native species of organism such as algae, barnacles, and small animals. These non-native species usually grow on the surface of any material that is immersed in water for a long period of time such as a ship's hull. The growth of this biofouling species on a ship's hull will increase this ship's drag and physical resistance which then leads to reduction of ship's speed and velocities. With this being said, the fuel consumption of the ship will also increase. Various methods were introduced to control or reduce the growth of marine biofouling such as antifouling paint. An antifouling paint possesses a few characteristics which help in reducing the marine biofouling growth such as self-cleaning properties. This same characteristic can also be found in superhydrophobic materials. A superhydrophobic surface is a surface which is very difficult to get wet and can be found in natural superhydrophobic materials such as lotus leaf, taro leaf and banana leaf. This paper studies the effect of natural superhydrophobic materials mixed with paint when subjected to various types of parameters. Different types of parameters were chosen such as several types of natural superhydrophobic materials as binders, thickness of paint, underwater depth, and sun-orientation were compared and recorded over 60 days. Results of the experiment were observed and discussed. The aim of this experiment is to determine the most efficient and suitable natural superhydrophobic material between lotus leaf, taro leaf, centella leaf, banana leaf, and rose petal.

Keywords Natural superhydrophobic material · Binders · Biofouling · Anti-fouling · Fiberglass

M. R. Zoolfakar (✉) · M. Z. Zahari
Marine and Electrical Engineering Technology, Universiti Kuala Lumpur Malaysian Institute of Marine Engineering Technology, Lumut, Malaysia
e-mail: redzuan@unikl.edu.my

F. Z. Z. Azaim
Student Development and Campus Lifestyle Section, Universiti Kuala Lumpur Malaysian Institute of Marine Engineering Technology, Lumut, Malaysia
e-mail: fatinzawani@unikl.edu.my

A. Ismail et al. (eds.), *Advanced Materials and Engineering Technologies*, Advanced Structured Materials 162,
https://doi.org/10.1007/978-3-030-92964-0_8

8.1 Introduction

The biodiversity of species that lives in the ocean is infinite. Biofoulings consist of fewer type of non-native species such as microorganism, plants, algae, or small animals. These non-native species will start accumulating on the submerged part of the ship on the very second it contacts the water [1]. The physical resistance of the vessel will increase generated by the accumulation of the biofouling and increases the drag of the vessel. These processes increase the frictional resistance, which will lead to the reduction of the vessel speeds and velocities [2]. The weight of the vessel will increase according to the accumulation of the biofouling, which causes more fuel consumption to cover up the speed loss. In return for this action, more harmful emission will spawn and will directly connect to another environmental problems [3]. An additional consequence, a rise of dry-docking operations will occur. The methods used during dry-docking, not only take a lot of time and energy, they also release a huge amount of toxic waste [3]. The first process of biofouling is the presence of the marine biofilm. Biofilm is a group of microorganisms with made up of cells that can bind to each other and can attach to a surface, especially a ship's hull. This biofilm is made of three major components that consist of extracellular polymeric substances (EPS), DNA, and protein. Microorganism life environment is represented by these major three components [4]. Biofilm can contain a single specie or several other microorganisms at one place together. This phenomenon usually takes place on a surface that they are attached among themselves, and they even share the nutrient among them and the birth of biofilm formation is on the very second the bacterium attached to the surface [4]. Besides all the problem from the marine biofouling, the prevention of this phenomenon also has long discovered by the Phoenicians in 1000 B.C.

Antifouling coatings are advanced paints that are applied to the hull of the ship to delay underwater marine development, which can impact the efficiency and stability of the vessels [5]. The hull coating will also serve as a shield against hull corrosion that can rust and weaken the metal, in addition to preventing marine growth. In both seawater and freshwater conditions, marine coating systems are applied to ships and offshore structures. They have the dual function of guarding against corrosion of the hull and maintaining the ships looking decent [6]. Moreover, the antifouling paints are divided into insoluble matrix paint and soluble matrix paint. The polymer matrix is insoluble in insoluble matrix paints, also referred to as touch leaching or constant contact, and did not wear down even after soaking it into the seawater. For example, using an elevated molecular weight, insoluble vinyl and acrylic are the regular polymers that were used. Right after the soluble pigments are dissolved, the organisms disperse by the film's penetrating seawater and must pass through the interlinking pores [6]. Soluble matrix paints were produced by adding a binder that could be absorbed in seawater to prevent any degradation of antifouling quality as a result. The classical film-forming material contains strong concentrations of rosin in these methods [6]. The next one is about tributyltin self-polishing co-polymer paints (TBT-SPC). Tributyltin self-polishing co-polymer paints (TBT-SPC paints) are the most successful in mitigating biofouling on ships, among the various solutions proposed in navigation history. Controlled hydrolysis makes it possible to remove the organotin biocide from the film and hence provides an antifouling action. The partly reacting brittle polymer backbone can be quickly eroded by moving seawater and exposure to a constantly renewed bioactive surface until enough TBT molecules have been released. Unfortunately, the TBT-SPC applications have a detrimental effect on the environment [7]. The substances that have been used in antifouling paint have a lot of chemical mixture that can be harmful to the marine biodiversity. Hence, the idea of using natural Superhydrophobic materials came out.

The first of superhydrophobicity was found in the vast nature on a lotus leaf and few other plant or leaf (e.g., lotus leaf and taro leaf) which did not get wet by the water. The lotus leaf has its unique surface structure and the presence of low surface energy material which is becoming the reason of this phenomenon. The surface must have a hierarchical micro- and nano-roughness and low surface energy at the same time to implement a superhydrophobic surface or coating [8]. Air that is trapped on the surface of the lotus leaf by the hierarchical micro- and nanoscale roughness is allowing the rise of the water contact angle and low surface energy can limit water's ability to bind with the leaf surface. The natural and exciting phenomenon of the lotus leaf has been given a special name: "the lotus effect". Superhydrophobic surfaces show outrageously high-water repellency, where the droplet of water rain down on the surface of the leaf, rolling with a minimal force of application and bouncing if dropping from a height on the surface of the lotus leaf [8]. A huge amount of attention has been given to the incredible water repellency, superhydrophobicity of a certain plant. There are various species of plant that contain the superhydrophobic characteristic, including Nelumbo nucifera (lotus), Brassica oleracea, Colocasia esculenta, and several others, as well as the wings of butterflies and the legs of water striders. The wettability of superhydrophobic material can be evaluated based on the contact angle. Most of the superhydrophobic surface has identical contact angle of around 160°, the lotus leaves rise above any other plants in state of improved stability and perfection as water repellent. If the hydrophobic effect of superhydrophobic materials is to be dynamically assessed, the rolling angle will be considered. The rolling angle at which the liquid falls on the solid surface corresponds to the critical angle α at which the solid surface tilts as, under the influence of gravity, the liquid falls on the solid surface [9]. The applications of superhydrophobic surfaces can be divided into several different parts such as oil water separation, corrosion resistance surfaces, self-cleaning properties, anti-bacterial properties, and drag reduction [10].

8.2 Methodology

8.2.1 Procedures

This section of the paper shows and explains all the flow of this experiment, including the steps and procedures to conduct this experiment to achieve the expected outcome as shown in Fig. 8.1. The flowchart is purposely made for the reader to understand the steps for the experiment to be performed perfectly. Parameters that lead to field testing, materials used, and the selection of equipment chosen to conduct a field test are based on ASTM D3626-78a (Standard Test Method for Testing Antifouling Panels in Shallow Submergence). Sensitive data analysis was used in this experiment to gain precise data collection. The planning, preparation, data collection, and data analyzing will be shown to make the methodology complete.

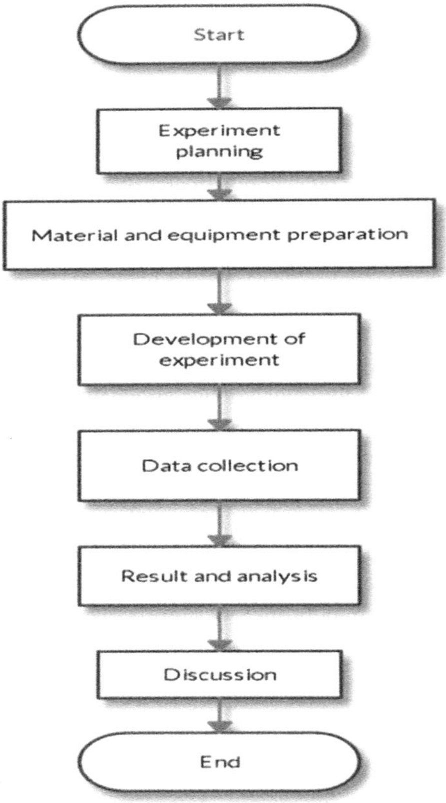

Fig. 8.1 Flowchart of experiment

8.2.2 Experimental Planning

In this section, it shows the planning of the whole experiment starting from the parameters, the equipment, and the devices or tools selection for the experiment. The material chosen for the test specimen is fiberglass. Three different materials underwent a preliminary test to determine which is one of these materials attracts the most marine biofouling growth and the result is that the fiberglass has the most percentage of marine biofouling growth. Fiberglass can be produced with fiberglass mat, catalyst, and raisin. Then, the fiberglass is dried under the sun for at least 6 h for the fiberglass to completely dry and harden.

Next, the powder of the natural superhydrophobic material as shown in Fig. 8.2 will be mixed with the Kossan marine paint as shown in Fig. 8.3 and stirred until it is

Fig. 8.2 Kossan marine paint

Fig. 8.3 Powder of natural superhydrophobic material

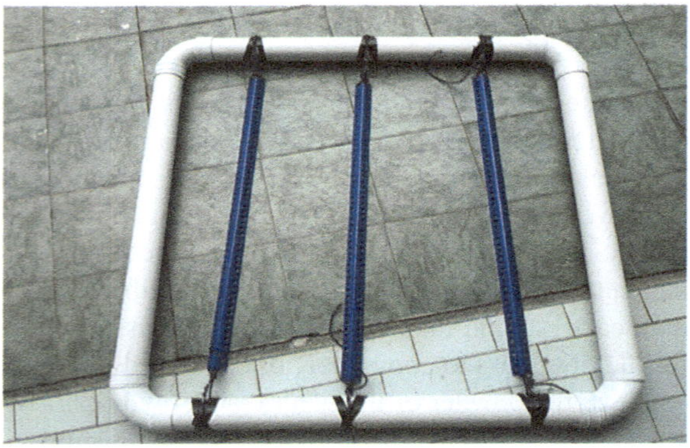

Fig. 8.4 Floating frame

fully mixed which each other. The amount of the natural superhydrophobic material was measured 10% from the volume of the paint. The painted plate will be left to dry about 8–12 h according to the paint manufacturer's guide.

Besides that, the floating frame as shown in Fig. 8.4 is used to hang the test specimen underwater and was made from 5 cm diameter of PVC pipes. Four PVC pipes and four PVC elbows were used to connect and made the floating frame. The floating frame will be tested for floating ability and every end of the elbows will be covered with the special PVC glue.

8.2.3 Parameters of the Experiment

Parameters, also known as variables, were chosen and used for the experiment. The parameters will be described below for further understanding. Table 8.1 is the list of the parameters in this experiment.

8.2.3.1 Binders

Types of natural superhydrophobic materials that have been chosen are Lotus leaf, Taro leaf, Centella leaf, Banana leaf, and rose petal. These five mains natural superhydrophobic materials play an important rule for the experiment. All five natural superhydrophobic materials will be dried out and will be grinded until they become a powder. The powder then will be mixed with the Kossan marine paint with accurate measurement of volume.

Table 8.1 Table of parameter

Binders	Material	Thickness	Depth	Sun-orientation	DAYS	F.R	A.F	O.P
Lotus	Fibreglass	1 layer	1	Yes	1	100	100	100
Lotus	Fibreglass	1 layer	1.5	Yes	1	100	100	100
Lotus	Fibreglass	1 layer	2	Yes	1	100	100	100
Lotus	Fibreglass	1 layer	1	No	1	100	100	100
Lotus	Fibreglass	1 layer	1.5	No	1	100	100	100
Lotus	Fibreglass	1 layer	2	No	1	100	100	100
Lotus	Fibreglass	2 layers	1	Yes	1	100	100	100
Lotus	Fibreglass	2 layers	1.5	Yes	1	100	100	100
Lotus	Fibreglass	2 layers	2	Yes	1	100	100	100
Lotus	Fibreglass	2 layers	1	No	1	100	100	100
Lotus	Fibreglass	2 layers	1.5	No	1	100	100	100
Lotus	Fibreglass	2 layers	2	No	1	100	100	100
Lotus	Fibreglass	3 layers	1	Yes	1	100	100	100
Lotus	Fibreglass	3 layers	1.5	Yes	1	100	100	100
Lotus	Fibreglass	3 layers	2	Yes	1	100	100	100
Lotus	Fibreglass	3 layers	1	No	1	100	100	100
Lotus	Fibreglass	3 layers	1.5	No	1	100	100	100
Lotus	Fibreglass	3 layers	2	No	1	100	100	100

8.2.3.2 Thickness of Paint

The mixture of Kossan marine paint with the natural superhydrophobic material will be painted with different thicknesses and will be used to evaluate the effectiveness of the mixture for a given thickness. The thickness or the layer of the paint will be based on ASTM D3623-78a and paint manufacturer guidelines.

8.2.3.3 Underwater Depth

The test specimens will undergo the immersion at various underwater depths. The justification for this action is to evaluate the efficiency of the natural superhydrophobic material at different level of depths. The depth of the immersion is also based on ASTM D3623-78a which is 1, 1.5, and 2 m.

8.2.3.4 Sun-Orientation

The immersed specimens will be placed and divided into 2 parts. One side of the test specimen will be exposed to the sunlight while the other side of the specimen

Table 8.2 Data table

No.	Parameter	Number of variables
1	Binders	5
2	Thickness	3
3	Depth underwater	3
4	Sun-orientation	2

will remain away from the sunlight. The purpose of adding sun-orientation in this parameter is to verify whether the presence of sunlight helps the growth of biofouling or does not have any effect for the growth of biofouling.

8.2.4 Data Table

Test specimens were evaluated and calculated for surface fouling physical condition of the paint for an interval of three days using the data table in Table 8.2.

Fouling Rating (F.R.) is the fouling presence on the test specimen which is intact during inspection. The procedure to calculation the F.R. is described in the following. Award each test specimen which is free from any biofouling a percentage rating of 100. Reduce the rating to 95 when there is any biofouling occurring. If mature forms of fouling are present, obtain the rating by subtracting from 95 the sum of the number of individuals' present and percent surface covered by.

Antifouling rating (A.F.) is the condition of the antifouling coating and it records qualitative descriptions of the coating. The procedure to calculate the A.F. is described in the following. Award an antifouling coating test specimen having no defect a percentage rating of 100. Subtract the percent surface affected by the coating from 100 to obtain the rating of imperfect films.

Overall performance (O.P.) is the lowest percent rating of the two preceding values.

8.2.5 Method of Collecting Data

The process of collecting data in this experiment is referring to the flowchart as shown in Fig. 8.5. Total period of the immersion is 60 days, and the specimen will be taken out from the water every 3 days to calculate the marine biofouling growth on the plate. Figure 8.6 acts as the tool to calculate the marine biofouling growth on the surface of the specimen. The tool will be placed on top of the specimen and visually can calculate the bow that has marine biofouling on it. The data calculation will be following the ASTM D3623-78a which consists of the fouling rating (F.R.), antifouling rating (A.F.), and overall performance.

Fig. 8.5 Flowchart of data collection

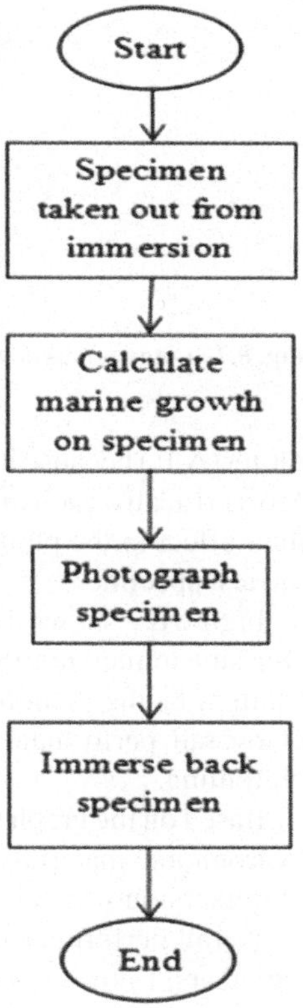

Fig. 8.6 Tools for calculating marine biofouling

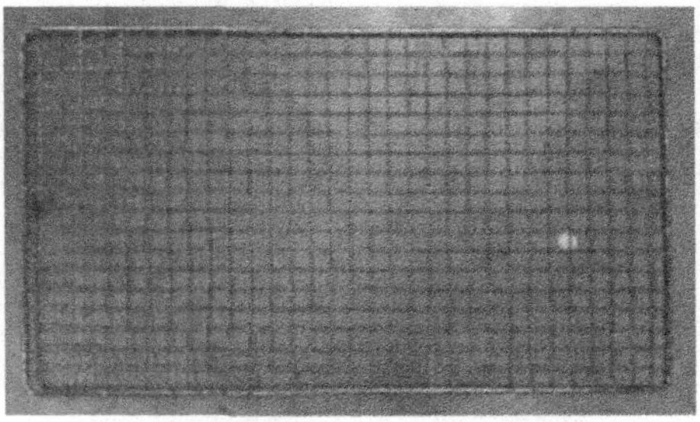

8.3 Results and Discussion

Over the period of 60 days, data of the test specimen were calculated and tabulated using the sensitive analysis. Then, the whole data was narrowed down to a specific target so that a graph can be easier plotted. The graph will be plotted based on the chosen parameters and will be compared between days and the overall performance. The overall performance was chosen instead of fouling rating (F.R.) and anti-fouling

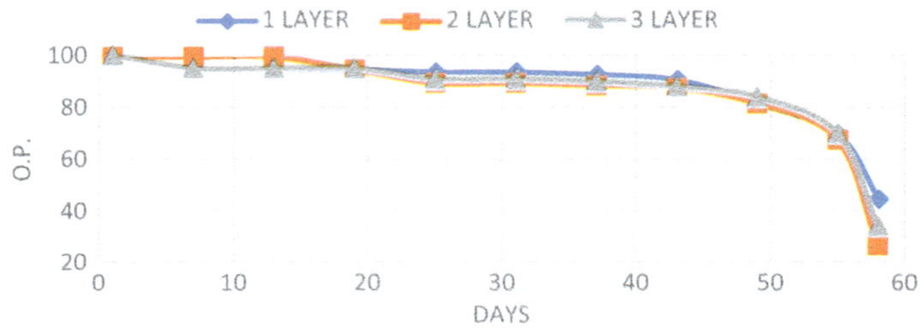

Fig. 8.7 Lotus versus days with layers while facing sun at 2 m depth underwater

rating (A.F.) because it is the lowest percent rating of the two others preceding values. Most critically, each of the parameters brings out different outcome and clearly shown their effect in the plotted graph based on the growth of marine biofouling on top of the test specimens.

Figure 8.7 shows the lotus leaf as binders with 3 different layers of thickness and constant immersion depth at 2 m underwater. The specimens were all having the sunlight facing them for 60 days. Two layers of coating show the worst degradation of overall performance and one layer of coating has the least growth of marine biofouling.

Based on the graph shown in Fig. 8.8, it shows five different types of natural superhydrophobic materials as binders with 3 layers thickness of paint and constant depth of immersion which is at 2 m underwater. Cantella leaf shows the worst degradation of overall performance and lotus leaf shows the most effective binders compared to the others. Lotus leaves have the highest contact angle among others which exceeds 160°and the graph proves that lotus is the suitable binder to avoid marine biofouling growth. The result shows taro leaf, centella leaf, banana leaf, and rose petal almost have the same outcome at the end of the experiment which is all below 20%.

Figure 8.9 shows the Taro as binder with three different depths or immersion underwater with two layers of paint thickness and the specimen were facing the sunlight. 2 m depth of immersion shows the worst degradation of overall performance

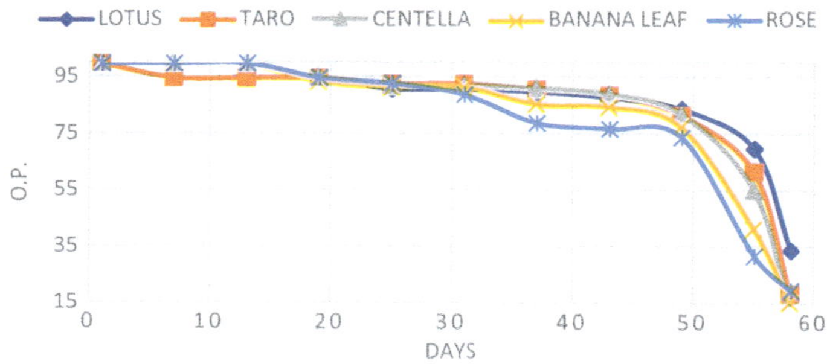

Fig. 8.8 Binders versus days with three layers thickness of paint while facing sun at 2 m depth underwater

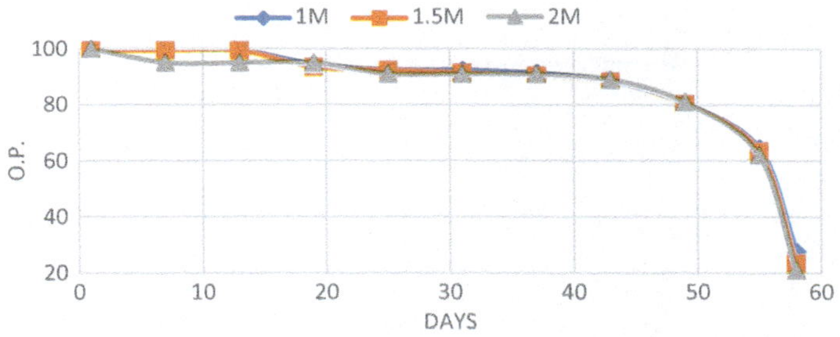

Fig. 8.9 Taro versus day with two layers of thickness while facing sun

and 1 m depth of immersion has the least growth of marine biofouling. The result shows that, the deeper the depth of immersion, the growth of marine biofouling will increase compared to others. The graph is the evidence that 2 m of immersion attract more marine biofouling growth than 1 and 1.5 m of immersion.

Based on the graph shown in Fig. 8.10, it shows five different types of natural superhydrophobic materials as binders with 2 layers of paint thickness and a constant depth of immersion which is at 1.5 m underwater. Rose shows the worst degradation of overall performance and lotus leaf shows the most effective binders compared to the others. Based on the graph, all the binders have different effect on the marine biofouling growth and at 1.5 m of immersion the differential marine biofouling growth is not that far compared to each other especially on the last day of experiment.

Figure 8.11 shows that Rose as binder on different sun-orientation with 1 layer of paint thickness at constant depth which is 1.5 m underwater. The graph shows that the different sun-orientations have only in a slight difference. The factor that affects this result is due to the condition of the surrounding where the specimen is left. The specimen is left alongside a pontoon that might block the sun from getting in contact and affecting the specimen. Another possible factor is the sizing of the plate, the size of the plate might be too small for the sun to take effect.

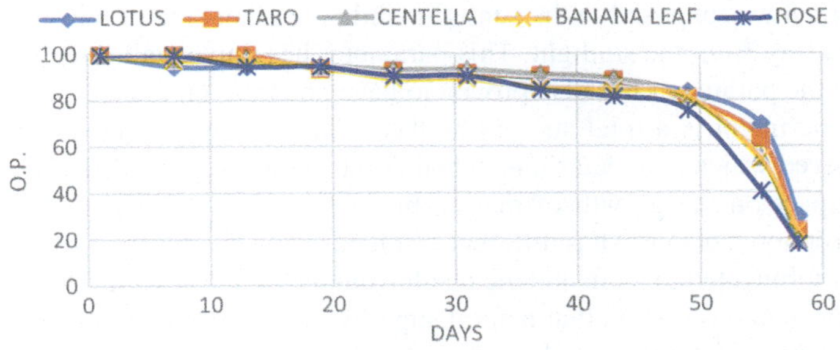

Fig. 8.10 Binders versus days with two layers thickness of paint while facing sun at 1.5 m depth underwater

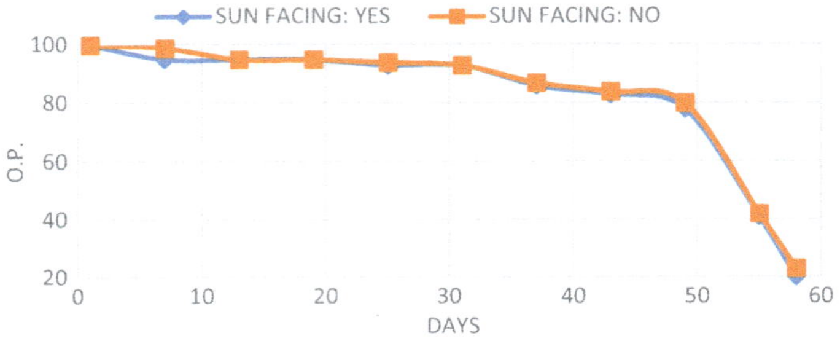

Fig. 8.11 Rose versus days with one layer thickness of paint at 1.5 m depth underwater

8.4 Conclusion

This paper reported on the research investigations to determine the effect of five different natural superhydrophobic materials on marine biofouling growth with four types of parameters. The growth of marine biofouling can be very troublesome especially on ship's owner and even can be dangerous to the sea ecosystem. Based on the result of the experiment, between five different binders that have been chosen, lotus leaf has proven that it is the best binder which can control and reduce the marine biofouling growth on the fiberglass compared to taro leaf, centella leaf, banana leaf, and rose petal. Next, is the thickness of paint. The mixture of marine paint with the natural superhydrophobic material with three different thickness and was used to evaluate the effectiveness of the mixture for a given thickness and the result shows inconsistencies between one layer and three layers thickness of paint. Another parameter in this experiment is the depth of immersion, the test specimens underwent the immersion at various underwater depths. The justification for this action is to evaluate the efficiency of the natural superhydrophobic material at different levels of depth. Based on the result accumulated, 2 m depth of underwater immersion has the worst degradation compared to the depth of 1 and 1.5 m. Lastly is the sun-orientation. The immersed specimens were placed and divided into two parts. One side of the test specimen was exposed to the sunlight while the other side of the specimen was remained away from the sunlight. This parameter has only small differences on the result and the parameter has also proven to have an effect on the test specimens.

The experiment is a total success as the main objective of the experiment has been achieved which is to determine whether the natural superhydrophobic material has an impact on the growth of marine biofouling. The efficiency of the natural superhydrophobic material has different result based on the different parameters that have been subjugated to them and the result from the field testing has been analyzed. This experiment has proven that natural superhydrophobic material can control and reduce marine biofouling growth on fiberglass.

Acknowledgements This research is conducted as one of the requirements that must be completed by the researcher that supported by UNIKL MIMET with guidance from experts and for this, we would like to express our gratitude to anyone who was deliberately or inadvertently involved in the completion of this study. We also thank the readers for their positive feedback and helpful ideas in writing this paper

References

1. Yebra DM, Kiil S, Dam-Johansen K (2004) Antifouling technology—past, present and future steps towards efficient and environmentally friendly antifouling coatings, In: Prog Org Coatings 50(2):75–104. https://doi.org/10.1016/j.porgcoat.2003.06.001.
2. Demirel YK, Uzun D, Zhang Y et al (2017) Effect of barnacle fouling on ship resistance and powering. Biofouling 33(10):819–834. https://doi.org/10.1080/08927014.2017.1373279
3. Alghamdi SA, Quijada R (2019) The maritime commons: digital repository of the world maritime the impact of biofouling on marine environment: a qualitative review of the current antifouling technologies. Impact Biofouling Marine Environ 91
4. Callow ME, Callow JA (2002) Mariñe biofouling: a sticky problem. Biologist 49(1):10–14
5. NRL (1952) The history of the prevention of fouling. Mar Fouling Its Prev 580:211–222
6. Woods Hole Oceanographic Institute (1952) Characteristics of antifouling coatings. Mar Fouling Its Prev 580:313–322
7. Qureshi AS, Ramaswamy R (2018) Evolution of antifouling paints. Int J Sci Res 7(1):433–443. https://doi.org/10.21275/ART20179277
8. Darmanin T, Guittard F (2015) Superhydrophobic and superoleophobic properties in nature. Mater Today 18(5):273–285. https://doi.org/10.1016/j.mattod.2015.01.001
9. Barati DG, Aliofkhazraei M, Khorsand S et al (2020) Science and engineering of superhydrophobic surfaces: review of corrosion resistance, chemical and mechanical stability. Arab J Chem 13(1):1763–1802. https://doi.org/10.1016/j.arabjc.2018.01.013
10. Sayre N (2015) Investigation of boat hull performance with superhydrophobic coating. Mater Sci

Chapter 9
Analysis of Safety of Navigation Compliance among Flag of Convenience's Ships that Factors of Ship Collision and Grounding: A Review

Che Nur Ashman Che Anuar, Wardiah Mohd Dahalan, and Aminatul Hawa Yahaya

Abstract The nationality of a ship would be based on what flag that they are flying. The flag that a ship is flying to ensure that the vessel is in act with the conformity with the applicable rules of the international laws. The marine accidents in Malaysia can be divided into five categories which are collision, grounding, contact, oil spill, and others. In three period at Tanjung Pelepas port, there are 33 accidents that occurred as 61% are from the oil spills, 15% collisions, 9% contacts, 9% grounding, and the others are 6%. The flag of conveniences (FOCs) are prospering that it equally criticized as for turning blind eye to the world's maritime safety and security norm. The FOCs are notoriously known for having a substandard ship which the ITF had declared a list of 32 countries where they are operating in international registries were including the FOC countries. There are several international organizations which are ITF, IMO, and others are fighting against FOCs as it does not make much differences in their practices as they observed. There are several incidents of collision of FOC registered vessels, and it would cause a million of tons of oils spills. Most of non-compliance vessels that are flying flags are FOCs, but not all FOC registries are poor performers as not all the national registries had the perfect record.

Keywords Collison · FOCs · Grounding · IFT · Officer of the watch (OOW)

C. N. A. C. Anuar · W. M. Dahalan (✉)
Universiti Kuala Lumpur Malaysian Institute of Marine Engineering, Technology, 32200 Lumut, Perak, Malaysia
e-mail: wardiah@unikl.edu.my

C. N. A. C. Anuar
e-mail: ashman.anuar@s.unikl.edu.my

A. H. Yahaya
Universiti Kuala Lumpur Business School, Jalan Persiaran Gurney, Kampung Datuk Keramat, 54000 Kuala Lumpur, Malaysia
e-mail: aminatulhawa@unikl.edu.my

A. Ismail et al. (eds.), *Advanced Materials and Engineering Technologies*,
Advanced Structured Materials 162,
https://doi.org/10.1007/978-3-030-92964-0_9

9.1 Introduction

The nationality of ships is based on the flag that they fly, wherever the vessels are located the flag states are responsible in ensuring that their vessels are in act with the conformity with the applicable rules of the international laws. In ensuring the maritime security and safety and in protecting the marine environment, flag states play an important role. The flag that is flying over ships is required to exercise an effective jurisdiction and control in the administrative, technical, and social matters [1].

The report of the UN Secretary-General that being submitted to the 63rd session of the UN General Assembly has clearly indicated that the number of flag states regularly failed to comply with the United Nations Convention of the Law of the Seas (UNCLOS) and other relevant international instrument and agreements. In the reports it is stated that some flag states failed to ratify and implement the vital social and environmental instruments, while some flag states failed to exercise an effective jurisdiction over the vessel that is flying their flag. This report highlighted the failure of the current regime as in addition to the risks posed to seafarers and to the marine environment, the preference for anonymity over accountability, this highlights adequately delete the potential catastrophic which would breach the security and safety [1].

The Marine accidents in Malaysia can be divided into five categories which are collision, grounding, contact, oil spill, and others. In the past year from 2015 to 2017 accidents which occurred in Tanjung Pelepas Port Limit shown that in the 3 years period there are 33 accidents occurred which 61% are from the oil spills, 15% collisions, 9% contacts, 9% grounding, and 6% others [2]. Based on the data, the collision and grounding accidents should be prevented as these incidents may cause a total losses or un-seaworthiness to the ship which would bring to a negative consequence to personnel and environment as shown in Figs. 9.1 and 9.2.

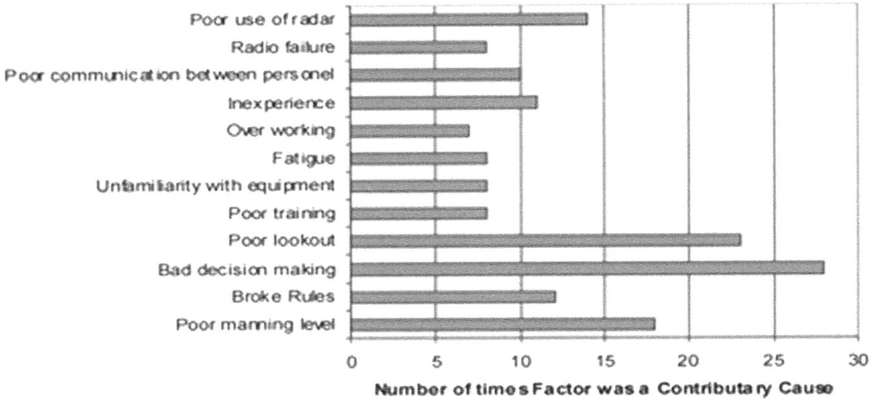

Fig. 9.1 Common factors in groundings

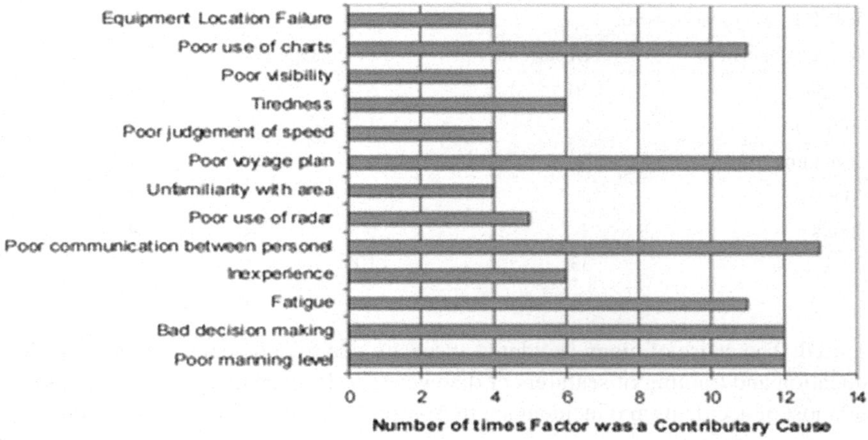

Fig. 9.2 Common factors in collisions

9.1.1 Flag of Convenience

Flag is a professional term that is used in describing the process of registration for a commercial ship at which a country would allow a ship to fly its flag and this is known as a flag state. The flag that flies legally would indicate the ships nationality. Once the ship is documented and registered it would take the nationality of the flag regardless from where originated and who owns the ship. This ship is also like a citizen that would hold a nationality by flying a flag. A nationality is essential for a ship to sail the high seas a ship with no nationality is known as a stateless ship which enjoys no protection on the high seas. Furthermore, they would be denied access to a foreign port which eventually this stateless ship will not be able to trade internationally. This stateless ship would suffer even more severe consequences by being a criminal's ship which would attempt to evade jurisdiction [3].

9.1.2 Major Cause of Accident at Sea

In a study the accident reports show that 85% of all accidents are directly due to human error or also associated with the human error which means due to inappropriate human response. 80% of accidents at sea are caused by human error as found at a recent paper IMO (2005) [4]. Collision is the most common type of accident that happens which can be confirmed by the latest data published by the Main Search and Rescue Coordination Centre of Turkey 2009. There are 60% of all accidents related to collision if grounding and contact are included in Table 9.1. Previous research indicates that the mistake is usually made not due to deficient or inadequate regulations, the mistake is due to the fact that regulations and standards are usually

Table 9.1 Accident occurred

Accident categories	Occurrences	Percentage (%)	Rank
Oil spill	20	61	1
Collision	5	15	2
Grounding	3	9	3
Contact	3	9	3
Others	2	6	4
Total	33	100	

ignored. The cause of many accidents at sea are due to deficiencies in the maritime education and training of seafarers or disregard for the current standards regulations and most of accidents and incident are mostly related to collision and grounding, [5].

9.1.2.1 Collision and Grounding

The main reason for a typical collision includes, according to the officer of watch (OOW), not following the COLREGs or the company's safety management system. These collisions are usually caused by a combination of inexperience and systematic issues in the organizations. It might seem to be very strange that the vessel can still collide and especially on the open sea as the two vessels might be the only ones on the horizon. This would manifest itself in the attitude that would become acceptable within the shore-based organization which would take unsanctioned risk and to make a short-cut [6].

9.1.2.2 FOCs Issues

The FOCs are prospering that is equally criticized as for turning a blind eye to the world's maritime safety and security norm. The FOCs are notoriously known for having a substandard ships at which the ITF had declared a list of 32 countries as shown in Table 9.2, where they are operating in international registries were including the FOC countries and top eleven FOC is shown in Fig. 9.3. The ITF list includes some of the world's top open registries such as Panama, the Marshall Islands, and Liberia. In January 2015, these three flags owned more than 41.8% of the shares in world shipping trade which in total makes over 1.75 billion dwt. The sheer size and involvement in the maritime transactions show that the list of FOCs is worth to review. There is concealment of ownership, insufficient regulation, poor working condition, and environmental effects are some of the maritime safety and security issues which are related to the FOCs [7].

Table 9.2 List of FOC Country

Antigua and Barbuda	Bahamas
Barbados	Belize
Bermuda (UK)	Bolivia
Cambodia	Cayman Islands
Comoros	Cyprus
Equatorial Guinea	Faroe Islands (FAS)
French International Ship Register (FIS)	German International Ship Register (GIS)
Georgia	Gibraltar (UK)
Honduras	Jamaica
Lebanon	Liberia
Malta	Madeira
Marshall Islands (USA)	Mauritius
Moldova	Mongolia
Myanmar	Netherlands Antilles
North Korea	Panama
Sao Tome and Príncipe	St Vincent
Sri Lanka	Tonga

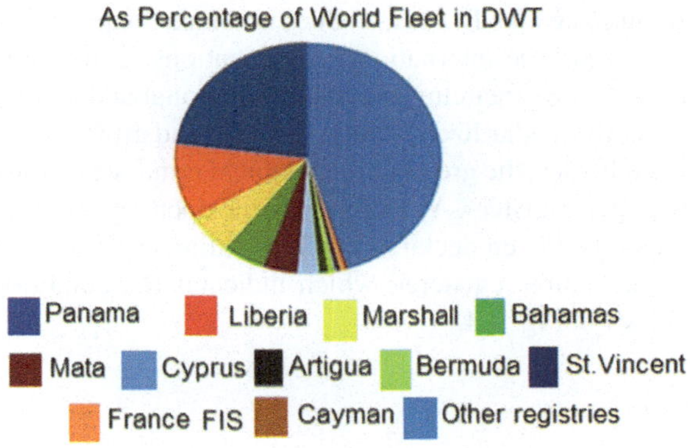

Fig. 9.3 Top eleven FOCs

9.2 Results and Discussion

The maritime safety is related to the safety of cargo, crew, ship, and environment. The FOCs are criticized for an unopened eye as the substandard and inefficient regulation as on the marine pollution and the safety of seafarer. This is due to the poor standards in regard of the safety, as there had been an instance boycotted from the ITF toward Panama flagged vessels. The FOCs had failed to enforce their own and other international requirement as this is one of the reasons for the large and widely scattered fleet that is becoming menacing. The FOCs are becoming incapable in enforcing and ensuring the international standards as the greedy and unethical shipowner which took this incapability for granted by taking an advantage toward unskilled seafarers by giving a lower salary in order to making money. This owner also did not care for the international regulation as the result in the FOC registered vessel would causing marine pollution such as oil spills dumping of sewage and waste, chemical spills, unmonitored exhaust gases, and others environmental problem. There are several incidents of collision of FOC registered vessel. It caused millions of tons of oil spills this is including the cases of rusty Erika, the prestige, and many more.

The advantage of incapability and substandard regulation of FOCs is that this ship became unsafe as the ship owners started exploiting the workers. These vessels that are being registered under FOC had become too degraded from the international requirement as even proper food and drinking facilities are not provided. These seafarers are paid lower than the market price which would lead to longer working hours [8].

9.3 Conclusion

The flag of convenience had become a great hurdle and threat in the maritime industry as the open registers were being brought into existing for genuine of a good reason. There are several international organizations such as the ITF, IMO, and others that are fighting against FOCs as it does not make many differences in their practices as they observed. The international organizations such as IMO, IACS, ILO, and others are nations are practicing several international and local conventions and regulations in order to standardized, safety, security, and growth of this maritime industry. After all the fallacies, the greedy, and unethical mind-set of shipowners, they are still going for the open registers. An example of the open registers of Panama, Liberia, and Marshal Island had been declared by ITF holding as the top three ranks of the world's as the largest ships registered which indicates the condition of the maritime industry as shown in Fig. 9.4.

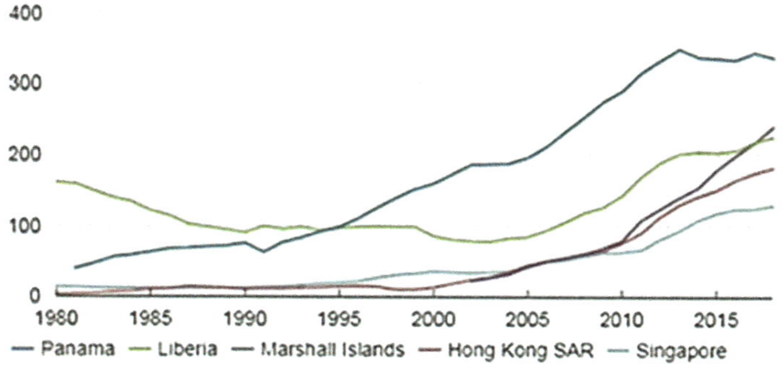

Fig. 9.4 FOCs values

There had been an increasing on the use of flags of convenience as both trade and high seas fisheries had grown internationally over decades. This area had been a concern in relation with the maritime security and a potential for the use of ships by terrorist operations and criminal organizations. Other than that, these open registers also would risk the safety of life at sea, the prevention of marine pollution, the conservation of fish stocks and biodiversity, and the protection of the marine environment. As stated, that the most of non-compliance vessels flying flags are FOCs, but not all FOC registries are poor performers as not all the national registries had a perfect record.

References

1. Hayashi M (2001) Toward the elimination of substandard shipping: the report of the international commission on shipping. Int J Mar Coast Law 16(3):501–513
2. Bolat F (2016) Analysis of ship accidents in Turkey from 2005 to 2015. Int J Eng Res Technol 5(12):478–481
3. Hamad BH (2016) An examination of bareboat charter registries and flag of convenience registries in international law. Ocean Dev Int Law 36(4):339–362
4. Uğurlu Ö, Yıldız S et al (2020) Analyzing collision, grounding, and sinking accidents occurring in the Black Sea utilizing HFACS and Bayesian networks. Risk Anal 40(12):2610–2638
5. Yu Z, Liu Z, Amdahl J (2019) Discussion of assumptions behind the external dynamic models in ship collisions and groundings. Ships Offshore Struct. 14(sup1):45–62
6. Lotilla RPM (1992) The efficacy of the anti-pollution legislation provisions of the 1982 law of the sea convention. Int Comp Law Q 41(1):137–151
7. Pinnell N (2020) Flag of convenience: substituting void-for-vagueness doctrine for equal protection analysis in manning v. Caldwell NCL Rev 99:1167
8. Timagenis GJ (1973) International control of dumping at sea. Anglo-AmL Rev 2(2):157–187

Chapter 10
A Portable Hydro Generator for Fishing Boats

Md Redzuan Zoolfakar, Muhammad Syahmi Syukri Abdul Shukur, and Nik Harnida Suhainai

Abstract The use of modern and renewable technology is on the rise in a variety of fields. This can be seen in the use of water energy, wind energy, geothermal energy, water energy, and other renewable energy sources in different aspects of life and industry. In Malaysia, hydroelectric power generation from dam is commonly used to produce electricity for consumers. As for marine vessels in Malaysia, they usually use generators onboard to produce electricity. By applying the hydroelectric method to the fishing boat, fuel consumption and gas emission can be reduced. This study focuses on zone A class small fishing boats which commonly have no power sources installed onboard the vessel. Without a fixed power source installed, small fishing boats use portable generators and batteries as sources of electrical power. However, some small boats are not installed with any power sources onboard due to limited space and high weight. The primary objective of this study is to analyze the electrical power output generated from the hydro generator onboard small fishing boats by using the portable method mounted on the boat body, a suitable method to carry out the experiment.

Keywords Compact power source · Power generation · Renewable energy · Alternative energy

M. R. Zoolfakar (✉) · M. S. S. A. Shukur · N. H. Suhainai
Marine and Electrical Engineering Technology, Universiti Kuala Lumpur Malaysian Institute of Marine Engineering Technology, Lumut, Malaysia
e-mail: redzuan@unikl.edu.my

M. S. S. A. Shukur
e-mail: syahmi.shukur@s.unikl.edu.my

N. H. Suhainai
e-mail: nikharnida@unikl.edu.my

A. Ismail et al. (eds.), *Advanced Materials and Engineering Technologies*, Advanced Structured Materials 162,
https://doi.org/10.1007/978-3-030-92964-0_10

10.1 Introduction

Energy is the most important sector for a country's growth. It is necessary for survival and essential for developmental activities such as education, health, transportation, and infrastructure in order to achieve a fair standard of living, as well as a critical factor for economic growth and jobs [1]. The three primary sources of energy are fossil fuels, nuclear power, and renewable energy. Renewable energy sources such as solar, wind, biomass, geothermal, and hydropower are used to reproduce energy and are therefore extremely useful in the fight against energy shortages [2]. Besides, non-renewable resources are facing depletion, strategic political control, and rising in costs which is causing renewable energy to become more competitive with non-renewable energy sources [3]. As a result of these issues, the use of renewable energy resources is becoming more common. Among the various sources of renewable energy, hydropower is the most important source of grid-connected energy in the world [4]. Hydropower alone provides about 16 percent of global electricity, with more hydropower projects planned to be implemented in the coming years due to the enormous benefits associated with it [5].

Hydropower is electrical energy generated by the flow of water which normally originates from stream, river, and ocean. This kind of energy has been harnessed since past centuries [6]. The efficiency of energy conversion in a hydropower plant varies depending on the flow rate and hydropower plant model [7]. Hydropower is a well-established and straightforward technology at which the water's potential energy is converted into kinetic energy, which spins a turbine that drives an electricity generator [8]. A hydro generator is also known as a tidal generator which makes it possible to take use of the kinetic energy that occurs around a ship hull to produce electricity. The principle is to capture and convert kinetic energy and flow-pressure energy of moving water to electrical energy at which the electrical energy can be supplied to the ships. The hydro generator is driven by the flow of water to produce power which does not rely in burning of fuel [9].

A portable hydroelectric generator consists of a flywheel that can be rotated from a flow of water, a drive shaft, and a generator. Taking advantages of water to generate electricity is a common practice and used globally such as in large-scale hydropower plants that use the potential energy of water stored in dams to run hydroelectric turbines [10].

10.2 Methodology

This part of the paper elaborates and explains about the steps, procedures, and details about the experiment that will be conducted. The process of experimental planning and execution to obtain the result and data is illustrated in the flowchart in Fig. 10.1. The methodology used to collect the primary data to method on processing the result and other related information which will be discussed for this research. It

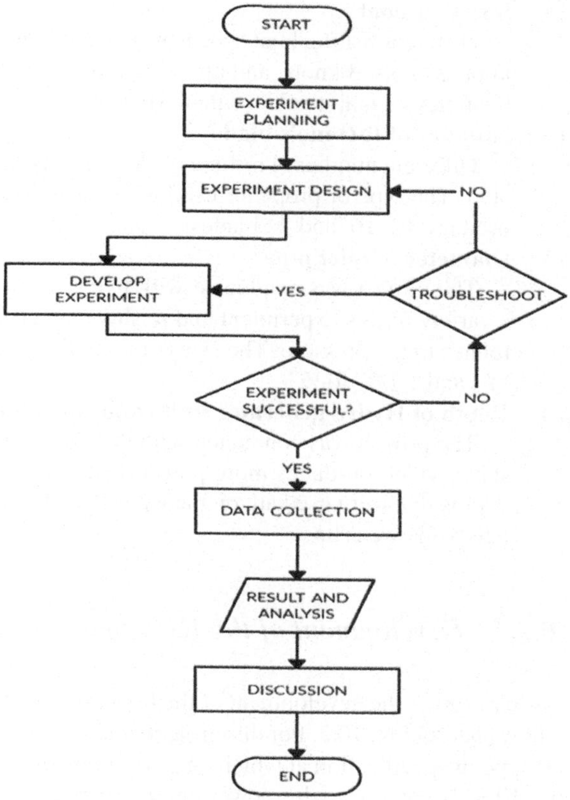

Fig. 10.1 Flow chart of experiment

also discusses on how the research is designed and what are the instruments used in conducting the study.

In general, this experiment is developed to obtain the data of power generation of a hydropower unit before analyzing it. Therefore, the methods used for this experiment are suitable to obtain the result for the project.

10.2.1 Experimental Parameters

Parameters or variables that will be used in this project, as well as their descriptions, will be described below. These parameters are required to differentiate and compare results from the experiment for analysis.

(a) **Speed of boat**
Different fishing boat speeds were used in this experiment and they were 1 knot, 2 knots, 3 knots, and up to 7 knots. This is to identify if the speed of the boat plays a major role for this experiment.

(b) **Number of turbine's blade**
Different number of turbine blades were used to determine the power generated. The type of propeller used was a crossflow type. The number of blades used are 12, 16, and 24 blades.

(c) **Diameter of inlet pipe**
The project was conducted with five sizes of inlet pipe diameters to act as a variety of this experiment and to show which will produce more power as it relates to the pressure. The sizes of pipe that were used are 1" to ½", 1 1/4" to ½", and 1 1/2" to ½".

(d) **Depth of Hydro generator unit from waterline**
The project was conducted with different depths of water generator unit to show which produces more power as it is related to the depth of water. The depths that were used are on the waterline, 0.5 m below the waterline, and 1 m below the waterline.

10.2.2 Development of the Experiment

This part shows the development of the flow of the experiment and the process refers to flow chart in Fig. 10.2. For this project, micro hydropower is used and there is only one type of specification at which only the number of blades is different as shown in Fig. 10.3. The micro-hydro power unit consists of a hard plastic casing and turbine that is connected to the winding coil by a shaft that will generate the electricity when rotating. The unit is mounted on a mild steel bracket with the boat hull. The complete experiment rig is shown in Fig. 10.4.

To conduct the experiment, there are several crucial steps that need to be done before conducting the experiment as shown in Fig. 10.2. It is also done to ensure that the experiment has taken precautious steps to avoid unwanted interference on the results. All the equipment and material used are assembled based on the experimental design and are tested for the functionality of the hydropower unit before continuing to begin with the experiment to make sure that the experiment is running smoothly without any possibilities of fault that will impact the results collected.

10.2.3 Method of Collecting Data

The voltage and current output of the hydropower unit are measured by a measuring device that is a multimeter for each of the variables, and the data is gathered and

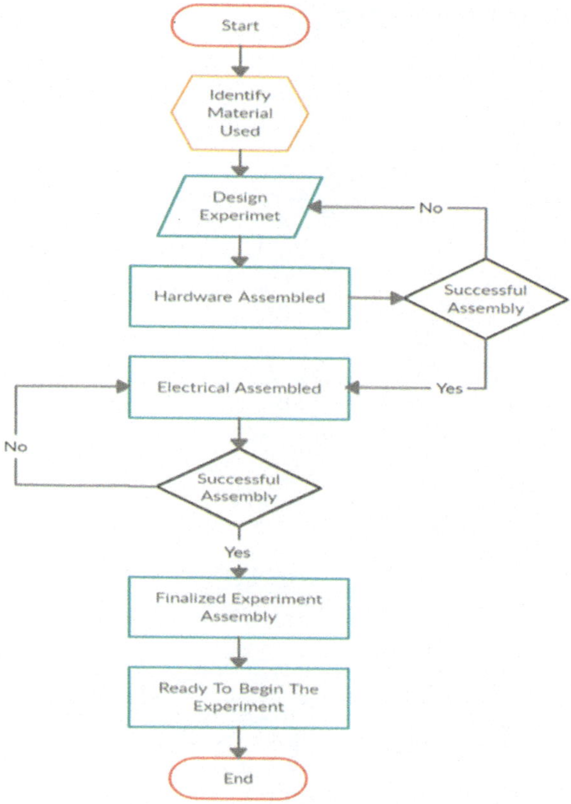

Fig. 10.2 Experiment development flow chart

Fig. 10.3 Hydro generator unit with three different number of blades

plotted into a graph for further analysis. The steps to take the result are shown in Fig. 10.5.

The flowchart in Fig. 10.5 shows the collecting data procedures of the experiment. The flowchart describes the process to gain the data for the project parameters till

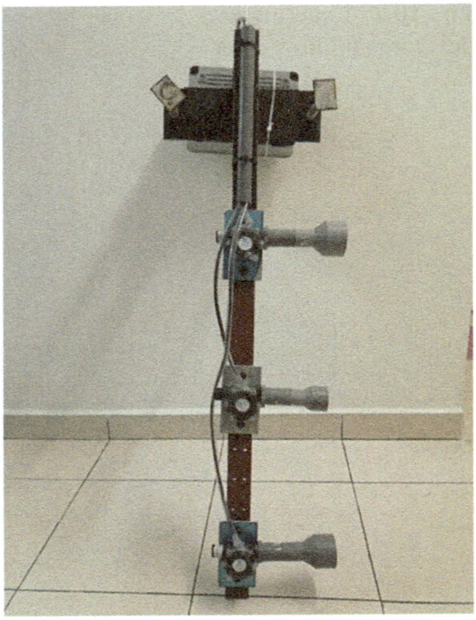

Fig. 10.4 Experiment rig

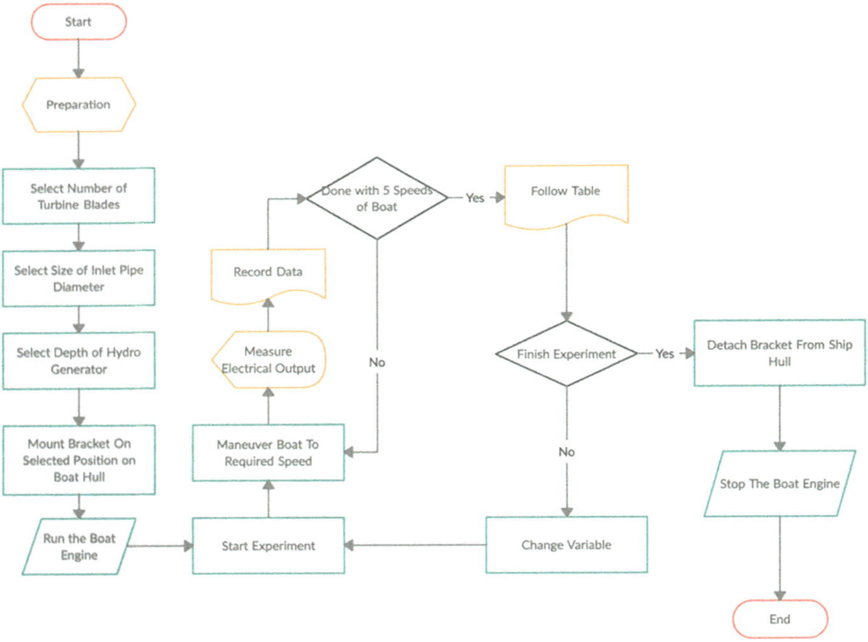

Fig. 10.5 Flow chart of experiment data collection

the data analysis obtained to be processed. Before the experiment starts, the turbine must be selected, attached the hydropower unit with the selected diameter of the inlet pipe, and fit the unit on the bracket at the required depth. Then, the bracket is mounted at the selected position on the boat hull according to Table 10.1. After all variables are selected, the boat engine is run and maneuvered to the required speed according to Table 10.1. The electrical output on the digital multimeter is recorded to take data according to Table 10.1. The experiment is continued by changing the variables that have not been done yet. The data are then transferred into a graph form

Table 10.1 Tabulation of data

No.	No. of turbine blades	Speed of boat (Knots)	Depth of waterline (m)	Diameter of inlet (inch)	mW
1	12	4	0	1"	
2	12	4	0	1–1/4"	
3	12	4	0	1–1/2"	
4	12	4	0.5	1"	
5	12	4	0.5	1–1/4"	
6	12	4	0.5	1–1/2"	
7	12	4	1	1"	
8	12	4	1	1–1/4"	
9	12	4	1	1–1/2"	
10	12	5	0	1"	
11	12	5	0	1–1/4"	
12	12	5	0	1–1/2"	
13	12	5	0.5	1"	
14	12	5	0.5	1–1/4"	
15	12	5	0.5	1–1/2"	
16	12	5	1	1"	
17	12	5	1	1–1/4"	
18	12	5	1	1–1/2"	
19	12	6	0	1"	
20	12	6	0	1–1/4"	
21	12	6	0	1–1/2"	
22	12	6	0.5	1"	
23	12	6	0.5	1–1/4"	
24	12	6	0.5	1–1/2"	
25	12	6	1	1"	
–	–	–	–	–	
108	24	7	1	1–1/2"	

for analysis purpose. Sensitive data analysis is used in collecting data and analysis for this experiment.

10.3 Results and Discussion

This part contains results and the discussion for the project experiment that has been conducted and completed successfully. This chapter is divided into several sections according to the variables to explain the data of the result. Each data is recorded into a table and plotted into a graph for better analysis and comparison throughout the experiment.

10.3.1 Sources of Data

The experiment has been conducted using the method shown in part 3 that used Table 10.2 to collect data and the result gathered to form a graph to show the result. The graph is plotted based on the parameters to show the comparison for each of the parameters.

As from the results, power output is created when water flows into the hydropower unit and makes the turbine turn. As the turbine rotor spins, the stator then produces the current as the magnetic field is cut. From there, the current is distributed by wire. Various parameters were used to identify the best situation to generate electricity. This experiment tells that each parameter shows a different power output. In theory, the greater the water flow, the greater the electrical current produced.

10.3.2 Result of Experiment

Figure 10.6 shows that the hydropower unit starts to produce a power output when the speed of the boat is greater than 3 knots. The speed of the boat below 4 knots cannot provide enough thrust to allow a high flow rate of water to flow through the turbine and to rotate the turbine blades fast enough to produce electricity. As shown in Fig. 10.6, the higher the speed of the boat which is causing a faster flow rate of the water to enter the hydro generator turbine inlet, the higher the power output produced from the hydro generator.

Based on Fig. 10.7, the hydro generator is mounted in three different depths which are 0, 0.5, and 1 m below the waterline of the boat. Result shows that the deeper the hydro generator, the higher the power output produced from the hydro generator. From a depth of 1 m, the hydro generator power output is the highest compared to 0.5 m and 0 m. This is due to a rise in hydrostatic pressure, which is the force of the liquid exerts on an object per unit area and this affects the rotation of the turbine

Table 10.2 Tabulation of result

No	No. of turbine blades	Speed of boat (Knots)	Depth of waterline (m)	Diameter of inlet (inch)	mW
1	12	4	0	1"	0
2	12	4	0	1–1/4"	0
3	12	4	0	1–1/2"	0.09
4	12	4	0.5	1"	0.06
5	12	4	0.5	1–1/4"	0.12
6	12	4	0.5	1–1/2"	0.45
7	12	4	1	1"	0.03
8	12	4	1	1–1/4"	0.03
9	12	4	1	1–1/2"	0.05
10	12	5	0	1"	0.02
11	12	5	0	1–1/4"	0.06
12	12	5	0	1–1/2"	0.3
13	12	5	0.5	1"	0.08
14	12	5	0.5	1–1/4"	4
15	12	5	0.5	1–1/2"	5.04
16	12	5	1	1"	0.12
17	12	5	1	1–1/4"	3.84
18	12	5	1	1–1/2"	5.1
19	12	6	0	1"	0.12
20	12	6	0	1–1/4"	0.35
21	12	6	0	1–1/2"	0.99
22	12	6	0.5	1"	0.2
23	12	6	0.5	1–1/4"	12.24
24	12	6	0.5	1–1/2"	13.75
25	12	6	1	1"	0.7
–	–	–	–	–	–
108	24	7	1	1–1/2"	45.24

blades. The deeper the hydro generator is placed, the greater the weight of the water pressing down on the turbine blades. Meanwhile, placing the hydro generator on the waterline produces a low power output due to insufficient water flow into the hydro generator. The power output is also affected by the existence of the wave motion of water.

Meanwhile, Fig. 10.8 shows the power output of three hydro generators with a variation in the number of turbine blades. It can be seen that the same trend of increasing in power output is obtained. But the hydro generator with 24 turbine blades produces more power output than 16 and 12 turbine blades. This happens

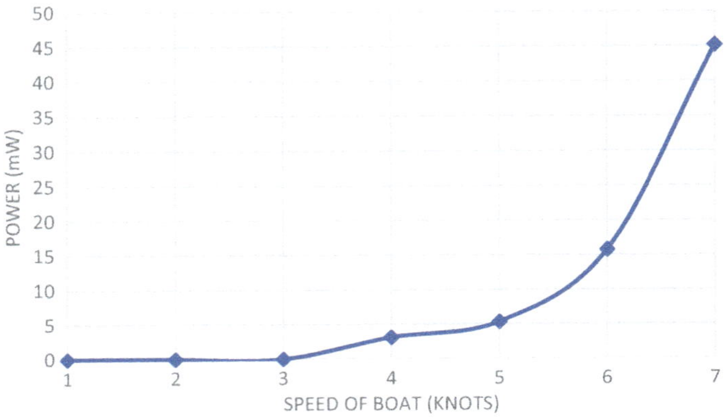

Fig. 10.6 Speed versus power output of hydro generator with 24 turbine blades and 1 ½" inlet diameter at 1 m depth

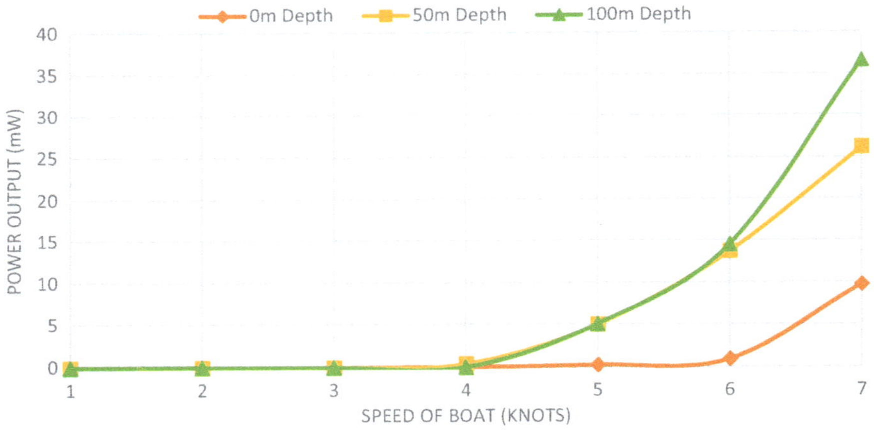

Fig. 10.7 Speed versus power output of hydro generator with 12 turbine blades and 1 ½" inlet diameter at variant depth of hydro generator

because in 24 blades turbine, more blades pick up the water stream when it enters the hydropower unit. As a result, it rotates faster than most to generate more electrical output.

Based on the graph in Fig. 10.9, the power output of a hydro generator with three different inlet scoop diameter sizes, which are 1,1 ¼, and 1 ½ inch, is shown. Result shows that the bigger the size of inlet scoop, the higher the power output produced by the hydro generator. This shows that the size of inlet diameter affects the power output produced. Bigger diameter of inlet allows more water to be scooped in and to flow through the hydro generator. Increasing the size of inlet diameter also allows bigger working pressure to act on the scoop and to produce higher velocity of water

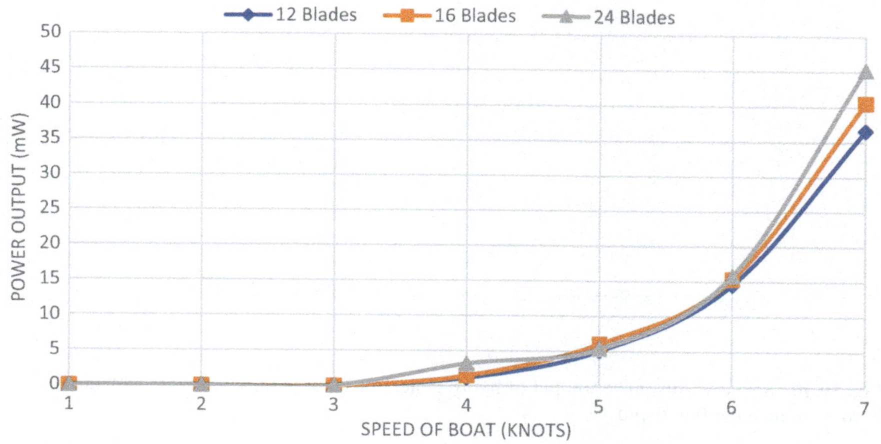

Fig. 10.8 Speed versus power output of hydro generator with inlet diameter size of 1 ½" and variant number of turbine blades at 1 m depth

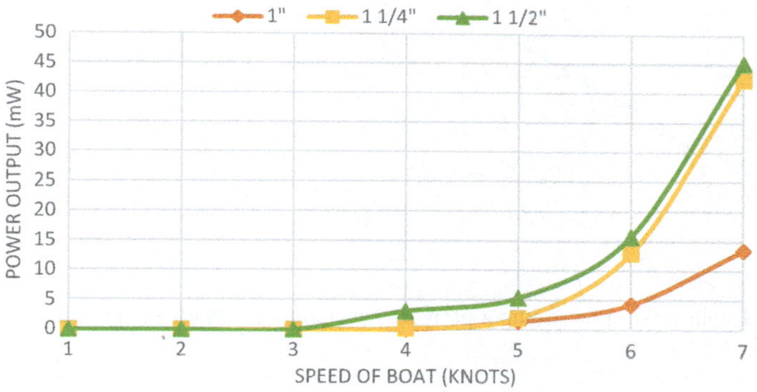

Fig. 10.9 Speed versus power output of hydro generator with 24 turbine blades and variant size of inlet diameter at 1 m depth

into the hydro generator and to push the hydro generator turbine. Meanwhile, in Fig. 10.10, the same hydro generator unit with a variant size of inlet diameter size is used on the waterline. Result shows that at the depth of 0 m, it cannot provide enough working pressure toward the inlet scoop to produce high velocity of water to push the hydro generator turbine blades.

Figure 10.11 shows the power output from the hydro generator unit with three different inlet scoop diameter sizes with different depth of water. The result shows that the performance of the hydro generator is affected by the size of inlet and the depth of the generator. The power output at 1 m depth has the highest value of the power output compared to 0 and 0.5 m. Thus, it shows that the bigger the diameter and deeper the depth of hydro generator, the more power output is produced by

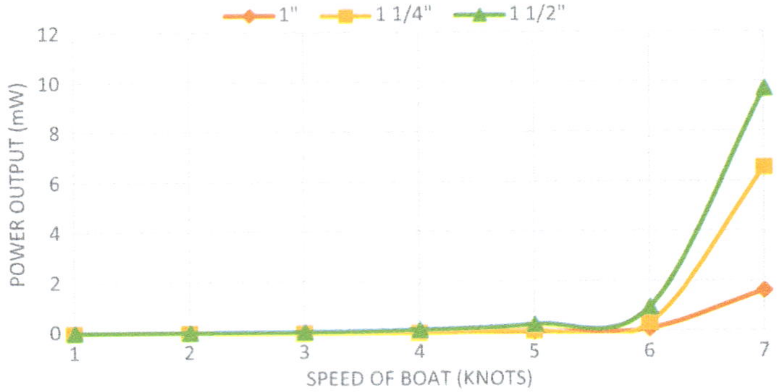

Fig. 10.10 Speed versus power output of hydro generator with 12 turbine blades and variant size of inlet diameter at 0 m depth

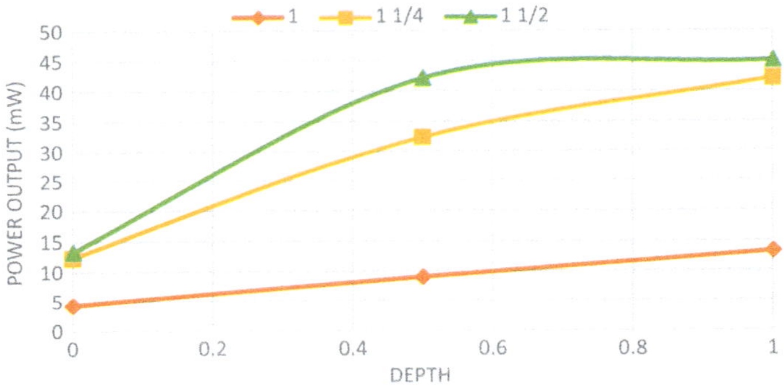

Fig. 10.11 Depth versus power output of hydro generator with 24 blades and variant size of inlet diameter

the hydro generator. Referring to Fig. 10.11, although the performance of the hydro generator is optimum when using the biggest diameter with deepest depth, the power will reach its peak at 45 mW. To achieve higher power, the diameter of the inlet and the depth of the hydro generator need to be increased.

10.4 Conclusion

In conclusion, this experiment was successful in achieving the objective which was to develop a portable hydropower for fishing boats, where this experiment was successfully fabricated using the materials and equipment selected in the methodology and adhering to the design to ensure this experiment works well. Next, it is to examine the hydropower unit's electrical power production. The results of this experiment can be found in the result and discussion part of paper, where graphs are plotted to demonstrate the contrast within the experiment's parameters and limitations. This experiment is carried out at a cost that is appropriate for all the chosen materials and design requirements.

Acknowledgements This research is conducted as one of the requirements that must be completed by the researcher that is supported by UniKL MIMET with guidance from experts, and for this, we would like to express our gratitude to anyone who was deliberately or inadvertently involved in the completion of this study. We also thank the readers for their positive feedback and helpful ideas in writing this paper.

References

1. Anaza SO, Abdulazeez MS, Yisah YA et al (2017) Micro hydro-electric energy generation—an overview. Am J Eng Res 2:5–12
2. Qazi A et al (2019) Towards sustainable energy. A systematic review of renewable energy sources, technologies, and public opinions. IEEE Access 7:63837–63851
3. Židonis A, Benzon DS, Aggidis GA (2015) Development of hydro impulse turbines and new opportunities. Renew Sustain Energy Rev 51:1624–1635
4. Hossain M et al (2017) (2018) A state-of-the-art review of hydropower in Malaysia as renewable energy: Current status and future prospects. Energy Strateg Rev 22:426–437
5. Yuebo X, Kabo-bah AT, Kabo-bah KJ et al (2018) Hydropower development—review of the successes and failures in the world. Elsevier Inc 2018. https://doi.org/10.1016/B978-0-12-813016-2.00004-6
6. Bagher AM, Vahid M, Mohsen M et al (2015) Hydroelectric energy advantages and disadvantages. Am J Energy Sci 2:17–20
7. Tammaruckwattana S, Reangkittakarn S, Rerkratn A (2018) Hydropower plant generator system. In: 2018 2018 3rd international conference on control and robotics engineering (ICCRE), pp 160–164
8. Varun BIK, Prakash R (2016) Hydropower as clean and renewable energy source for electricity generation. J Eng Res Appl Sci 1:359–369
9. Yutuc W, Jamal H, Afandi B (2018) Integrated full electric propulsion system for tanker ships with combined diesel and hydro generator drive. Mater Sci Forum 85:137–150
10. Blake M (2016) Portable hydroelectric generator. Pub No: US 2016/ 0271610 A1 Patent Application Publication, 19:1–5. https://patents.google.com/patent/US20160201643A1/en. Accessed 21 April 2021

Chapter 11
Natural Oleophilic Materials Study for Oil Pollution at Estuaries

Md Redzuan Zoolfakar, Abdullah Ahmad Fahmi Mohd Ruseli, and Sarah Nadiah Rashidi

Abstract Oil pollution or the introduction of oil into a body of water is a long occurring problem that affects the lives and growth of marine animals. In seas generally, oil pollution could be caused by either natural phenomenon or human error. Oil pollution may happen naturally through the natural oil seeps which occur when oil content underneath the surface of ocean floor escapes through cracks on the highly pressurized ocean floor. Even so, natural oil seeps are predictable and oil escapes in constant flow which allows easy containment. The collision or grounding of a ship is the example of oil pollution from human error. Collision or grounding of a ship may cause the hull to break open and releasing tons of oil which will harm the marine environment extensively. To combat oil pollution at sea, many laws and regulations have been applied by the International Maritime Organization. However, oil pollution at estuaries is often overlooked by many as it usually starts as a small-scale pollution and accumulates into a larger scale pollution. Main source of oil pollution in estuary is the fishing boat that accommodates the estuary area. There are various ways in dealing with oil pollution problems, and oleophilic material is one of them. An oleophilic material is a material which possesses a strong affinity toward oils instead of water. This materials could be in synthetic forms such as polystyrene or polyurethane and in natural forms which are kapok, rice husk, coconut coir, etc. This report will discuss and analyze the suitable and efficient natural materials that could be used to combat oil pollution at estuaries. The aim of this experiment that is conduct is to determine the oil sorption capacity of natural oleophilic materials in different types of water with the presence of several parameters.

Keywords Oleophilic · Natural materials · Oil pollution · Estuary water

---

M. R. Zoolfakar (✉) · A. A. F. Mohd Ruseli · S. N. Rashidi
Marine and Electrical Engineering Technology, Universiti Kuala Lumpur Malaysian Institute of Marine Engineering Technology, Lumut, Malaysia
e-mail: redzuan@unikl.edu.my

S. N. Rashidi
e-mail: sarahnadiah@unikl.edu.my

A. Ismail et al. (eds.), *Advanced Materials and Engineering Technologies*, Advanced Structured Materials 162,
https://doi.org/10.1007/978-3-030-92964-0_11

11.1 Introduction

One of the common sources of oil pollution is from marine vessels. Nowadays, most if not all ships possess oil separation equipment that treats oily wastewater before discharging it into the seas. Small recreational boats, however, do not possess any specific equipment for oily wastewater treatment. The amount of oil accumulated inside fishing boats is also fairly small compared to bigger ships for it to have an oil separation equipment, and it is hard to track as most of the oil spill from recreational boats goes unreported [1].

Those such a small amount of oil spill from recreational boats could accumulate and causes bigger problems if they are left unattended. The available space on the boats does not allow for any big oily water separation equipment, and cost is also an issue. The application of oleophilic materials could solve these problems for small recreational boats, and it does not require large space, and it is relatively cheap. According to [2], oleophilic materials refer to substances which possess the properties to attract oil. Oleophilic materials are available in both synthetic and natural.

Some of the examples of synthetic oleophilic materials are polypropylene, polyurethane, polystyrene, epoxy, calcium carbonate, and magnesium carbonate [3].

According to [4], an experiment has been conducted using open-cell foams from fibrillar blend of polypropylene and polytetrafluoroethylene. The data from the experiment show that the materials exhibit super hydrophobicity and possess the ability to selectively absorb various petroleum products.

According to [5], magnesium carbonate could be modified to create a superhydrophobic and oleophilic sorbent for oil spill clean-ups using a chemical process which includes the usage of palmitic acid. Based on the experiment, the modified magnesium carbonate is capable of absorbing oil up to about three times its weight and has high selectivity, buoyancy as well as rate of uptake of oil.

Other than that, authors of [6] executed an experiment using calcium carbonate treated with fatty acid as selective oil sorbent for oil separation purposes and the result from the experiment shows that the substance used could be a potential candidate for oil spill clean-ups due to its low-cost, effective, and environmentally friendly properties.

Polystyrene, a synthetic hydrophobic and oleophilic material, is fairly easy to be obtained. Authors of [7] conducted an experiment using the one-step method to develop both superhydrophobic and super oleophilic surfaces of polystyrene without involving chemical intervention. The end product from this experiment is proven to be able to be used as oil sorbent or in oily water separation processes.

Natural oleophilic materials are also easy to access as most of them are waste products from agriculture activities. Among them are kapok, rice straw, empty fruit bunch, etc.

According to [8], kapok's performance mainly depends on their packing densities. An experiment was conducted by them involving kapok in loose form and packed form. The results from the experiment show that its sorption capacity exceeds the sorption capacity of polypropylene by a significant value.

Rice straw is a waste byproduct from the rice harvesting process and usually being fed to livestock. According to [9], an investigation has been made involving the acetylation of rice straw with acetic anhydride at 100 degree Celsius with a few tertiary amine catalyst and without catalyst. In the end of the investigation, the acetylated straw shows a significant increases in hydrophobic characteristic, therefore offering a potential candidates for natural sorbents in oil cleanup.

Empty fruit bunch is residual materials that could be obtained from the palm oil extraction process. Empty fruit bunch or EFB could also be used as fuel by drying it first. EFB usually contains a high moisture which reduces combustion temperature and lowers the energy efficiency [10].

According to [11], empty fruit bunch was being modified using grafting reduced graphene oxide method. The modified EFB was then mixed together with mixture of oil and water before being dried for data analysis. The result from this study shows that the modified EFB demonstrated a hydrophilic characteristic by 131.6% of water absorption compared to the unmodified EFB which is 268.9%. The modified EFB also displays an oil–water separation increment by 17 times than the unmodified EFB fibers.

11.2 Methodology

11.2.1 Parameter

There are four main parameters that are used in this experiment. These parameters are constant on different types of raw materials throughout the experiment. The first parameter is the type of water which is used in the experiment. Types of water that is used are sea water, fresh water, and estuary water. These types of water will be collected from the same sources throughout the entire experiment.

The second parameters are the weight of materials. The materials used in this experiment possess various weight to volume ratio. For example, 5 g of kapok will be bigger in volume when compared to 5 g of rice husk. To make it equal for all materials, the amount of materials used will be determined using its weight instead of its volume. The weight of each materials used will be 5, 10, 15, and 20 g.

Next is the volume of oil. Only used engine oil (SAE-40) is used throughout this experiment to ensure the result of experiment is not being affected by the type of oil. However, the volume in each experiment is varied from 10, 15, and 20 ml.

The last parameter in the experiment is the time of reaction. The time allowed for the experiment to undergo varies from 12, 24, 36, and 48 h. Theoretically, any oil sorption will happen in early duration of experiment, but longer time will be provided to maximize the oil sorption process.

11.2.2 Raw Materials

Three oleophilic natural materials are used in this experiment: kapok, rice husk, and coconut husk. Materials will be gathered from only a source for each materials to prevent any disturbance that might affect the result from happening. Furthermore, only materials that are untreated with any form of treatment is used to make sure the absence of interference of any chemical during the experiment.

11.2.3 Preparation of Material

Based on Fig. 11.1, the natural oleophilic materials are soaked in hot water and left overnight to remove any contamination. The materials are then rinsed and dried before being weighed. After that, the materials will be packed into a meshed bag according to the parameters of experiment as shown in Figs. 11.2, 11.3, and 11.4.

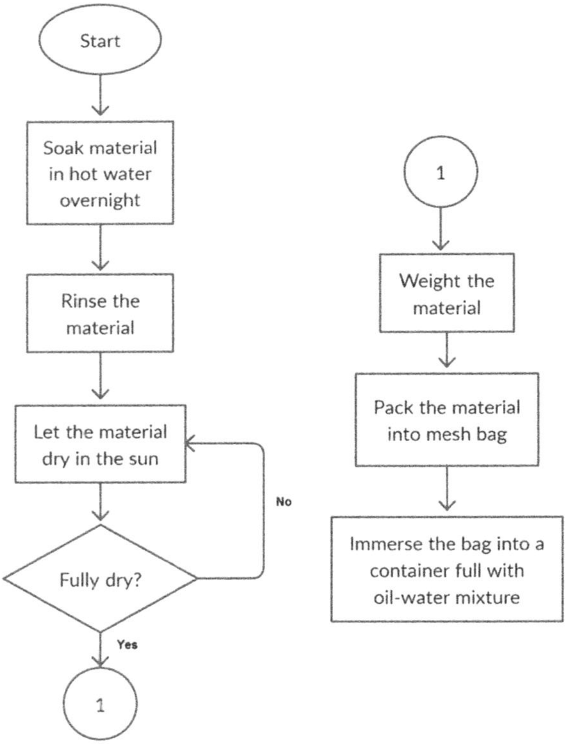

Fig. 11.1 Preparation of experiment flowchart

Fig. 11.2 15g of occonut coir packed in meshed bag

Fig. 11.3 15g of rice husk packed in meshed bag

Fig. 11.4 15g of kapok packed in meshed bag

11.2.4 Flowchart of Experiment

Referring to Fig. 11.5, the materials which have been packed into a meshed bag is immersed into a glass beaker containing 500 ml of water mixed together with a volume of oil based on the parameters. The specimen was left undisturbed for a period of time. After the period has ended, the specimen is taken out and left to dry. Once the specimen has fully dried, it is weighed and the data are recorded.

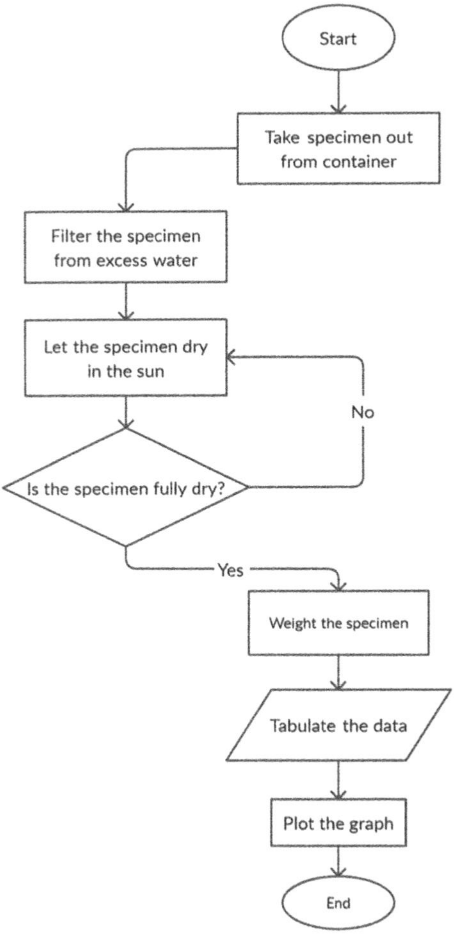

Fig. 11.5 Data collection flowchart

11.2.5 Method of Data Collection

The experiment is done based on Table 11.1 to minimize any error and overlooked disturbance during the experiment. Data obtained from the experiment are tabulated according to the table, and a graph is constructed.

Table 11.1 Sensitive data analysis

No.	Material	Type of water	Weight of material (g)	Volume of oil (ml)	Hours	OSC (g)
1	KAPOK	FW	5	10	12	8
2	KAPOK	FW	5	10	24	8
3	KAPOK	FW	5	10	36	9
4	KAPOK	FW	5	10	48	9
5	KAPOK	FW	5	15	12	9
6	KAPOK	FW	5	15	24	10
7	KAPOK	FW	5	15	36	12
8	KAPOK	FW	5	15	48	12
9	KAPOK	FW	5	20	12	15
10	KAPOK	FW	5	20	24	16
11	KAPOK	FW	5	20	36	17
12	KAPOK	FW	5	20	48	17
13	KAPOK	FW	10	10	12	7
14	KAPOK	FW	10	10	24	7
15	KAPOK	FW	10	10	36	8
16	KAPOK	FW	10	10	48	8
17	KAPOK	FW	10	15	12	11
18	KAPOK	FW	10	15	24	12
19	KAPOK	FW	10	15	36	12
20	KAPOK	FW	10	15	48	13
21	KAPOK	FW	10	20	12	15
22	KAPOK	FW	10	20	24	16
23	KAPOK	FW	10	20	36	17
–	–	–	–	–	–	–
324	Coconut COIR	EW	15	20	48	15

11.3 Results and Discussion

Result from the experiment has been observed and calculated. Based on the data obtained, a few graphs has been constructed.

Referring to Fig. 11.6, lines in the graph represent types of water used in the experiment as the parameters. The material used for the data in the graph is 5 g of rice husk with 15 ml of used engine oil. As shown in the graph, rice husks that are immersed in the sea water possesses the most oil sorption capacity in all period of the experiment. The reason for this to happen is due to the density of seawater is higher than fresh water and estuary water. By possessing higher density, sea water causes the rice husk to float higher than other water making it closer to the surface where the oil accumulated.

As the data in Fig. 11.7, the line in the graph shown represents the data from 5 g each of kapok, rice husk, and coconut coir in sea water mixed with 15 ml of used engine oil. As shown on the line of the graph, kapok and rice husk both are able to

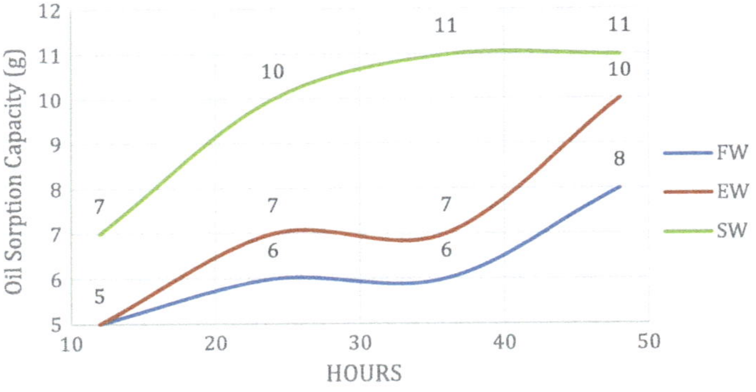

Fig. 11.6 Oil sorption capacity versus hours with 5 g of rice husk and 15 ml of oil

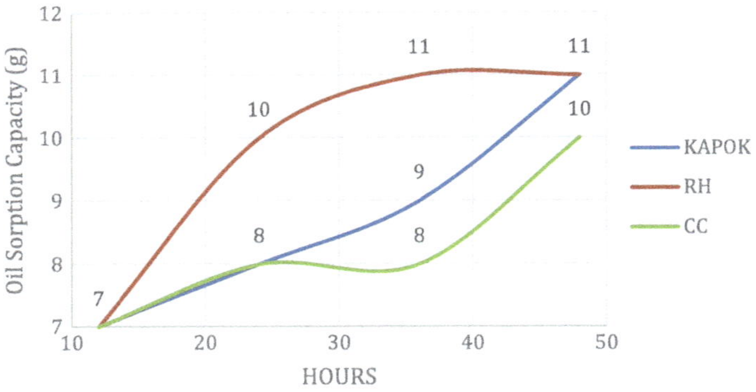

Fig. 11.7 Oil sorption capacity versus hours with 5 g of materials and 15 ml of oil in seawater

absorb 11 g of oil at 48 h while coconut coir is only able to absorb 7 g. Rice husk, however, absorbs more oil than kapok in 24 h and 36 h. The causes for this result are due to both the density and characteristic of the materials. Rice husk possesses higher density than kapok which allows it to float just below the surface. Kapok, however, is lighter in density with addition of it being in round shape after being packed in the meshed bag. Due to this, kapok floats above the oil surface with less surface contact with the oil. As times goes by, kapok starts to absorb more water and oil making it sink and increase the surface contact with the oil. The coconut coir absorbs the least amount of oil since it absorbs more water and at a faster rates making it sink below the surface of water.

In Fig. 11.8, the graph's lines represent the weight of the materials. Materials with the weight of 10 g are shown to be able to absorb more oil than both 5 g and 15 g. 5 g of materials is shown to absorb the least amount of oil due to its small size which results in small surface contact between materials and oil. As for 15 g of materials, it is the largest in and it possesses the biggest surface contact between materials and oil. However, 15 g of materials also sink faster than the rest of the weight of materials as it also absorbs the water the most. 10 g of materials has proven to be the most efficient in oil absorption, the amount of water it absorbs allows it to float just below the surface of water, and with the help of large surface contact, it is able to outperform other weight of materials.

In Fig. 11.9, the graph represents the materials being compared. Rice husk and kapok managed to absorb the highest amount of oil sorption capacity at 36 h and 48 h while coconut coir only managed to absorb 15g and 16 g. Kapok at 10 g of weight is able to absorb the most amount of oil, this is due to the amount of water it absorbs does not overweight the buoyance force of the materials. Thus, kapok is able to float in between the surface of oil and water with only half of it is submerged resulting it to stay in contact with the oil surface for a longer time compared to other materials. Rice husk and coconut coir both are able to absorb less amount of oil in earlier period of time and gradually increases as the time goes.

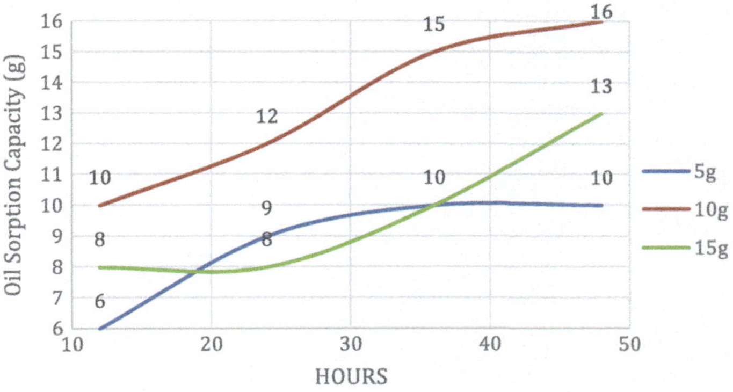

Fig. 11.8 Oil sorption capacity versus hours with coconut coir as the materials and 20 ml of oil in freshwater

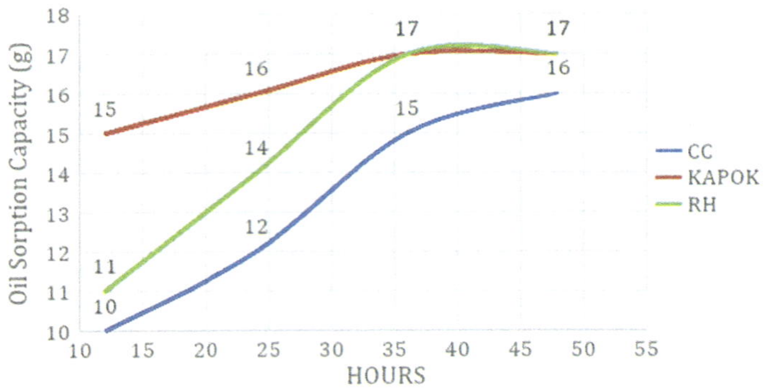

Fig. 11.9 Oil sorption capacity versus hours with 10 g of materials and 20 ml of oil in freshwater

In Fig. 11.10, the graph shown represents the various volume of oil in the same setting of experiment which is 15 g of kapok in estuary water. The result in the graph shows that 20 ml of oil produces the most oil sorption capacity for the materials. The explanation for this to happen is because of the volume of oil. 10 ml of oil is small in volume and once it accumulated in the surface, the area of surface it covers are small compared to other volume of oil. This results in smaller surface contact between materials and oil. The same theory could be applied for 15 ml and 20 ml of oil, once the oil accumulated on the surface of water, it covers a larger area, thus resulting in larger surface contact between materials and oil.

Referring to Fig. 11.11, an experiment has been conducted with a fixed setting of 20 ml oil and 15 g of materials in estuary water. Kapok has proven to be the best at absorbing oil with 16 g of oil sorption capacity at 48 h while rice husk and coconut oil only absorb 15 g of oil at the same time period. The affecting factor is the surface area of a 15 g kapok being larger than 15 g of rice husk and coconut coir. Larger

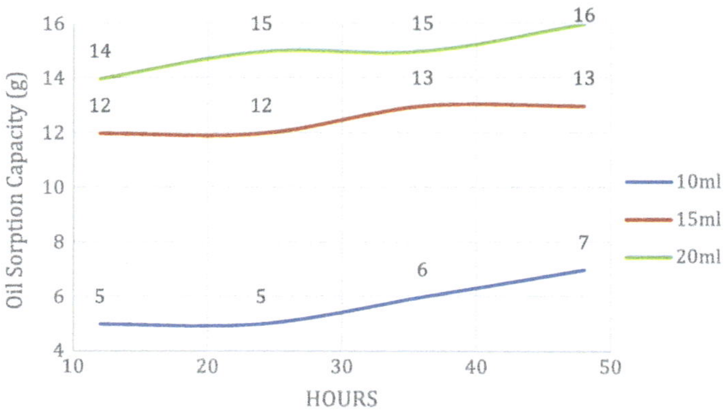

Fig. 11.10 Oil sorption capacity versus hours with 15 g of kapok in estuary water

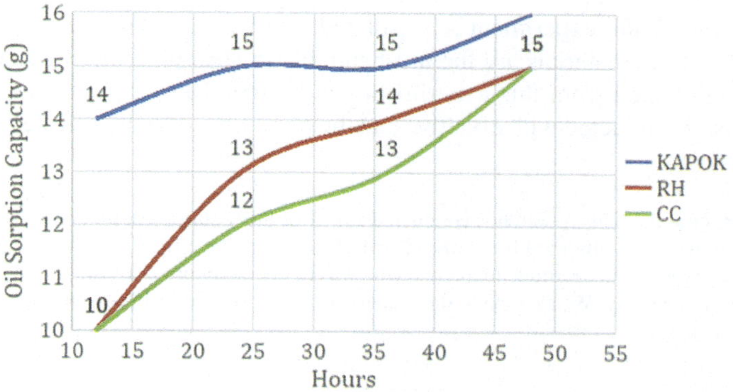

Fig. 11.11 Oil sorption capacity versus hours with 15 g of materials and 20 ml of oil in estuary water

surface area leads to larger surface contact between materials and oil which results in higher oil sorption capacity.

11.4 Conclusion

This paper reports on the research investigations to determine the oil sorption capacity of three natural oleophilic materials with four fixed parameters. Oil pollution has affecting lives of marine animals and if left unattended, humans lives too. The pollution of oil from big ship has been controlled as there are many laws and regulations been made to combat the pollution from big ship. In estuary, however, oil pollution is still a major problem due to fishing boats owner who place their boats there. The problem of oil pollution in estuary could be solved as there are many methods that exist today. The usage of oleophilic natural materials could also help to fight oil pollution using a simpler and less costly method compared to others.

For the preparation of the experiment, it is important that the materials gathered are not contaminated with any chemicals substances to prevent it from affecting the result of this experiments. As for the data collection, it is important to make sure the equipment used is the same throughout the entire experiment to avert any error.

The parameters used in this experiment show significant differences in each experiment as shown in the graphs in the result section. The data obtained from the experiment using the parameters are tabulated and compared in the form of graphs to make analyzing the data easier and more accurate. From comparing the graphs, we could conclude that the parameters which affect the result the most is the volume of the oil while the type of water and the weight of materials also show significant differences. The effective materials on lighter weight might not be effective in heavier weight. The same goes to the type of water and volume of oil.

In conclusion, this experiment is considered to be successful in determining which natural oleophilic materials are the most suitable in absorbing oil in estuary water. The result obtained from this experiment can be found in the result section. This experiment is conducted with effective and the optimum cost with selected materials and equipment.

Acknowledgements This research is conducted as one of the requirement that must be completed by the researcher that supported by UniKL MIMET with guidance from experts and for this, we would like to express our gratitude to anyone who was deliberately or inadvertently involved in the completion of this study. We also thank the readers for their positive feedback and helpful ideas in writing this paper.

References

1. Dell'amore C, Nunez C (2013) 3 Surprising sources of oil pollution in the ocean. https://www.nationalgeographic.com/science/article/140325-texas-pollution-oil-spills-animals-science. Accessed 20 Apr 2021
2. O'Lenick T (2018) Comparatively speaking: oleophilic vs. siliphilic. https://www.cosmeticsandtoiletries.com/research/chemistry/4823346.html. Accessed 23 Apr 2021
3. Yurgaitis D (2017) 10 Main types of sorbents for oil spills in water. https://blog.meltblowntechnologies.com/10-main-types-of-sorbents-for-oil-spills-in-water. Accessed 23 Apr 2021
4. Rizvi A et al (2014) Superhydrophobic and oleophilic open-cell foams from fibrillar blends of polypropylene and polytetrafluoroethylene. ACS Appl Mater Interfaces 6(23):21131–21140
5. Patowary M et al (2014) Chemical modification of hygroscopic magnesium carbonate into superhydrophobic and oleophilic sorbent suitable for removal of oil spill in water. Appl Surf Sci 320:294–300
6. Arbatan T et al (2011) Superhydrophobic and oleophilic calcium carbonate powder as a selective oil sorbent with potential use in oil spill clean-ups. Chem Eng J 166(2):787–791
7. Tu CW et al (2007) Fabrication of superhydrophobic and superoleophilic polystyrene surfaces by a facile one-step method. Macromol Rapid Commun 28(23):2262–2266
8. Lim TT, Huang X (2007) Evaluation of kapok (Ceiba pentandra (L.) Gaertn.) as a natural hollow hydrophobic-oleophilic fibrous sorbent for oil spill cleanup. Chemosphere 6(5):955–963
9. Sun XF et al (2002) Acetylation of rice straw with or without catalysts and its characterization as a natural sorbent in oil spill cleanup. J Agric Food Chem 50(22):6428–6433
10. Han J, Kim J (2018) Process simulation and optimization of 10-MW EFB power plant. Comput Aided Chem Eng 43:723–729
11. Sajab MS et al (2018) Oleophilicity and oil-water separation by reduced graphene oxide grafted oil palm empty fruit bunch fibres. Sains Malays 47(8):1891–1896

Chapter 12
Mushroom House Monitoring System at Pondok Seri Permai, Pasir Puteh, Kelantan

Sheikh Mohd Firdaus Sheikh Abdul Nasir, Hamid Yusoff, Halim Ghaffar, and Aliff Farhan Mohd Yamin

Abstract In Malaysia, most farmers depend on traditional agricultural practice. The adaptation of modern agricultural technology plays an important role in improving the overall efficiency and productivity. In modern agriculture, the internet of things (IoT) connects farmers to their farms through sensors to facilitate monitor of real-time conditions of their farms from anywhere. Oyster mushroom is widely cultivated among farmers in Malaysian. Although this crop is widely consumed and cultivated, it remains overshadowed by traditional cultivation approaches which result in low productivity, high labor efficiency, high cost and effort. Thus, this study aimed to develop a monitoring system based on IoT on environmental conditions of a mushroom farm, namely temperature, humidity, moisture and light intensity. Oyster mushroom requires an optimum temperature between 26 and 29 °C, humidity from 85 to 95% and carbon dioxide not exceeding 600 ppm. Sensors were placed at fixed locations in the farm to measure the parameter status transmitted to a remote monitoring station via low-power NodeMCU. The data obtained were stored on the cloud platform. The codes for the controller were written in the Arduino programming language, debugged, compiled and burned into the microcontroller using the Arduino integrated development environment. The results showed the success of monitoring environmental conditions through internet access from anywhere. This approach will reduce human efforts and also help to automate production, which benefits farmers in Malaysia.

S. M. F. Sheikh Abdul Nasir (✉) · H. Yusoff · H. Ghaffar · A. F. Mohd Yamin
Advanced Mechanic Research Group, School of Mechanical Engineering, College of Engineering, Universiti Teknologi Mara, Cawangan Pulau Pinang, Kampus Permatang Pauh, 13500 Permatang Pauh, Pulau Pinang, Malaysia
e-mail: sh.firdaus@uitm.edu.my

H. Yusoff
e-mail: hamidyusoff@uitm.edu.my

H. Ghaffar
e-mail: halim4346@uitm.edu.my

A. F. Mohd Yamin
e-mail: aliff.farhan6205@uitm.edu.my

A. Ismail et al. (eds.), *Advanced Materials and Engineering Technologies*, Advanced Structured Materials 162,
https://doi.org/10.1007/978-3-030-92964-0_12

Keywords Smart farming · Mushroom · Humidity · IOT · Monitoring system

12.1 Introduction

The purpose of this project is to monitor the environmental conditions such as temperature, humidity, carbon dioxide (CO_2) and lighting of the oyster mushroom farm at Pondok Seri Permai, Pasir Putih, Kelantan by using the internet of things (IoT). The IoT is a network of physical devices inserted with physics, software, sensors, actuators and properties that allows these objects to attach and exchange knowledge. Since such structures require refinement, a scientifically designed mushroom farm needs large investments which are consequently unaffordable by small and marginal mushroom farmers. In addition, mushroom units need to keep their air conditioners operating most of the year. For large oyster mushroom cultivation, we consider temperature, humidity, light and carbon dioxide. To reduce human effort and enhance yields, this system will provide a novel method of monitoring the farm. Food demand and space or land limitation for agro-economic activities highlight urban agricultural technology and it becomes one of the promising solutions to secure food supply. In addition, extreme weather and climate changes influence crop production, thereby increasing prices and lowering the crop quality produced [1].

Mushroom farming is time-consuming and requires precise methods. Mushrooms have a rough time reproducing if the conditions are not suitable. Most mushroom farmers in Malaysia use time-controlled humidifier, which is inconsistent in preserving humidity and temperature in their fields, resulting in lower productivity [2, 3, 10].

It grows best at an elevation of 800 m above sea level, with temperatures ranging from 16 to 22 °C and relative humidity levels of 60–90% [4, 5]. It is not difficult to cultivate mushrooms in the lowlands if the environment is built in accordance with the real mushroom ecosystem. Temperature, humidity, sun and fresh air are the four elements that make up the atmosphere setting [2]. The air flow of the mushroom growing chamber is critical as it directly influences the CO_2 content of the space [10].

A mushroom cultivation system that is integrated with technology that can detect and maintain environmental conditions that are required by the farmer. It is essential to implement a sensor-based environmental control system. The majority of mushroom farmers in Malaysia have small-scale operations with only few or no employees, and they do not operate 24 h a day [2]. The mushroom farm is almost always deserted. As a result, a remote monitoring system is required to monitor the condition of mushroom farms to avoid unfavorable conditions that could lead to mushroom death. To address the issues raised above, a system that can remotely monitor and regulate environmental conditions is needed [3, 10].

This study highlights the design of an interface circuit for agricultural sensors, namely light, carbon dioxide, temperature and humidity sensors. A remote monitoring system was also designed including a user interface to display accurate data from sensors and a closed-loop control system to maintain optimum mushroom growth conditions by controlling temperature, humidity and light. In addition, this paper presents an IoT-based monitoring and environmental condition control for oyster mushroom indoor cultivation. This is a smart urban farming system with less maintenance, less manpower and space saving [6, 7]. Moreover, this study intends to improve and enhance the conventional farming system in general. The use of IoT platform will improve the capability of current equipment for remote monitoring purpose and at the same time provide data logs for analysis and reference [8, 9].

12.2 Methodology

The oyster mushroom cultivation was carried out in a room with length of 30 m, wide of 15 m and height of 3 m inside a concrete building as depicted in Fig. 12.1. The wall was made of bricks and the roof of the room was constructed under the main concrete. Four rows of shelves were installed in the room; each shelf contained around 1000 blocks of mushrooms. A 3500 CMH ventilation system for each exhaust fan, 50 watts light for 3000 lumens and cooling pad were installed on the front building to bring air from outside to inside. All the devices installed in position outside the cultivation room were fixed by electrical and electronic devices to monitor temperature, humidity, CO_2 level and light intensity during the experimental procedure as seen in Fig. 12.2. Optimal ranges of temperature, light intensity, CO_2 and humidity for oyster mushroom cultivation were 26–29 °C, 200–500 lx level, 85–95% humidity and 600 ppm CO_2. The humidity and temperature were generated by the exhaust fan and cooling pad system as shown in Fig. 12.3. Light intensity was provided by fluorescent lamps. The ranges of normal conditions for oyster mushrooms indoor cultivation are summarized in Table 12.1.

Fig. 12.1 Inside the building

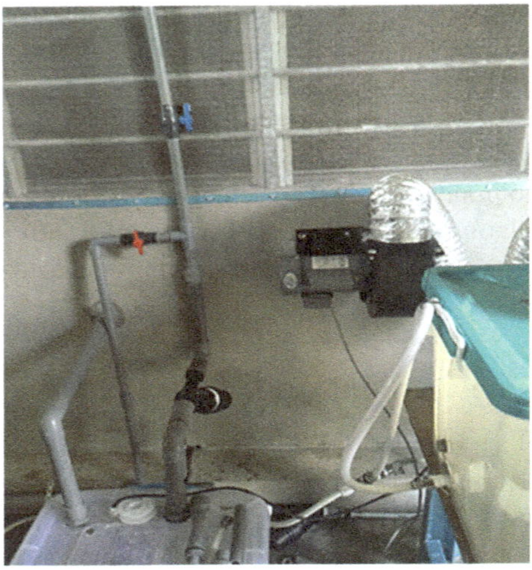

Fig. 12.2 Installation of irrigation and drainage systems

Fig. 12.3 Cooling system

Table 12.1 Parameters for oyster mushroom cultivation

Parameter	Range	Unit	Controlled equipment
Temperature	26–29	°C	Exhaust fan
Humidity	85–95	RH %	Cooling pad
CO_2	600<	ppm	Exhaust fan
Light	200–500	lux	Lamp

12.3 Software Implementation

In this project, the ESP32 controller was used as the main controller. It was a microcontroller with an integrated WIFI module. It was programmed using the C++ language with Arduino IDE software. All the sensor readings were captured by the ESP32 controller and displayed on the TFT display on the device and also sent to the server. The controller itself has a WIFI capability and was programmed to act as a WIFI client, thus enabling a direct connection to an available a WIFI Router with internet connection. An app known as Blynk was used as the IoT for this platform. The sensor data captured by the controller were sent to the Blynk server, and users can use the Blynk app to display the readings on an Android or iOS phone. Figure 4a, b show the algorithmic data transfer by ESP32 to the Blynk server and from Blynk server to the Blynk app, respectively. The data can be displayed in various types such as sensor value, graph, virtual meters and so on. The working block diagram of the whole system is displayed in Fig. 12.5.

Figure 12.6 depicts the schematic diagram of the greenhouse integrated with IoT. The main controller for this project was ESP32 DOIT DEVKIT V1. It was a 32-bit LX6 microprocessor with clock frequency of up to 240 MHz, supported Wi-Fi connectivity with 150 Mbps speed, supported both classic Bluetooth v4.2 and BLE specifications, 34 programmable GPIOs, up to 18 channels of 12-bit ADC and 2 channels of 8-bit DAC and last but not least serial connectivity including 4 × SPI, 2 × I2C, 2 × I2S, 3 × UART. The sensors used were DHT22 for temperature and humidity input, MH-Z14A infrared gas module for CO_2 reading and VEML7700 ambient light sensor to measure the lumen. The reading capability of the DHT22 sensor was in the range of 0–100% RH for humidity and −40–80 °C for temperature,

Fig. 12.4 Algorithm data transfer **a** ESP32 to server and **b** from server to application

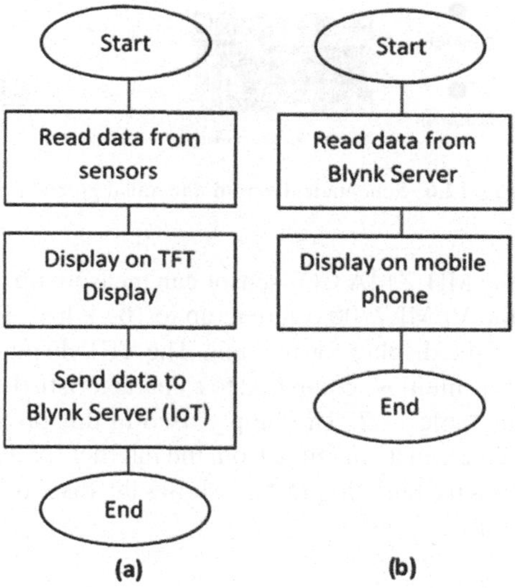

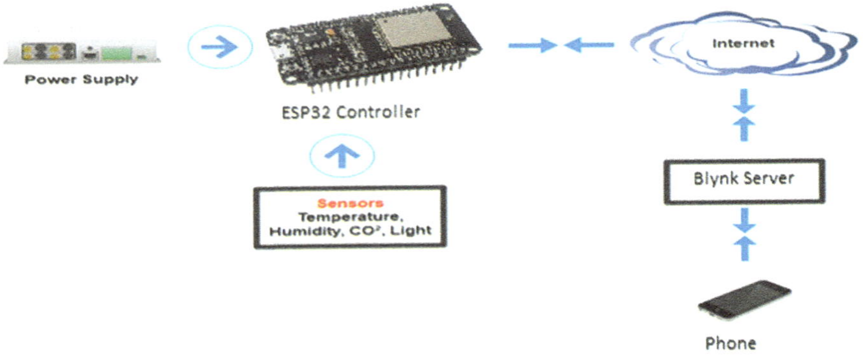

Fig. 12.5 System block diagram

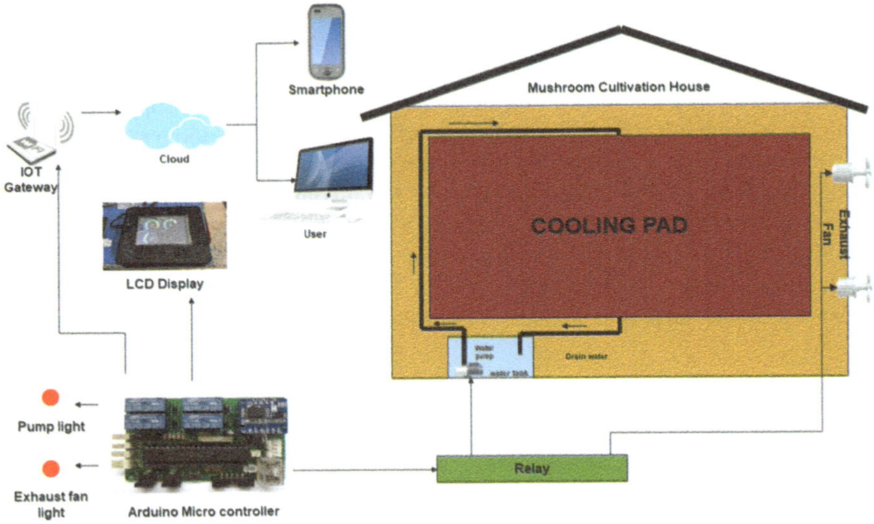

Fig. 12.6 Schematic diagram integrated green house

the MH-Z14A CO_2 sensor can measure up to 5000 ppm of carbon dioxide gas while the VEML7700 can read up to 167 k lux of ambient light. A 4″ TFT LCD was used as the display for the unit. The TFT display used graphical method, so the display was limitless compared to a characteristic LCD. All the main components are shown in Table 12.2. The supply used in this project was 5 V and 3.3 V. The system can obtain an actual time from the internet for timestamp but DS3231 RTC was also used as a backup. Figure 12.7 shows the display of the sensor readings and the controller unit.

Table 12.2 List of components

Main controller board and Wifi	ESP32 DOIT DEVKIT V1
Humidity and temperature sensor	DHT 22
CO_2 sensor	MH-Z14A infrared gas module
Light sensor	VEML7700 ambient light sensor
Display	4″ TFT LCD
RTC	DS3231

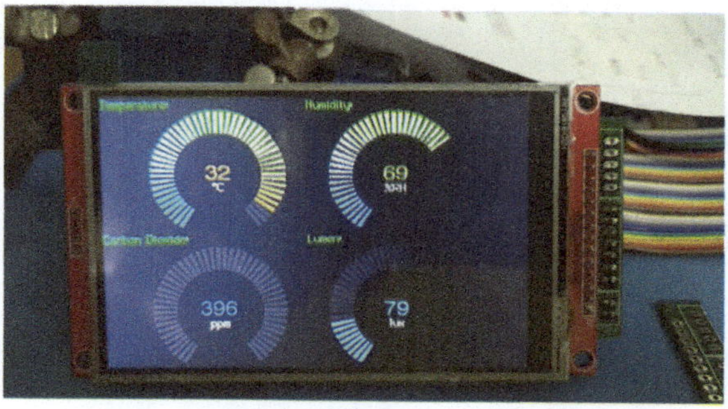

Fig. 12.7 System for environmental condition monitoring oyster mushroom indoor cultivation

12.4 Result and Discussion

This project reevaluates the concept among farms and agronomists. It is believed that IoT has revolutionized data monitoring, management and analysis. Temperature and humidity sensors were examined and the data obtained were sent to the IoT platforms for access and monitoring. Moreover, environmental parameters for mushroom farming require continuous monitoring. This system will play a crucial role in automation of mushroom cultivation. This project is beneficial for automation and monitoring of mushroom cultivation. It will also assist farmers in increasing agriculture yields and managing mushroom production efficiently. This system will also provide assistance to farmers to obtain proper live feed from ambient temperature and moisture with more accurate results with minimum manpower.

Figure 12.8 shows the graphical interface of the dashboard app used in the system. In this system, four parameters were monitored at a 10-min sampling rate, namely room temperature, relative humidity, light intensity and carbon dioxide. Figure 12.8 also shows the data with function of time which were obtained from the IoT server platform. Apart from using a browser, the data can be duplicated and monitored by users via third-party apps on android. This method provides flexible monitoring system to users. After implementing this system, the mushroom grew rapidly as shown in Fig. 12.9.

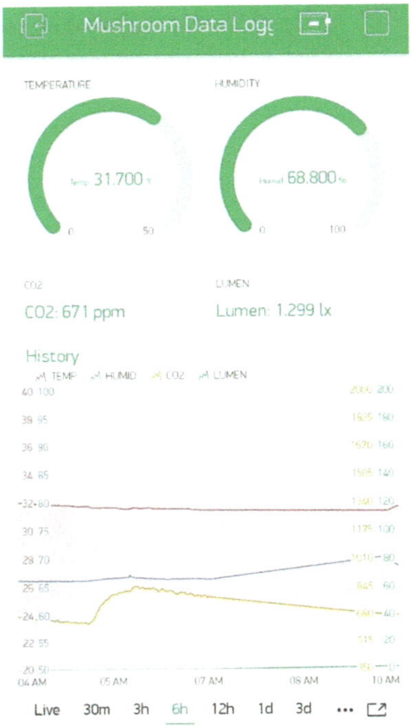

Fig. 12.8 User interface

Fig. 12.9 Oyster mushroom productivity after monitoring environmental conditions

12.5 Conclusion

In conclusion, monitoring and control of environmental conditions for oyster mushroom indoor cultivation were successfully conducted. The IoT-based monitoring system is an optimistic technology which can enhance the mushroom production based on the following reasons:

1. The system framework is based on open-source platform with low capital cost.
2. Easy and user-friendly for data migration.
3. The system is modular which can be expanded and easy to increase the number of data/sensors.

Nevertheless, this system has limitations that need to be considered, such as remote monitoring depending on the quality of internet connection which is currently not recommended. However, this can be resolved with the proposed system. In addition, hardware and internet data costs, operating costs (monthly/yearly) of cloud server platform should be considered as well. Many platforms are only free for personal or non-commercial use with limited access. Furthermore, data security on IoT is not undisclosed.

References

1. Talavera JM et al (2017) Review of IoT applications in agro-industrial and environmental fields. Comput Electron Agric 142:283–297
2. Grant JJ (2002) An investigation of the airflow in mushroom growing structures, the development of an improved, three-dimensional solution technique for fluid flow and its evaluation for the modelling of mushroom growing structures. Diss. Dublin City University
3. Jecko AA et al (2019) Design and implementation of wireless monitor and controlling system for the identification of water level. In: International conference on smart systems and inventive technology (ICSSIT). IEEE
4. Quratul AQA (2015) The transformation of agriculture based economy to an industrial sector through crowd sourcing in Malaysia. Int J Comput Sci Inf Technol Res 3(1):34–41
5. Rosmiza MZ et al (2016) Prospects for increasing commercial mushroom production in Malaysia: Challenges and opportunities. Mediterr J Soc Sci 7(1 S1):406–406
6. Akkaş MA, Radosveta S (20197) An IoT-based greenhouse monitoring system with Micaz motes. Procedia Comput Sci 113:603–608
7. Zhang Z et al (2017) Remote monitoring system for agricultural information based on wireless sensor network. J Chin Inst Eng Trans Chin Inst Eng 40(1):75–81
8. Terroso S, Fernando et al (2019) An open IoT platform for the management and analysis of energy data. Future Gener Comput Syst 92:1066–1079
9. Cambra C et al (2017) An IoT service-oriented system for agriculture monitoring. In: 2017 IEEE international conference on communications (ICC). IEEE
10. Zheyang H, Tengis T, Batminkh A (2020) A study of the incubator model for growing mushrooms. Int J Adv Culture Technol 8(1):19-25

Chapter 13
The Effect of Hilbert Curve Pattern Intensity in ASTM D638 Type III on Stress Concentration and Cyclic Application

Sheikh Mohd Firdaus Sheikh Abdul Nasir, Khairul Anuar Abd Wahid, Muhammad Nur Farhan Saniman, Wan Mansor Wan Muhammad, and Mohd Fahim Bin Ramlee

Abstract Various infill patterns are introduced in 3D printing to generate low-density objects that lead to reduced cost and fabrication time through mass reduction. However, as a trade-off, the strength of the 3D printed component is uncertain. Many works have been carried out to investigate the overall strength of the infill pattern but most of them do not address the specific study of the stress concentration region and cyclic loading durability. This paper focuses on investigating an effect of stress concentration region and cyclic loading durability of Hilbert curve pattern in a 3D printer sample that has been designed by following the ASTM D638 Type III standard. The density percentage of the infill pattern was varied from 20 to 60%. The analysis has been carried out by using the computational method. The results show that the stress concentration region is increased proportionally with infill density percentage probably due to the increasing of sharp edges amount in the Hilbert curve pattern. On the other hand, the results have shown that cyclic loading durability is inverse proportionally where the sample with 20% of infill density can withstand until 5000 times cycles under continuous cyclic loading and has been reduced consistently when the infill increases to 60% which the sample able to withstand up to 600 times cycles before the break.

S. M. F. Sheikh Abdul Nasir
Mechanical Department, Universiti Teknologi Mara, Permatang Pauh, Pulau Pinang, Malaysia
e-mail: sh.firdaus@uitm.edu.my

K. A. Abd Wahid (✉) · M. N. Farhan Saniman · W. M. Wan Muhammad · M. F. Bin Ramlee
Mechanical Engineering Section, Universiti Kuala Lumpur Malaysia France Institute, 43650 Bandar Baru Bangi, Selangor, Malaysia
e-mail: khairulanuarabdwahid@unikl.edu.my

M. N. Farhan Saniman
e-mail: mnfarhan@unikl.edu.my

W. M. Wan Muhammad
e-mail: drwmansor@unikl.edu.my

A. Ismail et al. (eds.), *Advanced Materials and Engineering Technologies*, Advanced Structured Materials 162,
https://doi.org/10.1007/978-3-030-92964-0_13

Keywords 3D printing · Mass reduction · Infill pattern · Stress concentration · Cyclic loading

13.1 Introduction

The 3D printing technology, which has been developed in the 1980s, is a manufacturing technology to produce three-dimensional products layer by layer [1, 2]. Since 3D printing is affordable, and easier to use compared to other quick prototyping technologies, the demand is increased to create complex geometries. The beauty of 3D printing technology is the capability to reduce the amount of material used during printing the components. In consequence, not only the cost for production could be reduced, but also speeding up the overall manufacturing processes [3, 4]. One of the approaches is to introduce an infill pattern inside the printed 3D structure rather than a solid fill. There are several types of infill patterns available such as rectilinear, concentric, honeycomb, octagram spiral, Archimedean chords, and Hilbert curve. Each of these infill patterns has their own characteristics. Marc et al. [5] stated that rectilinear at 50% infill density is the fastest type of infill pattern to achieve full coverage printing. However, finishing the quality of the 3D printed surface was not uniform as some side was thicker than another side. According to Miguel et al. [6], the honeycomb infill pattern could provide a good tensile strength compared to other patterns with a slights percentage of difference.

Furthermore, Grzegorz et al. [7] and Liseli et al. [8] stated that the honeycomb infill pattern offered the highest tensile strength compared to other patterns. On the contrary, Khan et al. [9, 10] mentioned that the rectilinear infill pattern could provide a higher tensile strength compared to honeycomb and concentric infill patterns. Despite such confusion, even though there are more than six infill patterns available, the most popular patterns that were being studied in 3D printing were only rectilinear [11] and honeycomb [12] infill patterns. Consequently, there is very little information on rarely used infill patterns such as the Hilbert curve. In addition, to the authors' knowledge, there is no specific study that investigates the stress concentration and cyclic loading effect to the various percentage of infill density of the Hilbert curve in 3D structure.

13.2 Methodology

The first stage of this study started with a design geometry of the dogbone specimen which is the standard design for tensile testing by following the ASTM D638 standard. PLA material properties have been defined as shown in [13]. The dogbone shape has been designed by using the Solidworks software as shown in Fig. 13.1.

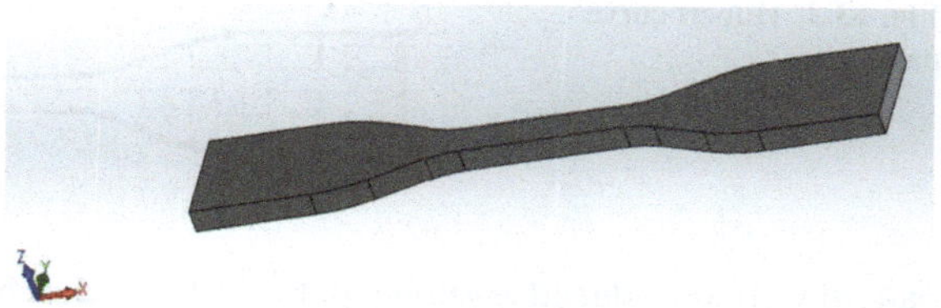

Fig. 13.1 Model of ASTM D638 specimen (unit: mm)

Afterward, the dogbone shape specimen model was then exported in Standard Triangulation Language (STL) file format, which then was further processed by a slicing software to implement various density infill patterns as shown in Fig. 13.2.

To study the influence of infill patterns on the tensile properties of 3D printed parts, the specimens were designed based on ASTM D638 Type III standards, as shown in Fig. 13.1. The value of gage length L_O, a width of narrow section W, and thickness T were 50, 19, and 9.6 mm, respectively. The resulted volume is 34,943.69 mm^3. The 3D model was then exported in Standard Triangulation Language (STL) file format, which then was further processed by a slicing software to implement various low-density infill patterns. After that, the G-code files for each 3D model with specific infill patterns were generated by the slicing software. The G-code files, which translated the 3D design into position coordinates, were then used in a FFF 3D printer to fabricate the specimens' layer by layer. The various densities of the Hilbert curve

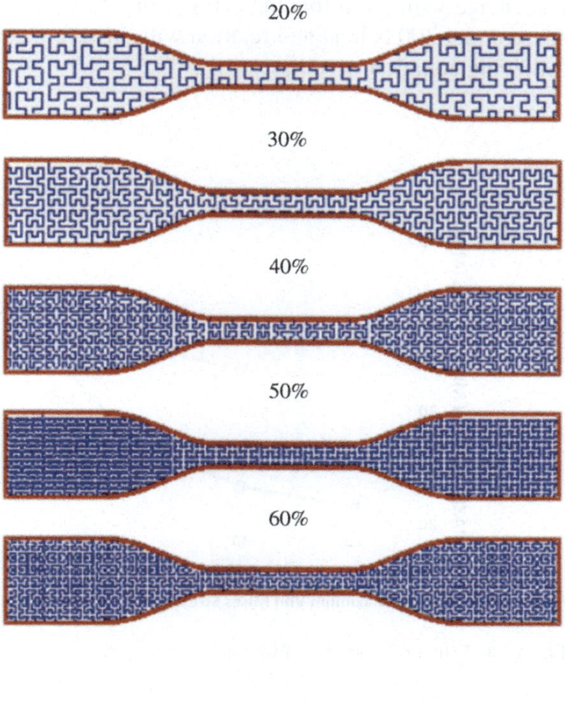

Fig. 13.2 Different infill density at 0 orientations

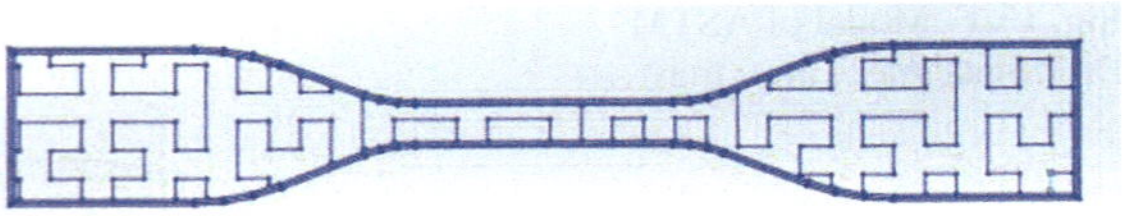

Fig. 13.3 Hilbert curve pattern

pattern were introduced as shown in Fig. 13.2. Reference [13] shows the mechanical properties of ASTM D638 by using PLA material.

This software using after the specimen was set all parameters in slic3r software. The function of Gcode to XYZ software is to generate the coordinate from Gcode which has been exported using the slic3r software. After that, the coordinate file will be opened using the SolidWorks software to design the solid parts. Data will be collected using the SolidWorks simulation to determine the stress concentration and fatigue occurs at the specimen with different loads (Fig. 13.3).

13.3 Results and Discussion

Static simulation was carried out to determine the maximum von Mises stress before concentration data will be calculated. Based on the static simulation, the value of force is averagely 300 N due to the average limit load that can be applied before the specimen breaks. 300 N then is applied to each specimen with various infill density percentage with point loading setting. Figure 13.4 shows the maximum von Mises stress under 300 N load and relation with stress concentration factor, *K* by using the following equation.

$$K = \frac{\sigma_{\max}}{\sigma_{\text{nom}}} \tag{13.1}$$

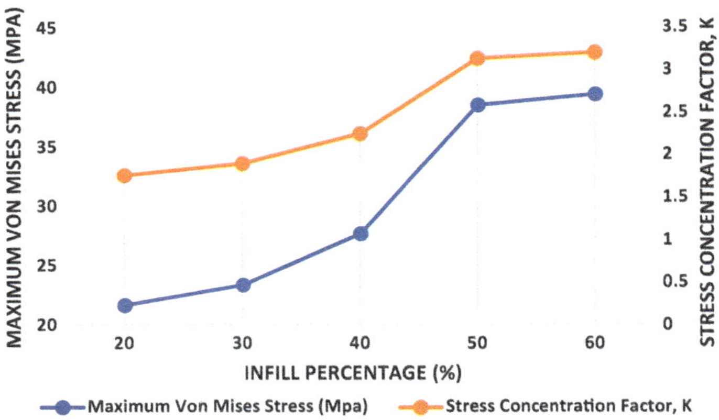

Fig. 13.4 Effect of load onto different infill density

$$\sigma_{nom} = \frac{F}{A} \tag{13.2}$$

where K is stress concentration factor, σ_{max} is the maximum von Mises stress, σ_{nom} is the nominal stress, and A is an area of specimen shape surface.

As shown in Fig. 13.4, the maximum von Mises stress proportionally increases with the percentage of the infill density along with stress concentration factor, K. The result shows that the specimen with 20% infill density experiences the maximum von Mises stress at 21.7 MPa which contributes to 1.77 stress concentration. It was shown that the increment of percentage infill density accumulates more stresses as the maximum von Mises stress and linearly increases up to 38.5 MPa and 3.14 for stress concentration factor for a specimen with 50% density. However, the maximum von Mises stress and stress concentration factors are almost consistent when the infill density increases by up to 60%. Based on this trend it indicates that a higher percentage of infill density leads to accumulating more stress as well as stress concentration factor. This is probably due to the fact that the total number of sharp edges in a Hilbert curve pattern increases proportionally with the amount of infill density. It was a mechanic phenomenon where the stress concentration is increased as the flow of stress is disrupted due to irregularities in geometry which in this case is the sharp edges of the Hilbert curve pattern.

Cyclic loading of 50 N and increases up to 320 N is applied continuously onto the designed specimen to investigate the limit of loading the specimen can be embraced as shown in Fig. 13.5. Figure 13.6 shows the result of fatigue result of applied cyclic loading. The results show that all specimens experience a similar S-N pattern as the loading increases. The specimen with 20% infill density is able to sustain 5184 cycles compared with specimen 60% infill which only can sustain 634 cycles. Less infill density seems able to sustain longer cycle probably due to plastic region properties as the brittle effect decreased. On the contrary, the specimen with the highest infill density which is 60% can sustain the highest stress compared to others as shown zoom graph. This is because of higher infill density can increase the overall tensile strength due to percentage density inside the specimen.

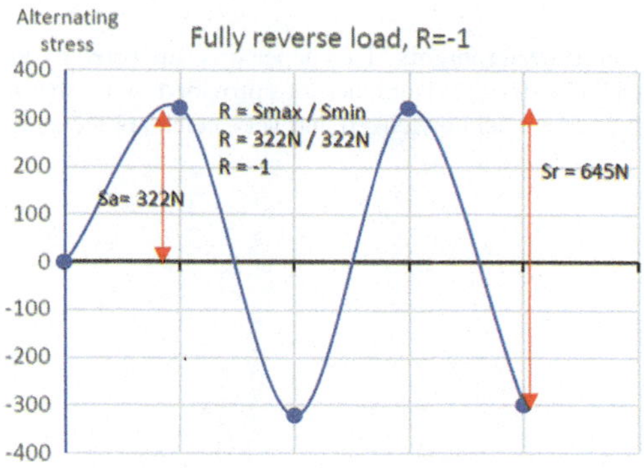

Fig. 13.5 Fully reverse cyclic loading on specimen

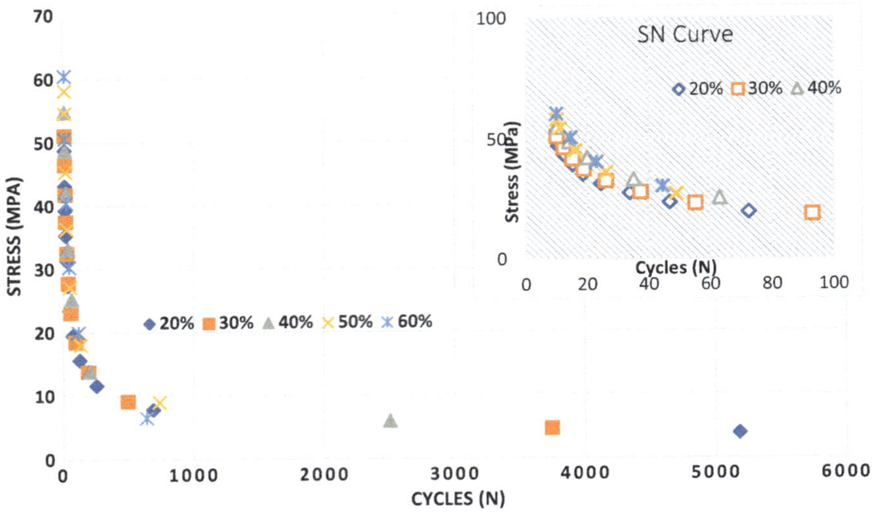

Fig. 13.6 S-N curve of stress and numbers of cyclic loading

13.4 Conclusion

An effect of stress concentration region and cyclic loading durability of Hilbert curve pattern in 3D printer sample that has been designed by following the ASTM D638 Type III standard and has been carried out by using the computational method. The density percentage of the infill pattern was varied from 20 to 60%. The results show that the stress concentration region is increased proportionally with infill density percentage probably due to the increasing of sharp edges amount in Hilbert curve pattern. On the other hand, the results have shown that cyclic loading durability is inverse proportionally where the sample with 20% of infill density can withstand until 5184 times cycles under continuous cyclic loading and has been reduced consistently when the infill increases to 60% at which the sample is able to withstand up to 634 times cycles before the break.

Acknowledgements This paper was supported by financial aid under Short Term Research Grant (STRG: str19024) and facilities provided by Universiti Kuala Lumpur (UniKL) and UiTM Penang branch are acknowledged and very much appreciated.

References

1. Xin W, Man J, Zuowan Z et al (2017) 3D printing of polymer matrix composites: a review and prospective. Compos B Eng 110:442–458
2. Thamizh TTM, Ishan K, Shubham A et al (2020) 3D printer operation using a design model. Int J Adv Sci Technol 29(6s):127–134
3. Vojislav P, Juan VHG, Olga JF et al (2010) Additive layered manufacturing: Sectors of industrial application shown through case studies. Int J Prod Res 49(4):1061–1079
4. Gulla SP, Krishnan VK, Kolli KK (2020) Exploration analysis of 3D printer. Int J Adv Sci Technol 29(7):1249–1254
5. Marc DV, John T, Russell T (2015) Effect of infill patterns on print quality of dispenser-printed electronic ink. Electron Lett 15:1186–1187
6. Miguel FV, Wilson C, Santiago F (2016) Effect of infill parameters on tensile mechanical behavior in desktop 3D printing. 3D Print Addit Manuf 3(3):183–192
7. Grzegorz Ć, Cezary G, Krzysztof K et al (2017) The influence of printing parameters on selected mechanical properties of FDM/FFF 3D-printed parts. In: IOP conference series materials science engineering, vol 227, no 012033, pp 1–10
8. Liseli B, Guha M, Hazel M (2015) Study of infill print design on production cost-time of 3D printed ABS parts. IJRapidM 5(3/4):308
9. Shaheryar AK, Bilal AS, Muhammad F et al (2017) Evaluation of the effect of infill pattern on mechanical strength of additively manufactured specimen. Mater Sci Forum 887:128–132
10. Shah F K, Hafizawati Z, Yee L C et al (2018) Effect of infill on tensile and flexural strength of 3D printed PLA parts. In: IOP conference series materials science engineering, vol 429, no 012101, pp 1–6
11. Tianyun Y, Juan Y, Zichen D et al (2020) Tensile failure strength and separation angle of fdm 3D printing pla material: experimental and theoretical analyses. Compos B Eng 188(107894):1–13
12. Miguel FV, Wilson C, Santiago F et al (2016) Effect of infill parameters on tensile mechanical behavior in desktop 3D printing. 3D Print Addit Manuf 3(3):183–192
13. Technical Data Sheet: https://www.sd3d.com/wp-content/uploads/2017/06/MaterialTDS-PLA_01.pdf

Chapter 14
Sound Vibration Signal Enhancement for Bearing Fault Detection by Using Adaptive Filter: Adaptive Noise Canceling and Adaptive Line Enhancer

Sheikh Mohd Firdaus Sheikh Abdul Nasir, Khairul Anuar Abd Wahid, Muhammad Nur Farhan Saniman, Wan Mansor Wan Muhammad, and Irfan Abdul Rahim

Abstract This paper investigates the effectiveness of the adaptive filter, ANC, and ALE to improve vibration and sound signals. These signals have been used to detect the natural development of bearing defects for machine diagnosis applications. However, during measurement, these signals have been corrupted by the noise that was coming from other machine parts. In this work, the noise has been successfully removed by using an adaptive filter. Two types of adaptive filters will be compared which are the adaptive noise canceling (ANC) and the adaptive line enhancer (ALE). This investigation is carried out by collecting the vibration and sound signal from a bearing that has been loaded with 20 kg mass and rotated with fixed 1500 rpm. This bearing is continuously rotated for 40 h. It was shown that the ANC filter is more efficient compared to ALE with the least mean square error.

Keywords Machine diagnosis · Bearing fault detection · Adaptive filter

---

S. M. F. Sheikh Abdul Nasir
Mechanical Department, Universiti Teknologi Mara, Permatang Pauh, Pulau Pinang, Malaysia
e-mail: sh.firdaus@uitm.edu.my

K. A. Abd Wahid (✉) · M. N. Farhan Saniman · W. M. Wan Muhammad
Mechanical Engineering Section, Universiti Kuala Lumpur Malaysia France Institute, 43650 Bandar Baru Bangi, Selangor, Malaysia
e-mail: khairulanuarabdwahid@unikl.edu.my

M. N. Farhan Saniman
e-mail: mnfarhan@unikl.edu.my

W. M. Wan Muhammad
e-mail: drwmansor@unikl.edu.my

I. Abdul Rahim
School of Manufacturing Engineering, Universiti Malaysia Perlis, Kampus Tetap Pauh Putra, 02600 Arau, Perlis, Malaysia
e-mail: irfanrhim@unimap.edu.my

A. Ismail et al. (eds.), *Advanced Materials and Engineering Technologies*, Advanced Structured Materials 162,
https://doi.org/10.1007/978-3-030-92964-0_14

14.1 Introduction

The defect machine components can be detected based on the changing of their signal characteristics [1, 2]. However, the change of signal characteristics from the defect machine component is difficult to be detected and evaluated since the defect machine signal usually will be corrupted by other signals which are called as noise that came from another machine components [3, 4]. Therefore, this noise signal needs to be removed so that the true signal from the defect machine component easily can be detected.

Several available methods can be used to remove the unwanted signal. One of the well-known methods was the fixed filter such as low pass, high pass, bandpass, and band stop filter [5]. These fixed filters are applied to remove the unwanted signal based on frequency range [6]. By setting the cut off, the unwanted frequency range, these unwanted signals can be filtered. As a result, only the signal of interest has remained. However, the fixed filter has several drawbacks where ranges of frequency components that need to be removed must be determined first. Otherwise, the signal of interest will be removed unintentionally. The challenge is to determine accurately the range of the cut-off frequency.

To overcome the drawbacks of these filters, the blind source separation (BSS) algorithm has been introduced. BSS is a technique that can be used to separate the mixed signals into individual signals without the need to know the frequency range of the unwanted signal. Instead of that, the BSS separates the mixed signal based on the reference signal. These reference signals can be obtained by attaching sensors to each source of signal in the machine system that contributes to the mixed-signal [7, 8]. During the separation process, the BSS algorithm assumed that each source signal is statistically independent or de-correlated. However, in the actual manufacturing system, all the component or machine part are connected hence the signals will influence each other. As a result, the reference signal will be no more independent. Therefore, such a situation will contribute to fault detection accuracy.

Adaptive filters use as the same as the BSS principle to remove unwanted signals but with improved algorithms. Due to that, this technique has known to be applied in diverse applications such as telecommunication [9, 10], sound recognition [11, 12], image processing [13], medical fields [14, 15], and machine diagnostics [16, 17]. However, the current practice in machine diagnosis applications typically used a single source signal as a reference signal. The single source signal may provide only partial information that can be obtained due to inherent limitations of sensor that have varying degrees of uncertainty. Hence multiple sensors may improve the signal enhancement.

In this paper, vibration and sound signals are used to detect the fault generate from bearing while the relationship and their effectiveness are investigated. The tested bearing is run to failure for 40 h continuously. Along with this work, a growth index is proposed to indicate the growth level of bearing deterioration. The noise signal that compounded the vibration and sound will be cleaned by using an adaptive filter algorithm.

14.2 Methodology

14.2.1 Theory

The adaptive filter is a well-known and reliable technique that is able to remove unwanted signals without the need to know the characteristics of unwanted signals. This is because the adaptive filter automatically adjusts the filter coefficients based on the reference signal that highly correlates with the unwanted signal in a mixed signal. The coefficient correlation is based on the minimum square error (MSE) of the adaptive filter output. The lower MSE value, the unwanted signal will be closer to the reference signal. Therefore, the adaptive filter will reiterate to adjust the coefficients until the MSE is minimal. Once the MSE is minimal, the adaptive filter structure is in optimal design. Least mean square (LMS) and recursive least square (RLS) algorithms are the common algorithms employed by the adaptive filter to dictate the filter of how to adjust the coefficients. Due to its simplicity, the LMS algorithm is used in this paper. Figure 14.1 shows the operation of the adaptive filter.

From Fig. 14.1, the adaptive filter is started once the mixed-signal and reference signal is driven into an adaptive filter. The mixed-signal that is denoted as $u(n)$ contains two-component signals which are $s(n)$ and $v(n)$ where each of them represents as machine defect signal and unwanted signal, respectively. While the reference signal is denoted as $v_o(n)$ is highly correlated with the unwanted signal $v(n)$ in the mixed signal.

$$u(n) = s(n) + v(n) \tag{14.1}$$

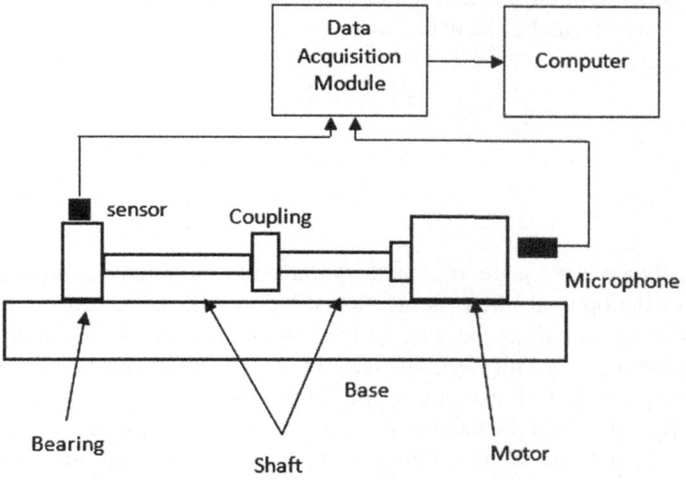

Fig. 14.1 Fault simulator device

$$v_o(n) \sim v(n) \tag{14.2}$$

With an initial set of filter coefficients, the filter will adjust the reference signal characteristics and force as possible to match with the unwanted signal in mixed signals. This filtered reference signal is denoted as $y(n)$ and can be described as following:

$$y(n) = \sum_{i=0}^{M-1} \vec{w_i}(n)u(n-i) \tag{14.3}$$

where M is the filter length, $\vec{w_i}(n)$ is the filter coefficients, and $u(n-i)$ is the primary input with a previous adjustment step.

After the $y(n)$ is obtained where its beliefs have been adaptively filtered to match with an unwanted signal component in the mixed-signal component, the subtraction process will be performed between the mixed-signal with $y(n)$ and can be described as the following equation:

$$e(n) = u(n) - y(n) \tag{14.4}$$

$$e(n) = [s(n) + v(n)] - y(n) \tag{14.5}$$

where $y(n) \approx v(n)$.

The error signal $e(n)$ then can be used to indicate how well the noise is eliminated from the mixed signal by using a statistical performance function known as a mean square error which can be calculated as follows:

$$J\left[|e(n)|^2\right]_{\min} \tag{14.6}$$

$$\varepsilon = \frac{J_{\min}}{\sigma_{\mathrm{ref}}^2} \tag{14.7}$$

where $J_{\min}$ is a mean square, $\sigma_{\mathrm{ref}}^2 i$ is the variance of the reference signal, and ε is the scale of the optimal filter. As mentioned before, the mean square needs to be minimal since it will affect the scale of the optimal filter design and indicates how close $y(n)$ the signal is with $v(n)$. The scale of the optimal filter is known in between the range $0 \le \varepsilon \le 1$. If ε is zero, the adaptive filter with the optimum filter coefficient has perfectly eliminated the unwanted signal from the mixed signal whereas on the other hand if ε is 1, the adaptive filter does not eliminate the unwanted signal at all. Therefore, if ε is not minimal, that means that the y (n) signal is not close enough with v (n), thus a new set of adaptive filter coefficient needs to be adjusted. The filter coefficient can be readjusted by updating the filter weight based on the following Eqs.

(14.8) and (14.9). In this work, two types of an algorithm will be used which are the adaptive noise canceling (ANC) and adaptive line enhancer (ALE). The filter weight coefficient for both algorithms can be found at the following equation, respectively.

$$\vec{w}(n+1) = \vec{w}(n) + \mu.e(n).\vec{\mu}(n) \tag{14.8}$$

$$\vec{w}(n+1) = \vec{w}(n) + \mu.e(n).u(n-i) \tag{14.9}$$

where μ is the step size of the adaptive filter, $\vec{w}(n)$ is the filter coefficient vector, $e(n)$ is an error signal, and $\vec{\mu}(n)$ is the filter input vector. Based on this weight filter coefficient, the adaptive filter will remove the unwanted signal to reduce the residual of random variables by using the following equation.

$$x(n) = \frac{u(n) - \overline{u}}{\sigma} \tag{14.10}$$

where $x(n)$ is the pre-processed signal, $\overline{u}$ is the mean value of $u(n)$, and σ is the standard deviation of cleaned $u(n)$.

14.2.2 Simulation Work

The simulation of the mixed signal is carried out to test the effectiveness of the adaptive filter to remove the unwanted signal that has been mixed with an actual signal. Since there are two types of signals used in this work, which are vibration and sound, two synthetic mixed signals will be generated by using the command function in MATLAB.

The synthetic mixed signal that represents the vibration signal contained two-components signal where each of them has frequency component of 5 Hz and 20 Hz respectively and can be modeled as the following equation:

$$u(n)_1 = s(n) + v(n) \tag{14.11}$$

$$u(n)_1 = 2\sin(2\pi 5t) + \sin(2\pi 20t) \tag{14.12}$$

While for the second synthetic mixed signal that represents the sound signal, two component signals are mixed where each of them has frequency component of 50 Hz and 200 Hz respectively and they can be modeled by using the following equations:

$$u(n)_1 = s(n) + v(n) \tag{14.13}$$

$$u(n)_1 = 2\sin(2\pi 50t) + \sin(2\pi 200t) \tag{14.14}$$

Both have been compounded with the random signal that represents a noise signal and has a standard deviation of 0.5006 to make it more realistic.

The simulation of the mixed signal has been carried out to test the effectiveness of the adaptive filter for the classification process before the adaptive filter is applied to classify the real mixed signal.

The first mixed signal is simulated using Eq. (14.16) where $s(n)$ represents a signal coming from the defect machine component signal and $v(n)$ represents the signal coming from other components in the same machine. These two signals are mixed to become $u(n)_1$ signals. Both sinusoidal signals are sampled 1000 times in 1 s

$$u(n)_1 = s(n) + v(n) \tag{14.15}$$

$$u(n)_1 = 2\sin(2\pi 120t) + \sin(2\pi 50t) \tag{14.16}$$

The second mixed signal is simulated from Eq. (14.18) where $s(n)$ represents a signal coming from the defect machine component and $\text{rand}(n)$ represents a signal coming from a random noise signal.

$$u(n)_2 = s(n) + \text{rand}(n) \tag{14.17}$$

$$u(n)_2 = 2\sin(2\pi 120t) + \text{rand}(n) \tag{14.18}$$

These $s(n)$ and $\text{rand}(n)$ signals are mixed to produce a distorted $u(n)_2$ signal. The sinusoidal signal is sampled 1000 times in 1 s while the random noise is characterized by a standard deviation of 0.5006 and the number of data points is 10001. Both mixed signals $u(n)_1$ and $u(n)_2$ are taken as primary input and driven into adaptive filters for classification process.

14.2.3 Experimental Works

The real mixed signal has been generated from the simulator device as illustrated in Fig. 14.1. This test simulator device consists of a DC geared motor driving at a speed of 200 rpm. This motor is connected to a bearing through a shaft and a coupling.

In the first test, the vibration signal from the defect bearing component is taken where it is mixed with the vibration signals coming from the other parts of the test simulator device. The Vibration signal coming from a defect-free similar bearing from the test simulator device is taken as the reference signal while the delay version of the mixed signal is used as the reference signal.

For the second test, the sound signal from the defect motor component is taken where it is mixed with the sound signals coming from the other parts of the test simulator device. Similarly, the sound signal coming from the defect-free motor in the test simulator device during rotation is used for the reference signal for the ANC filter while the delay version of mixed-signal is used as the reference signal for the ALE filter.

In this experiment, the vibration signal is captured by a MEMS accelerometer type ADXL202 while the sound signal is captured by the microphone. Both signals were then transferred into the data acquisition module using the National Instrument USB6210 model for the digitizing process before the filtering process can be carried out by the computer.

In the first test, the vibration signal from the defect bearing component is taken where it is mixed with the vibration signals coming from the other parts of the test simulator device. The Vibration signal coming from defect-free similar bearing from the test simulator device is taken as the reference signal for the ANC filter while the delay version of the mixed signal is used as the reference signal for the ALE filter. For the second test, the sound signal from the defect motor component is taken where it will be mixed with the sound signals coming from the other parts of the test simulator device. Similarly, the sound signal coming from the defect-free motor in the test simulator device during rotation is used for the reference signal for the ANC filter while the delay version of mixed-signal is used as the reference signal for the ALE filter.

14.3 Results and Discussion

14.3.1 Simulation Results

This section discusses the classification result for the simulation of the mixed vibration signals. Figures 14.2 and 14.3 show the result of signal classification for mixed signal $u(n)_1$ that has changed the signal pattern by using the ANC and ALE filter respectively. The result shown in Fig. 14.3c is the classified signal from the mixed signal by using the ANC filter while Fig. 14.4c shows the result of the classified signal $s(n)$ by using the ALE filter. Generally, both filters succeed to classify the unknown signal of interest from the mixed signal, however, the classified signal by using the ANC filter is more accurate than the ALE filter since the optimal design for the ANC filter is $\varepsilon = 0.0014$ while $\varepsilon = 0.0666$ for ALE. Therefore, in this simulation result, the ANC filter is more accurate than the ALE filter for classification of the mixed signal.

Figures 14.4 and 14.5 show the results of signal classification from the simulation of the mixed sound signal $u(n)_2$ by using ANC and ALE, respectively. The result shown in Fig. 14.4c shows the classified signal $s(n)$ from mixed signal by using the ANC filter while Fig. 14.5c shows the classified signal $s(n)$ from mixed signals by

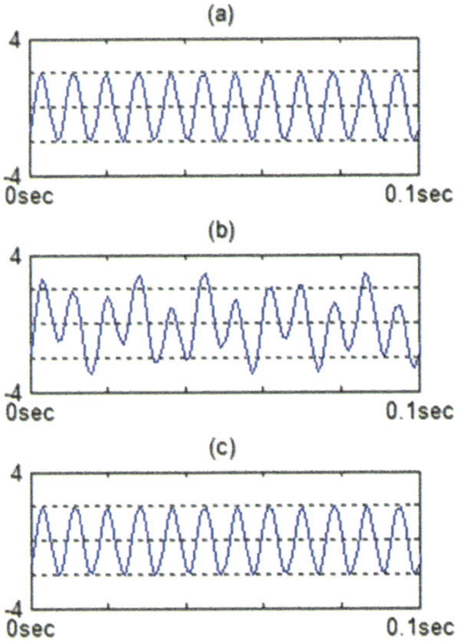

Fig. 14.2 Classification of mixed simulated vibration signal from noise by using ANC filter, **a** original defect signal, **b** mixed signal, **c** classified defect signal by using ANC

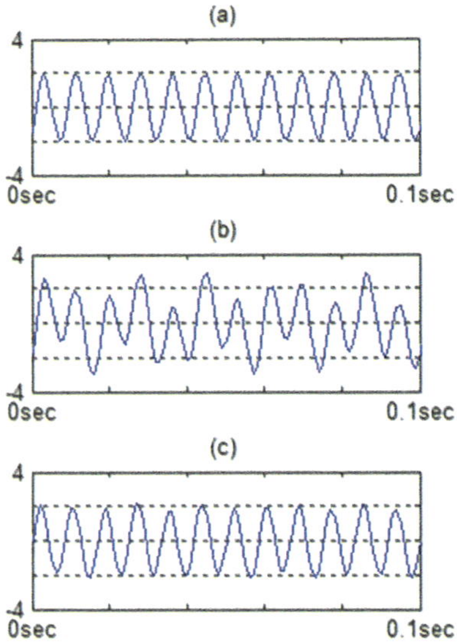

Fig. 14.3 Classification of the mixed simulated sound signal from noise by using ALE filter, **a** original defect signal, **b** mixed signal, **c** classified defect signal by using ALE

using the ALE filter. Generally, both ANC and ALE filter succeeded to classify the signal $s(n)$ from being distorted. However, the classified results by using ANC show more accuracy than ALE since the normalized least square error is $\varepsilon = 0.0151$ for the ANC adaptive filter while $\varepsilon = 0.1382$ for the ALE adaptive filter. Therefore, in this simulation result, the ANC filter is more accurate than the ALE filter to classify mixed signals.

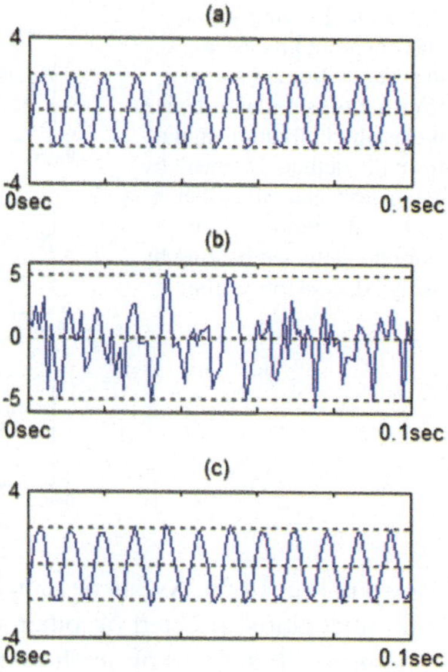

Fig. 14.4 Classification of defect signal from mixed simulation sound signal by using ANC filter **a** original defect signal, **b** mixed signal, **c** classified defect signal by using ANC

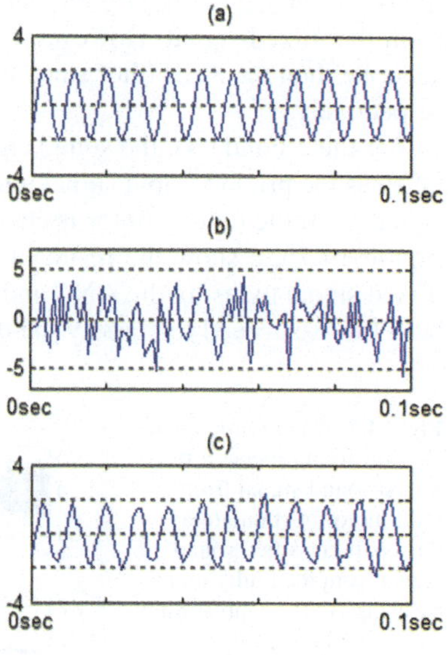

Fig. 14.5 Classification of defect signal from mixed simulation sound signal by using ALE filter, **a** original defect signal, **b** mixed signal, **c** classified defect signal by using ALE

14.3.2 Experimental Results

This section discusses the classification result by using a real mixed signal which is obtained from the fault simulator device. For the first test, the mixed vibration signal from the defected bearing is taken as the primary input for ANC and ALE filter as

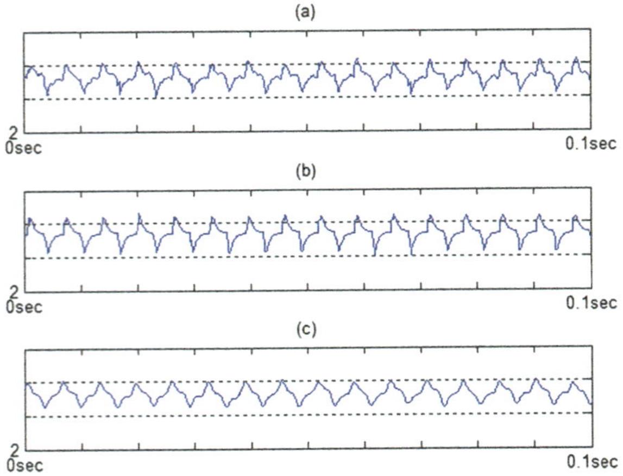

Fig. 14.6 Bearing signal classification process **a** distorted vibration signal from a fault simulator device, **b** classified vibration signal from the defected bearing by using ANC adaptive filter, **c** classified vibration signal from the defected bearing by using ALE adaptive filter

shown in Fig. 14.6a. As can be seen the defected vibration signal Fig. 14.6a is having little fluctuation affected by other vibration signals that are coming from another component. Figure 14.6b, c show the clean result of the classified bearing signal obtained from the ANC and the ALE adaptive filter, respectively. Generally, both filters succeeded to classify the defect signal from the defected bearing component from the mixed signal. However, the classified signal from the ANC filter is more accurate than the ALE filter since the least square error shown is 0.0109 and 0.109, respectively.

For the second test, the sound signal from defect motor as shown in Fig. 14.7a is taken as the primary input signal into the ANC and the ALE filter. The pattern sound signal from the defect motor is changed when it is mixed with other sound signals. Figure 14.7b, c show the result of the classifying defected motor signal from the mixed signal by using the ANC and the ALE adaptive filter, respectively. Generally, both filters succeed to classify the defect signal from the defected motor. However,

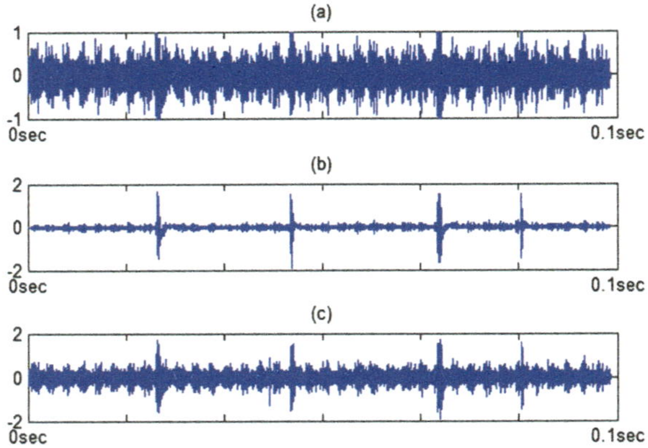

Fig. 14.7 The motor signal classification process **a** the mixed sound signal from fault motor in simulator device, **b** classified sound signal from the faulty motor by using ANC adaptive filter, **c** classified sound signal from the faulty motor by using ALE adaptive filter

ANC adaptive filter shows more accurate than ALE adaptive filter to classify the defect motor signal from mixed signal since the least square error of classifying signal by ANC filter is $\varepsilon = 0.0427$ while ALE filter is $\varepsilon = 0.0441$.

14.4 Conclusion

Two types of adaptive filters known as the adaptive noise canceller (ANC) and adaptive line enhancer (ALE) have been used to classify the signal of interest from the mixed signal. Generally, the simulation and experimental work showed that both adaptive filters succeed to classify the defect signal from the mixed signal. However, based on the least mean square error, the ANC filter is more accurate than ALE. This is probably due to the fact that reference input used in ANC has many signal components correlated in mixed signals due to noise source at the machine body part. However, even though the ALE filter is less accurate than ANC, the ALE filter is more practical to apply in a real environment for fault detection application since the ALE filter does not need the reference signal which is hard to obtain in an actual complex machine.

Acknowledgements This paper was supported by Short Term Research Grant (STRG)-strl 19204 awarded by Universiti Kuala Lumpur, Center of Research and Excellence (CORI) and acknowledge financial aid under UiTM Cawangan Pulau Pinang, as well as facilities provided.

References

1. Goyal PBS (2015) The vibration monitoring methods and signal processing techniques for structural health monitoring: a review. Arch Comput Methods Eng 23:585–594
2. Zhiwei G, Carlo C, Steven D (2015) A survey of fault diagnosis and fault tolerant techniques—part 1: fault diagnosis with model 0 based and signal-based approach. IEEE Trans Ind Electron 62(6):3757–3767
3. Yanxue W, Jiawei X, Richard M et al (2016) Spectral kurtosis for fault detection, diagnosis, and prognostics of rotating machines: a review with applications. MSSP 66067:679–698
4. Jinglong C, Zipeng L, Jun GC et al (2016) Wavelet transform based on inner product in fault diagnosis of rotating machinery: a review. MSSP 70–71:1–35
5. Tao Y, Gabriel R (2017) Bandpass to bandstop reconfigurable tunable filters with frequency and bandwidth controls. IEEE Trans Microw Theory Tech 65(7):2288–2297
6. Xiaole L, Houguang L, Jianhua Y et al (2017) Improving the bearing fault diagnosis efficiency by the adaptive stochastic resonance in a new nonlinear system. MSSP 97:58–76
7. Tarak B, Zerhouni N, Rechak S (2018) Tool wear condition monitoring based on continuous wavelet transform and blind source separation. Int J Adv Manuf Technol 97:3311–3323
8. Xia-Zhou L, Chi X, Yi-Qing N (2019) Wayside detection of wheel minor defects in high-speed trains by a Bayesian blind source separation method. Smart sensors for structural health monitoring. Sensors 19(18), 3981:1–16
9. Swathi N, Indira VBSSD, Sasibhushana GR (2015) An adaptive filter approach for GPS multipath error estimation and mitigation. Proc ICMEET 2015:539–546

10. Mohamed AA, Toufik L, Aladdine A (2016) Adaptive filters for direct path and multipath interference cancellation: application to FM-RTL-SDR based passive bistatic radar. In: International conference on sciences of electronics, technologies of information and telecommunications, pp 461–465
11. Rajesh K, Pradeep CR (2016) A framework for sign gesture recognition using improved genetic algorithm and adaptive filter. Cogent Eng 3(1):1–9
12. Rachel EB (2017) In-ear microphone speech quality enhancement via adaptive filtering and artificial bandwidth extension. J Acoust Soc Am 141:1321–1331
13. Delian L, Zhaohui L, Xiaorui W (2015) Moving target detection by nonlinear adaptive filtering on temporal profiles in infrared image sequences. Infrared Phys Technol 73:41–48
14. Lu L, Zhenhong J, Jie Y et al (2015) A medical image enhancement method using adaptive thresholding in NSCT domain combined unsharp masking. Int J Imaging Syst Technol 25(3):199–205
15. Feng J, Aleksei G, Seungmin R et al (2018) Medical image semantic segmentation based on deep learning. Neural Comput Appl 29:1257–1265
16. Siliang L, Qingbo H, Tao Y et al (2016) Online fault diagnostics of motor nearing via stochastic resonance based adaptive filter in an embedded system. IEEE Trans Syst Man Cybern Syst 47(7):1111–1122
17. Faris E, David M, Cristobal RC et al (2014) Effectivenes of adaptive filter algorithms and spectral kurtosis in bearing fault detection in a gearbox. J Vib Eng Technol 219–229

Chapter 15
Effect of Needle Size and Needle Height to Substrate in Encapsulation Process of LED Packaging

Md. Abdul Alim, Mohd Zulkifly Abdullah, Mohd Sharizal Abdul Aziz, Roslan Kamarudin, Mohd Syakirin Rusdi, Muhammad Hafifi Hafiz Ishak, and Prem Gunnasegaran

Abstract The purpose of this study is to investigate the effect of needle size and substrate to needle height by measuring the droplet volume and epoxy covered area in encapsulation process of LED packaging. Five needle gages (16G, 18G, 21G, 22G, and 23G) were used in this experiment. Average volume of epoxy drops was determined for each size of needle. Experimental data shows that droplet volumes are directly related with the needle diameter. Optimum covered area by the epoxy encapsulation was found at the needle tip to substrate distance at 3.240 mm for the 16G needle. These findings are relevant and helpful for controlling the amount of epoxy, encapsulation area, and the encapsulant shape in encapsulation process of LED packaging.

Md. A. Alim · M. Z. Abdullah (✉) · M. S. A. Aziz · R. Kamarudin · M. S. Rusdi
School of Mechanical Engineering, Universiti Sains Malaysia, Engineering Campus, 14300 Nibong Tebal, Penang, Malaysia
e-mail: mezul@usm.my

Md. A. Alim
e-mail: abdulalim@student.usm.my

M. S. A. Aziz
e-mail: msharizal@usm.my

R. Kamarudin
e-mail: roslan_k@usm.my

M. S. Rusdi
e-mail: syakirin@usm.my

M. H. H. Ishak
School of Aerospace Engineering, Universiti Sains Malaysia, Engineering Campus, 14300 Nibong Tebal, Penang, Malaysia
e-mail: mhafifihafiz@usm.my

P. Gunnasegaran
Department of Mechanical Engineering, College of Engineering, Universiti Tenaga Nasional, Putrajaya Campus, Jalan IKRAM-UNITEN, 43000 Kajang, Selangor, Malaysia
e-mail: prem@uniten.edu.my

A. Ismail et al. (eds.), *Advanced Materials and Engineering Technologies*, Advanced Structured Materials 162,
https://doi.org/10.1007/978-3-030-92964-0_15

Keywords Droplet volume · Epoxy drop · Needle size · Encapsulation · LED packaging

15.1 Introduction

LED encapsulation and lens formation are important to LED success. It has a significant impact on the light extraction performance of LED chips. The efficiency of light extraction is not only related to the phosphor coating but also closely related to the geometry of the interface and the lens shape [1, 2]. It is difficult to develop an encapsulation process that is cheap, dependable, and repeatable. Generally, permanent molds with cavities and compression molding systems are used in hemispherical lens forming. Then, the curing is performed in an oven and LEDs are separated from the cavities. The management of the tool's parasite adhesion on LEDs is a huge concern in these systems [3, 4]. The injection molding process can make several different forms of lens, but these molds are dedicated to a specific type and are not for rapid production. Developing a mold-free injection-based encapsulation process on large surfaces is very essential to establish the rapid industrial production. Many studies have been carried out to develop the moldless encapsulation. Researchers are trying to develop the best effective way [5, 6]. A needle-based dispensing system is convenient to perform the encapsulation process in laboratory-based research. Epoxy is dispensed over the substrate by a needle from a specific height to make the lens. Epoxy drops are needed to control to form a suitable shaped lens [7]. So it is very important to know the droplet size of epoxy and the epoxy covered area by a single drop. The droplet size delivered from different sized needles can be determined by many techniques. Droplets from different needles can be measured by means of measuring beaker and precision weigh scale. The weight of the certain drops of fluid can be recorded in milligrams, which is equivalent to volume in microliters, and the mean volume of an individual drop can be calculated [8].

The pendant drop method has been developed to measure the droplet volume of different drop configurations. Different sizes of needles, a camera, and a light source are required for this method. The image dimension can be measured based on resolution. By this way, the diameter of the droplet sphere is measured and the volume is calculated [9]. Recently, an open-sourced fully featured software written in the Python language named "Open Drop" is also used to measure droplet volumes [10]. The epoxy droplet is dispensed over the substrate for encapsulation. Several image-based methods are used to assess the maximum spreading of epoxy during encapsulation. The open-source image processing software image-J is commonly used in the scientific community [11, 12]. The objective of this work is to observe the effect of needle size and substrate to needle height in encapsulation process of LED packaging by measuring the droplet volume of the epoxy using different needle size and determination of covered area by epoxy for different needle tip to substrate distance. Another objective of this work is to find the optimal technical condition for the suitable dispensing height (needle tip to substrate distance) for this specific substrate.

15.2 Methodology

Five standard blunt needle gages—16G, 18G, 21G, 22G, and 23G were used in the experiment. These sizes were used as they are most commonly available in the market. The dimensions of the needles used in the experiment are shown in Table 15.1. These needles were well fit with the syringe. A goniometer was used to ensure accurate vertical positioning of the syringe with the experimental setup.

The experimental setup was constructed using a micrometer, syringe, light source, a laboratory stand, microscope, and a monitor to obtain standardized results with minimal error. A measuring beaker was placed vertically under the needle to measure the weights of epoxy drops. The same experimental setup was used to observe the epoxy covered area after dispensing over the substrate (shown in Fig. 15.1). The distance between needle tip and substrate is controlled by using a control knob.

The encapsulant used in the tests is a product of Penchem Technologies Sdn. Bhd, Malaysia. It has two parts. Part-A contains an epoxy ($C_{21}H_{25}ClO_5$) and part-B is the hardener (Alicyclic anhydrides). It is mandatory to mix part-A and part-B with equal volume % before application as encapsulant. This transparent epoxy system is especially suitable for LED encapsulation. This epoxy is needed to be cured at 120 °C/1 h and 125 °C/2 h. Part-A and B are taken in a beaker with the same volume % and mixed by a mechanical stirrer. The solution is then kept in a vacuum chamber to remove the bubbles inside. After that, it is taken into syringe for dispensing. Micrometer spindle was placed in such a way that every turn in thimble scale and main scale can create pressure on the plunger of the syringe. A certain plunger pressure is required to dispense the epoxy droplet.

Three approaches are applied to determine the mean drop volume for each size needle (shown in Fig. 15.2). For measuring through a measuring beaker, twenty drops are taken in the beaker which was previously been tared on a weighing scale and it was calculated the average droplet volume. Droplet images were taken by microscope for each type of needle, and image processing system was used manually to calculate the droplet volume. An open-source software—"Open Drop" was applied to calculate

Table 15.1 The dimensions of different needles used in experiments

Needle size	OD (mm)	ID (mm)	Needle length (mm)
16G	1.62	1.19	12.5
18G	1.28	0.84	12.5
21G	0.81	0.51	12.5
22G	0.71	0.41	12.5
23G	0.63	0.33	12.5

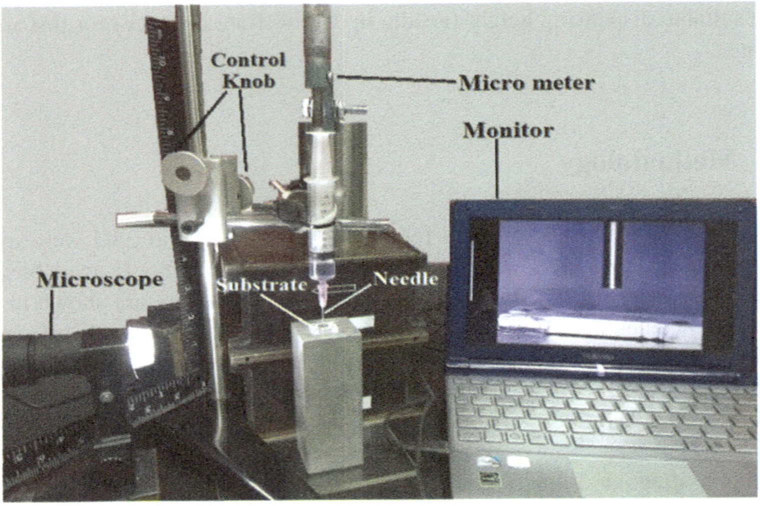

Fig. 15.1 Experimental setup for dispensing the epoxy

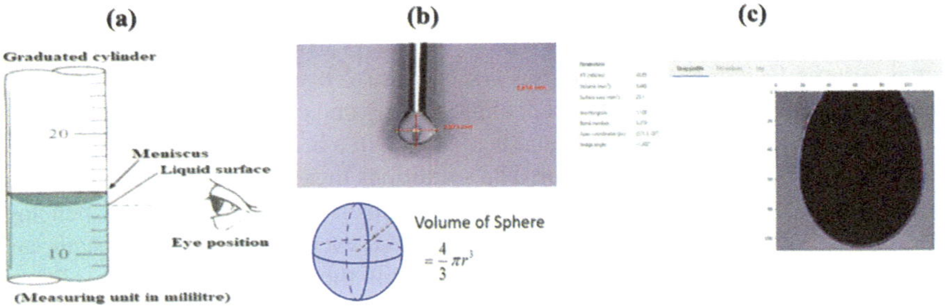

Fig. 15.2 Different approach of measuring droplet volume—**a** measuring cylinder, **b** image system (manually), and **c** open drop software

the droplet volume depending on the image taken for each drop. After calculating the maximum droplet volume, the suitable needle size was chosen. The droplet of epoxy was dispensed from five different height to substrate (1.620, 2.430, 3.240, 4.050 and 4.860 mm) for the specific needle. All the experiments before curing were maintained at room temperature. The curing was then performed at 120 °C for one hour and then checking the sample once, the curing was been continued for 2 h at 125 °C. The best epoxy covered area was determined after curing the epoxy over substrate by the open-sourced software "Image-J".

15.3 Results and Discussion

The mean droplet volume of 20 drops dispensed from blunt needles of five needle sizes was measured by three approaches. Maximum value of droplet volumes was found by measuring cylinder method for 16G, 18G, 21G, 22G, and 23G were 17.241, 13.333, 11.364, 10.417, and 9.524 mm^3, respectively. There was a significant correlation between needle gage and drop volume with larger needles dispensing larger drop volumes. The volume measuring by the manual image system and open drop software are not significantly different compared to the measuring cylinder. The average drop volumes are shown as graph in Fig. 15.3. Comparing results from repeated testing of individual needle gages documented little variation with no correlation found in drop volume. According to the data found from the experiment, the 16G needle dispenses the highest droplet volume and the lowest is from 23G. From the graph, it is observed that the droplet volume is directly related to the needle size at normal temperature and pressure.

Epoxy dispensing from different height was observed. 16G needle was used in this approach. The needle was vertically hold to the stand and ensured that the droplet was falling to the target position of the substrate. The substrate was kept over a flat type of vice and maintained on a perfect horizontal plane. Uniform pressure was applied over the syringe plunger to maintain equivalent droplet size. For each height, 20 samples were taken and cured at oven at 120 °C for 1 h and later again at 125 °C for 2 h. The epoxy covered area for five different height was measured for each sample. Image-J software was used to measure the area. The randomly taken images of epoxy covered area (green colored) over the substrate for each height are shown in Fig. 15.4.

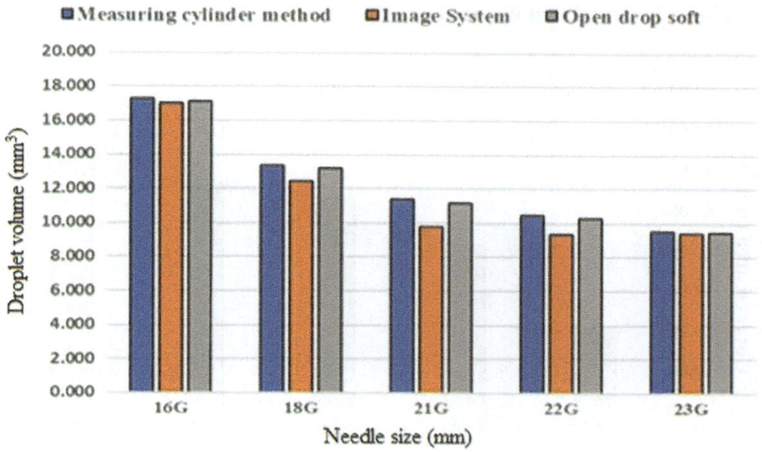

Fig. 15.3 Needle size versus droplet volume graph

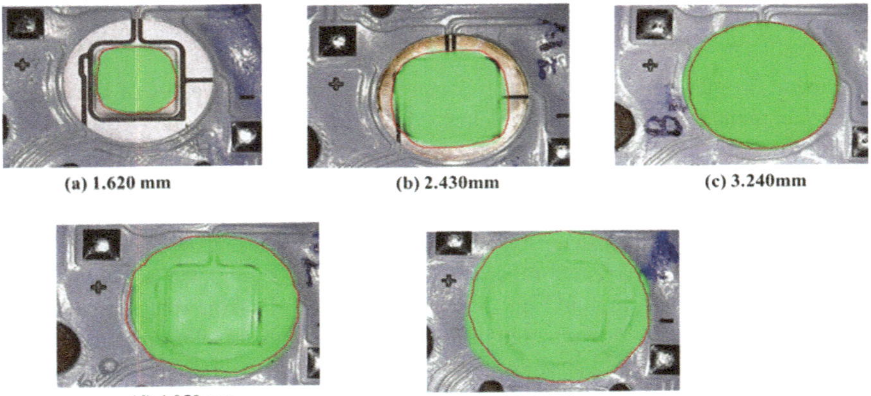

Fig. 15.4 Covered area measurement by Image-J software for different dispensing height

The average area was taken for each height, which is shown as graph in Fig. 15.5. The focusing area over the substrate is circular of 50.24 mm^2. For the dispensing height 1.620, 2.430, 3.240, 4.050, and 4.860 mm, the average covered area for 16G needle are 20.69, 41.31, 52.45, 55.75, and 61.74 mm^2, respectively.

From the data, we can determine that 41.18% of target area is covered by 1.620 mm height. Furthermore, 82.20%, 104.40%, 110.97%, and 122.89% area are covered by 2.430, 3.240, 4.050, and 4.860 mm dispensing height, respectively. We note that this study examined the effect of needle size on drop volume as well as encapsulation covered area. In this study, the shape of encapsulation is not considered. For the single drop of epoxy spreading over the substrate at normal temperature and pressure was observed. It is observed that the 16G needle is dispensing maximum droplet volume

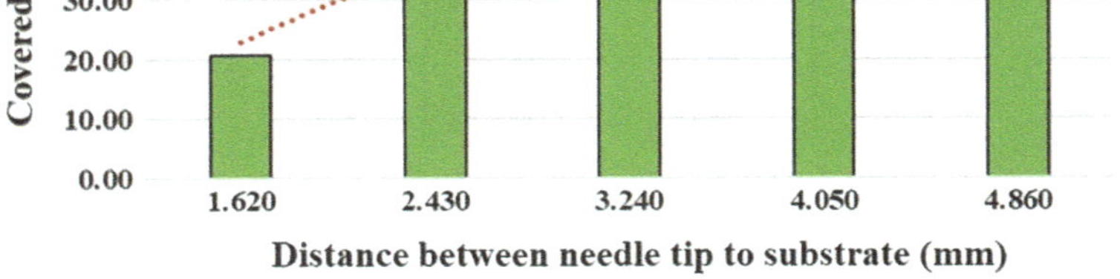

Fig. 15.5 Trendline with increase of encapsulation area with increase in distance between needle tip and substrate

and with this needle the optimum epoxy covered area is found for the 3.240 mm dispensing height.

15.4 Conclusion

The needle size significantly affects epoxy drop volume. The smallest volume is dispensed by the smallest gage needle 23G. Highest droplet volume was found with the needle of biggest diameter. Needle size 16G on 3.240 mm distance from substrate provides optimum epoxy covered area by single drop of epoxy. The described study shows that it is possible to perform the encapsulation process without the help of a mold. It will be easy to maintain proper shape of encapsulation by the right amount of epoxy dispensing from calculated height. This study will be helpful to adjust the automatic equipment of encapsulation process in LED packaging.

Acknowledgements The authors acknowledge the technical and financial support from the Universiti Tenaga Nasional (Grant no: J510050002/2021069), Collaborative Research in Engineering, Science and Technology (CREST), Malaysia and Universiti Sains Malaysia (USM).

References

1. Hu R, Luo X, Zheng H et al (2012) Design of a novel freeform lens for LED uniform illumination and conformal phosphor coating. Opt Express 20(13):13727. https://doi.org/10.1364/oe.20.013727
2. Alim MA, Abdullah MZ, Aziz MA et al (2020) Experimental study on luminous intensity of white LEDs of different configurations. IOP Conf Ser Mater Sci Eng 1007(012145). http://doi.org/10.1088/1757-899X/1007/1/012145
3. Wang L, Weelden TV, Yuan C, Wei J (2015) LED encapsulation and integration with film assisted molding technology. In: 2013 10th China international forum on solid state lighting, ChinaSSL 2013, pp 89–94. http://doi.org/10.1109/SSLCHINA.2013.7177321
4. Mani AA, Moulin G, Gasse A, Mendizabal L, Consonni M, Henry D (2016) Innovative process development for LEDs dome shaping without molding. In: 2016 6th electronic system technology conference, ESTC 2016. http://doi.org/10.1109/ESTC.2016.7764503
5. Chen K, Zhang R, Lee SWR (2010) Integration of phosphor printing and encapsulant dispensing processes for wafer level LED array packaging. In: Proceedings—2010 11th international conference on electronic packaging technology and high density packaging, ICEPT-HDP 2010, pp 1386–1392. http://doi.org/10.1109/ICEPT.2010.5582822
6. Chu Y, Chen C, Tsou C (2015) A silicon-based LED packaging substrate with an island structure for phosphor encapsulation shaping. IEEE Trans Compon Packag Manuf Technol 5(2):155–162. http://doi.org/10.1109/TCPMT.2014.2379255
7. Chang CCL, Huang YC, Hu SF et al (2012) LED package with dome/side-emitting-enhancement silicone lens achieved by dispensing and geometry transferring. In: Twelfth international conference on solid state lighting and fourth international conference on white LEDs for solid state lighting, vol 8484, p 84841B. http://doi.org/10.1117/12.928280
8. Tripp GK, Good KL, Motta MJ, Kass PH, Murphy CJ (2016) The effect of needle gauge, needle type, and needle orientation on the volume of a drop. Vet Ophthalmol 19(1):38–42. https://doi.org/10.1111/vop.12253

9. Hoorfar M, Neumann AW (2006) Recent progress in axisymmetric drop shape analysis (ADSA). Adv Colloid Interface Sci 121(1–3):25–49. https://doi.org/10.1016/j.cis.2006.06.001
10. Berry JD, Neeson MJ, Dagastine RR, Chan DYC, Tabor RF (2015) Measurement of surface and interfacial tension using pendant drop tensiometry. J Colloid Interface Sci 454:226–237. https://doi.org/10.1016/j.jcis.2015.05.012
11. Papadopulos F, Spinelli M, Valente S et al (2007) Common tasks in microscopic and ultrastructural image analysis using ImageJ. Ultrastruct Pathol 31(6):401–407. https://doi.org/10.1080/01913120701719189
12. Lü C, Ren H, Zhang Y, Shen Y (2010) Leaf area measurement based on image processing. In: 2010 international conference on measuring technology and mechatronics automation, ICMTMA 2010, vol 2, pp 580–582. http://doi.org/10.1109/ICMTMA.2010.141

Chapter 16
Microstructure Observation and Microhardness Study of Friction Stir Welded Blank of Aluminum to Steel

Muhamad Zulkhairi Rizlan, Ahmad Baharuddin Abdullah, and Zuhailawati Hussain

Abstract Friction stir welding is commonly used to join dissimilar materials that are difficult to weld when using conventional methods. Weld joints of dissimilar aluminum alloy 6061 and mild steel were obtained by friction stir welding using a variation of parameter values. Macroscopic examination indicated successful bonding between the two materials although some defects may result in decreased joint strength. Further microstructure analysis revealed grain refinement from stirring action of tool pin with steel particles randomly scattered in plasticized aluminum. Chemical composition analysis identified Al-rich intermetallic compound Fe_2Al_3 at the weld interface. Thicker intermetallic layer was influenced by higher heat input from increased rotation speed and decreased welding speed. Maximum hardness was measured at the interface, indicating formation of a brittle intermetallic compound.

Keywords Friction stir welding · Aluminum · Steel · Microstructure · Hardness

16.1 Introduction

Aluminum alloy and steel joint offers weight reduction and energy conservation [1]. The joint has the potential in developing industrial products such as in automotive due to the high strength-to-weight ratio which leads to improved fuel efficiency [2]. However, joining dissimilar materials like aluminum and steel provide a challenge owing to their difference in thermal and mechanical properties [3]. When using fusion welding, difference in properties especially density and melting point results in

M. Z. Rizlan (✉) · A. B. Abdullah
Metal Forming Research Lab, School of Mechanical Engineering, USM Engineering Campus, 14300 Nibong Tebal, Penang, Malaysia

A. B. Abdullah
e-mail: mebaha@usm.my

Z. Hussain
School of Material and Mineral Resources Engineering, USM Engineering Campus, 14300 Nibong Tebal, Penang, Malaysia
e-mail: zuhaila@usm.my

A. Ismail et al. (eds.), *Advanced Materials and Engineering Technologies*, Advanced Structured Materials 162,
https://doi.org/10.1007/978-3-030-92964-0_16

solidification segregation. Residual stresses as a result of the difference in expansion coefficient could distort the structure and generate cracks [1]. Moreover, the limited solubility of aluminum (Al) and iron (Fe) in each other forms brittle intermetallic compounds (IMCs) at the interface [4]. These IMCs could initiate cracks and reduce the joint strength [1]. Thicker intermetallic layer also leads to joint failure under stressed conditions [5]. Friction stir welding (FSW) is a solid-state joining process known for its energy efficiency, environmental friendliness, and versatility [4]. FSW has been used to join dissimilar metals such as Al and steel as it is a solid-state process which can avoid problems related to solidification such as shrinkage, oxidization, and porosity [2]. Compared to fusion welding, FSW provides better mechanical properties and flexibility to control heat input and limiting IMC growth [1].

To improve the weld quality of FSW process, heat generation and plastic deformation of materials should be controlled. Kar et al. [6] studied the effect of process inputs such as rotation speed, welding speed, probe plunge depth, and tool design on joint structure and properties of aluminum alloy 6061 and mild steel. They reported that the rate of heat generation and peak temperature are directly proportional to rotation speed. They also stated that rotation speed controls the interface structure by severe thermo-mechanical deformation. Yasui et al. [7] in their high-speed FSW of aluminum alloy 6063 and S45C steel found that heat input decreased with increasing welding speed. Therefore, high frictional heat is generated from the combination of low welding speed and high rotation speed [8]. Habibnia et al. [9] from their findings in friction stir welding of aluminum alloy 5050 and stainless steel 304 suggested reduction in heat generation by decreasing the rotation speed and increasing the welding speed to get a strong and sound weld joint. According to Tanaka et al. [10], weld joint strength of aluminum alloy 7075 and mild steel in their research is inversely proportional to IMC layer thickness.

Decrease in welding speed and increase in rotation speed result in mixed layered structure of aluminum and steel-rich layers and IMC formation in the stir zone which significantly reduced the joint strength [11]. When welding speed is decreased, more diffusion time and heat input were provided for the growth of IMC along and across the joint interface and non-continuous IMCs became sufficiently dense to form a continuous layer [8]. Liu et al. [12] in their work on friction stir welded joint of aluminum alloy 6061-T6 and high strength steel however stated that welding speed did not have significant effect on IMC formation. Instead, IMC formation is caused by increase in rotation speed and tool offset. Their finding was supported by Sameer and Birru [13] in their research on friction stir welding of aluminum alloy 6061 and interstitial-free (IF) steel. Apart from rotation speed, welding speed, and tool offset, Kar et al. [6] also mentioned that the probe plunge depth into steel significantly affected joint morphology at the interface, IMC layer formation, and mechanical properties. Meanwhile, from Movahedi et al. [14] observation on friction stir welded joint of aluminum alloy 5083 and St-12 steel, IMC layer with thickness less than 2 μm will not degrade joint quality. The IMC layer is responsible for high hardness of joint as well. As mentioned by Balamagendiravarman et al. [15] in their study on friction stir welding of commercially pure aluminum and 304 stainless steel, maximum hardness was measured at stir zone due to the intermetallic layer. The same result was reported by Wan and Huang [8] who measured maximum hardness at the interface of aluminum alloy 6082-T6 and Q235A steel.

16.2 Methodology

Aluminum alloy 6061 (AA6061) and mild steel plates both with the same thickness and dimensions of 1.5 mm and 200 mm × 80 mm, respectively, were butt joined using the FSW process by utilizing a conventional milling machine. Mild steel plates were placed on the advancing side (AS) while aluminum plates on the retreating side (RS) as shown in Fig. 16.1. This was done to avoid incomplete bonding between aluminum and steel if steel was placed on retreating side [16]. The tool used was made of H13 tool steel with straight cylindrical pin 1 mm in length and 5 mm in diameter, while the tool shoulder was 10 mm in both length and diameter. The tool tilt angle was 3° with 1.0 mm offset toward aluminum. The tool offset was to avoid tool wear and breakage when plunging at interface due to higher strength of steel compared to aluminum and to avoid superheating of aluminum on account of the difference in material flow behavior [4].

In order to investigate the effect of welding parameters on the weld joint quality, a set of experiments was designed using the Taguchi method with L9 orthogonal array selected. Table 16.1 shows the details for the parameters used in the experiments while Fig. 16.2 shows the experiment in progress. The weld joints of AA 6061 and mild steel were then prepared for post-weld analysis. Specimens for microstructure and hardness analysis were polished using hand grinder with 240, 320, 400, and 600 grit sandpapers and rotary polisher grinder with 1500 grit sand paper, diamond compound 6 μm, and alumina 0.3 μm. For microscopic observations, the specimens were etched with Keller's reagent on aluminum side and Nital reagent on steel side. Scanning electron microscope (SEM) was used for metallography examination with energy-dispersive spectroscopy (EDS) for chemical composition of interface.

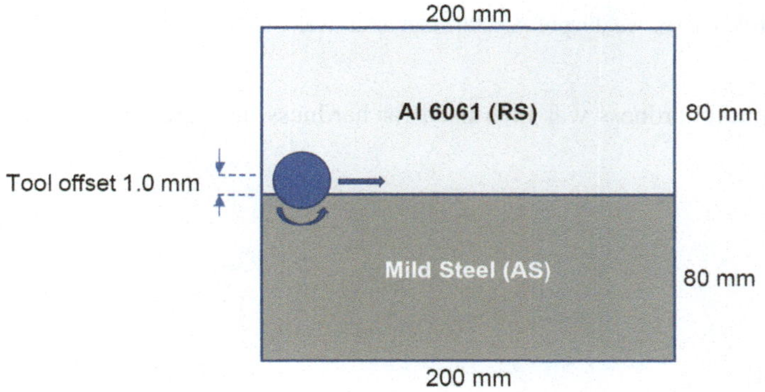

Fig. 16.1 AA6061 and mild steel dimensions and placement

Table 16.1 Experiments designed using Taguchi method

Experiment No	Rotation speed (rpm)	Welding speed (mm/min)	Plunge depth (mm)
1	600	22	0.2
2	600	45	0.275
3	600	90	0.35
4	865	22	0.275
5	865	45	0.35
6	865	90	0.2
7	410	22	0.35
8	410	45	0.2
9	410	90	0.275

Fig. 16.2 Friction stir welding of AA6061 and mild steel

Vicker's microhardness was then used for hardness measurements across the weld line.

16.3 Results and Discussion

16.3.1 Macrostructure

A few defects can be observed on the surface of weld line as shown in Fig. 16.3. Material reduction and keyhole at both ends of weld line were unavoidable as a result of FSW process. Flashes were formed at higher rotation speed due to greater heat generation. Root-flow defect is caused by surface lack of fill and can be controlled by proper selection of process parameters. From SEM analysis result shown in Fig. 16.4, the four weld zones can be observed. The four zones are the stir zone or nugget zone (NZ), thermo-mechanically affected zone (TMAZ), heat affected zone (HAZ), and base metal (BM). The stir zone shows basin-like shape that widens toward the upper surface.

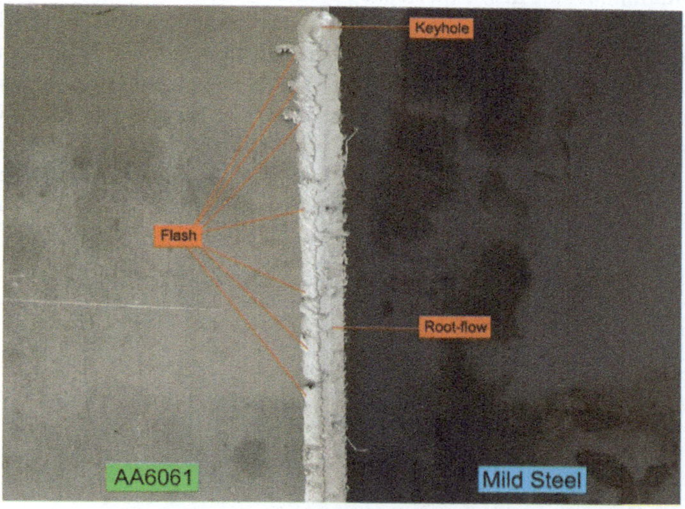

Fig. 16.3 Defects on weld joint

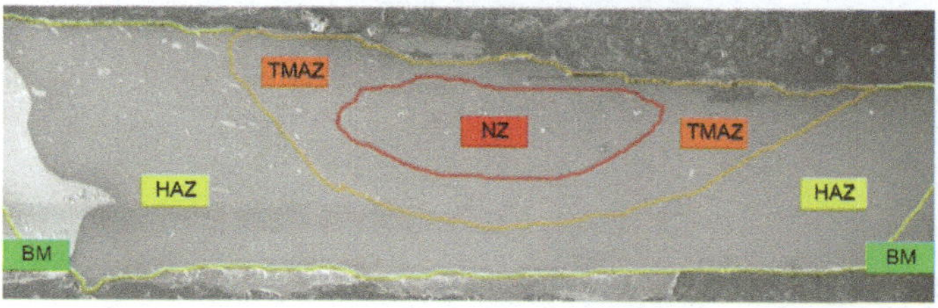

Fig. 16.4 Weld zones of AA6061 and mild steel joint

16.3.2 Microstructure

Some steel particles with irregular shapes are scattered in aluminum side as can be seen in Fig. 16.5. The stirring action of tool pin caused abrasion and shearing which broke these plastic deformed steel fragments and mixed them with the plasticized aluminum alloy, creating a steel reinforced aluminum composite structure. However, these steel pieces produce large IMC around them which decreases the tensile strength. Contrary to steel particles in aluminum side, aluminum elements are minimal in steel side. A few voids were visible near the joint interface at both aluminum and steel sides as shown in Fig. 16.6. Voids may be caused by insufficient heat input or lack of material flow during friction stir welding process.

Near the interface, the grains are finer and equiaxed due to severe plastic deformation and recrystallization. Compared to aluminum, the microstructural change in steel side was minimal owing to the difference in mechanical and thermal properties. The aluminum side also experienced greater plastic deformation caused by the tool offset toward aluminum. As a result of stirring action from the tool pin, stir zone consisted of refined and equiaxed grains attributed to dynamic recrystallization from deformation and heat input [17]. In TMAZ, grains are elongated owing to the deformation from thermo-mechanical effect. The microstructure in HAZ consists of coarser grain structure as the zone was only affected by welding heat. As for the steel side of HAZ, the microstructure does not show much differences compared to base material. Figure 16.7 shows the grain structure near the interface, TMAZ, and HAZ.

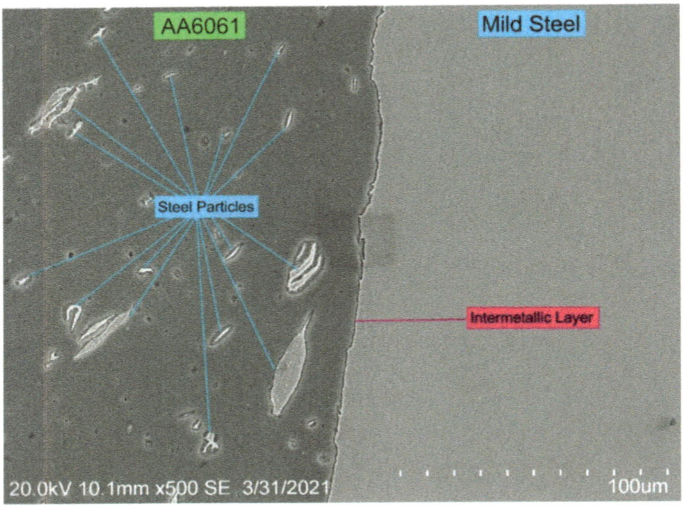

Fig. 16.5 Steel fragments in aluminum side

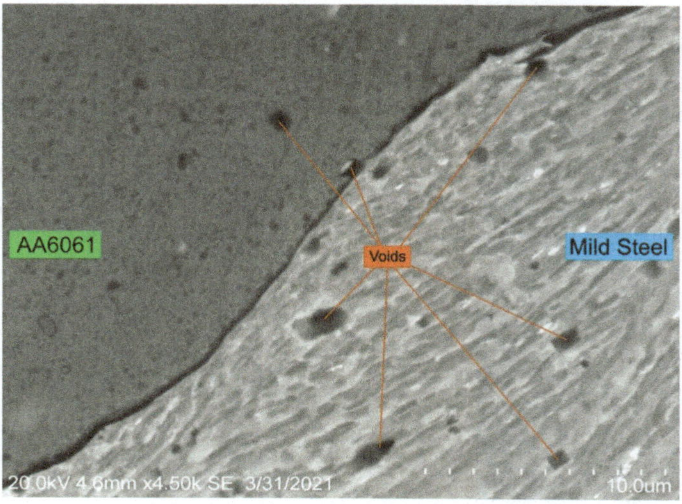

Fig. 16.6 Voids near the interface on both aluminum and steel sides

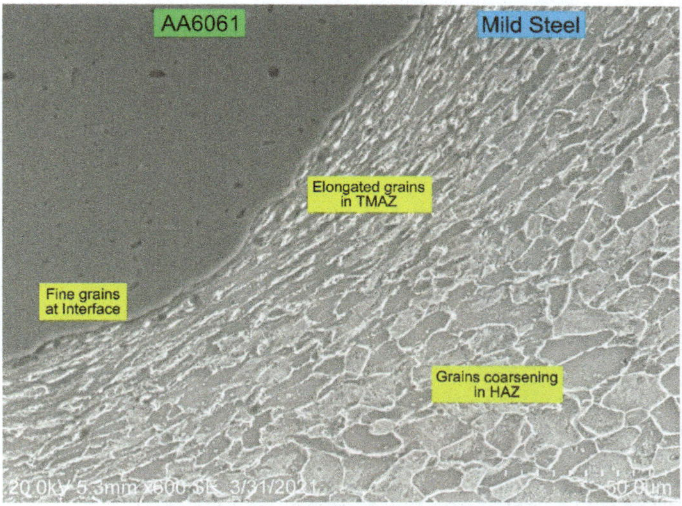

Fig. 16.7 Grain structure near the interface, TMAZ, and HAZ

16.3.3 Intermetallic Compound

Intermetallic compounds formed at stir zone due to thermal and mechanical action from pin stirring. The formation of brittle intermetallics especially Al-rich IMCs such as Fe_2Al_3, $FeAl_2$, Fe_2Al_5, and $FeAl_3$ are reported to form at low temperature which can reduce joint strength [18]. From SEM analysis results, thin continuous intermetallic compound layers were formed at the interface with thickness less than 2 μm. The intermetallic compound at Al/Fe interface was identified as $FeAl_3$ and Fe_2Al_5 from EDS analysis result as shown in Fig. 16.8. Apart from the interface, IMC layers with different thickness formed around the steel fragments that were stirred into aluminum as can be seen in Fig. 16.9. Al-rich Fe_2Al_3 intermetallic was formed at higher rotation speed due to severe plastic deformation and dynamic recrystallization from tool stirring action and material mixing. Thickness of IMC layer increases with

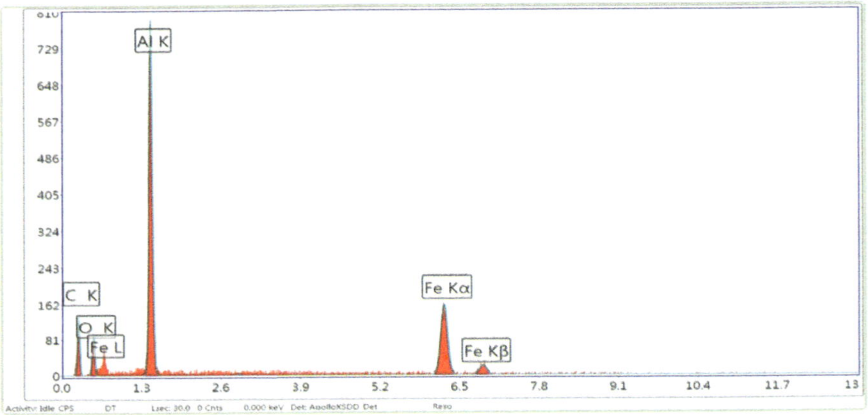

Fig. 16.8 EDS analysis identifying Fe_2Al_5 IMC

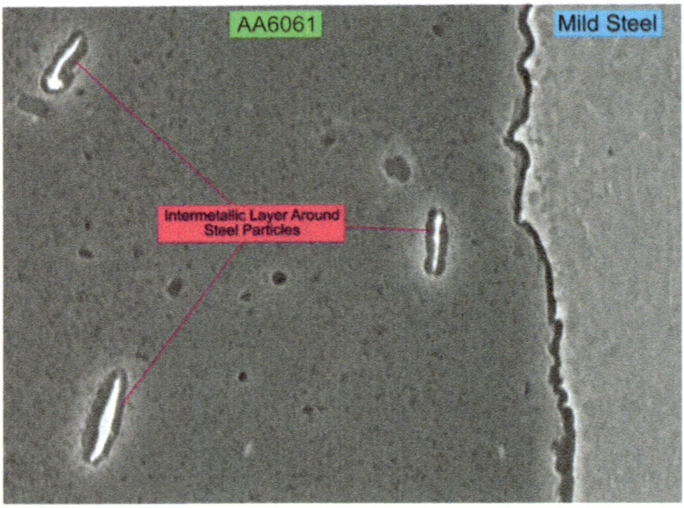

Fig. 16.9 IMC layer around steel particles

higher rotation speed on account of higher heat input and decreases with lower welding speed. Higher heat input causes formation of thicker IMC which makes the joint more brittle and decreases the joint strength.

16.3.4 Microhardness

Grain size and brittle intermetallics influenced the hardness value. Hardness is lowest across AA6061, highest at weld interface, and intermediate across mild steel. Maximum microhardness was recorded at the interface in the stir zone due to materials mixing, plastic deformation, and recrystallization from pin stirring action as well as formation of brittle intermetallic compound. Hardness increases near interface of steel side due to high deformation. At the aluminum side, sudden increase in microhardness near the interface is attributed to the existence of small steel particles and IMC.

High hardness measured in stir zone compared to base metal, HAZ, and TMAZ due to the presence of brittle IMC. At NZ next to TMAZ, the hardness decreased as materials experienced highest temperature and most precipitates dissolved. Hardness gradually rose at TMAZ in response to working hardening caused by deformation. In TMAZ of aluminum side, hardness decreased slightly as a result of second phase particle dissolution and coarsening from thermo-mechanical effect owing to dynamic recovery and recrystallization. Minimum hardness value in HAZ was recorded on both sides. The hardness is minimum in HAZ since the zone was only subjected to thermal cycle under which grain and precipitate coarsening occurred. The hardness in HAZ is further decreased and gradually rose to hardness of BM at the boundaries. Figure 16.10 shows the hardness distribution across the aluminum and steel interface.

16.4 Conclusion

In this study, FSW of aluminum alloy 6061 and mild steel has been investigated and the conclusions drawn are as follows:

1. Grains are finer and equiaxed at the weld interface from severe plastic deformation and recrystallization as a result of tool pin stirring action. However, the stirring action distributed broken steel particles into the aluminum side which initiated IMC layer formation around them.
2. IMC layers of less than 2 μm were formed which is favorable as thicker IMC layer will degrade the joint quality. Al-rich IMCs were formed at higher heat input from increase in rotation speed and decrease in welding speed.
3. The highest hardness was measured at the weld interface, followed by TMAZ, base metal, and HAZ. With most of the measurements exceeded 200 HV, the high interface hardness indicated the formation of brittle intermetallic compound.

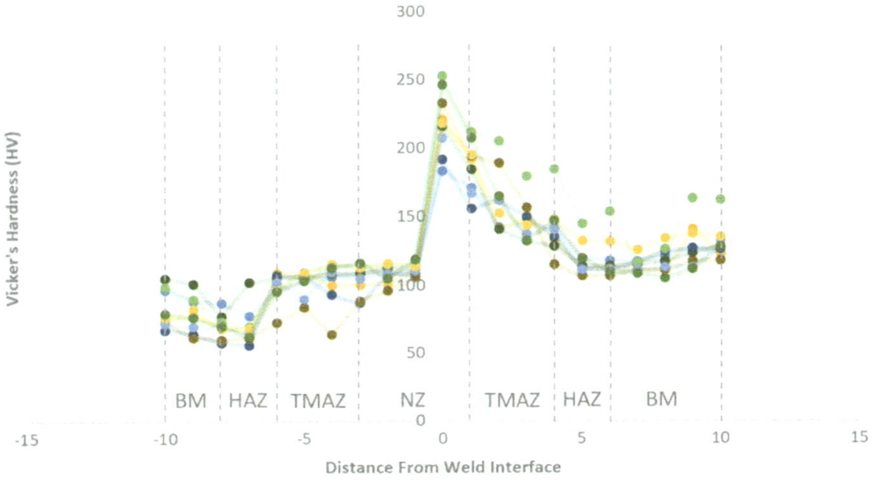

Fig. 16.10 Hardness distribution across Al/Fe interface

4. Steel particles distribution in aluminum, formation of Al-rich IMC and IMC layer thickness were all affected by the heat input and influenced the weld joint quality. Therefore, in order to obtain a strong weld joint, the heat input needs to be properly controlled by careful selection of rotation speed and welding speed.

Acknowledgements The author would like to acknowledge the support from Universiti Sains Malaysia (USM) thru RUI Grant (1001/PMEKANIK/8014031).

References

1. Li P et al (2020) Interfacial microstructure and mechanical properties of dissimilar aluminum/steel joint fabricated via refilled friction stir spot welding. J Manuf Process 49:385–396
2. Wei Y, Xiong J, Li J, Zhang F, Liang S (2017) Microstructure and enhanced atomic diffusion of friction stir welding aluminium/steel joints. Mater Sci Technol 33:1208–1214
3. Elnabi MA, Osman T, El-Mokadem AE, Elshalakany AB (2018) Mechanical properties and microstructure of dissimilar friction stir welding of pure aluminum to low carbon steel. Adv J Grad Res 4:47–58
4. Murugan B, Thirunavukarasu G, Kundu S, Kailas SV, Chatterjee S (2018) Interfacial microstructure and mechanical properties of friction stir welded joints of commercially pure aluminum and 304 stainless steel. J Mater Eng Perform 27:2921–2931
5. Ogura T et al (2012) Partitioning evaluation of mechanical properties and the interfacial microstructure in a friction stir welded aluminum alloy/stainless steel lap joint. Scr Mater 66:531–534
6. Kar A, Vicharapu B, Morisada Y, Fujii H (2020) Elucidation of interfacial microstructure and properties in friction stir lap welding of aluminium alloy and mild steel. Mater Charact. https://doi.org/10.1016/j.matchar.2020.110572

7. Yasui T, Tsubaki M, Fukumoto M, Shimoda Y, Ishii T (2006) High-speed weldability between 6063 and S45C by friction stir welding. Study of welding of dissimilar metals by friction stir welding (1st report). Weld Int 20:284–289
8. Wan L, Huang Y (2017) Microstructure and mechanical properties of Al/steel friction stir lap weld. Metals 7:1–14
9. Habibnia M, Shakeri M, Nourouzi S, Givi MKB (2015) Microstructural and mechanical properties of friction stir welded 5050 Al alloy and 304 stainless steel plates. Int J Adv Manuf Technol 76:819–829
10. Tanaka T, Morishige T, Hirata T (2009) Comprehensive analysis of joint strength for dissimilar friction stir welds of mild steel to aluminum alloys. Scr Mater 61:756–759
11. Chen YC, Komazaki T, Kim YG, Tsumura T, Nakata K (2008) Interface microstructure study of friction stir lap joint of AC4C cast aluminum alloy and zinc-coated steel. Mater Chem Phys 111:375–380
12. Liu X, Lan S, Ni J (2014) Analysis of process parameters effects on friction stir welding of dissimilar aluminum alloy to advanced high strength steel. Mater Des 59:50–62
13. Sameer MD, Birru AK (2019) Investigations on microstructural evolutions and mechanical properties of dual-phase 600 steel and AA6082-T6 aluminum alloy dissimilar joints fabricated by friction stir welding. Trans Indian Inst Met 72:353–367
14. Movahedi M, Kokabi AH, Seyed RSM, Najafi H (2012) Effect of tool travel and rotation speeds on weld zone defects and joint strength of aluminium steel lap joints made by friction stir welding. Sci Technol Weld Joint 17:162–167
15. Balamagendiravarman M, Kundu S, Chatterjee S (2017) An analysis of microstructure and mechanical properties on friction stir welded joint of dissimilar 304 stainless steel and commercially pure aluminium. Arch Metall Mater 62:1813–1817
16. Kimapong K, Watanabe T (2004) Friction stir welding of aluminum alloy to steel. Weld J 83:277–282
17. Zhou L et al (2020) Microstructure and mechanical properties of Al/steel dissimilar welds fabricated by friction surfacing assisted friction stir lap welding. J Mater Res Technol 9:212–221
18. Anaman SY, Cho HH, Das H, Lee JS, Hong ST (2019) Microstructure and mechanical/electrochemical properties of friction stir butt welded joint of dissimilar aluminum and steel alloys. Mater Charact 154:67–79

Chapter 17
A Study on the Electromagnetic Radiation in Human Head Tissues on 5G Mobile Exposure

Ooi Zhi Kai and Nur Hidayah Mansor

Abstract The 5th generation mobile network, also known as 5G, is rapidly developed and expected to be widely implemented soon. This situation raises concern on the effect of millimetre-wave (mm-wave) utilised by 5G devices to impact human health negatively. While mm-wave carries no risk of causing genetic mutation for being non-ionising, the main concern is the heating effect on human skin, measured by measuring the specific absorption rate (SAR) of skin exposed to a 5G antenna. This research sets up to simulate the 5G mobile phone antenna effect on human skin-heads. Three antennas are designed and optimised close to the human head model for an operating frequency of 28 GHz, 30 GHz, and 33 GHz, respectively. The characteristic of the three antennas was examined for their performance then a 3D SAR plot is generated. All three antennas designed have acceptable performance. The SAR plots of 3 antennas show the maximum SAR decrease as the operating frequency increases with similar SAR distribution.

Keywords 5G mobile antenna · 5G network · Specific absorption rate · Electromagnetic radiation

17.1 Introduction

Modern growth in communication technology has introduced mobile handphones beginning from the 1st generation (1G) mobile phone towards the 5th generation (5G) smartphones. The 5G of mobile development promises much faster speed and capacity than the previous generation, with data bandwidth up to 10 Gbps, ultra-low latency with about 1 ms, better coverage, and performance. 5G has much complete application, especially in industry 4.0, with the integration of IoT, analytics, additive

O. Z. Kai · N. H. Mansor (✉)
School of Mechanical Engineering, Universiti Sains Malaysia, 14300 Nibong Tebal, Penang, Malaysia
e-mail: menurhidayah@usm.my

O. Z. Kai
e-mail: zhikaiooi@student.usm.my

A. Ismail et al. (eds.), *Advanced Materials and Engineering Technologies*, Advanced Structured Materials 162,
https://doi.org/10.1007/978-3-030-92964-0_17

manufacturing, robotics, artificial intelligence, advanced materials, and augmented realities into the industry, which are much dependent on speed and capacity that 5G brings to the table. To achieve such speed and capacity, 5G utilises a higher frequency than its predecessor, exploiting a higher frequency higher than 30 GHz, which is considered a millimetre-wave. Such development has an impact on more significant exposure to electromagnetic radiation as well as heat-induced in human tissues. Realizing human health is at stake, more researchers put effort into studying the impact of electromagnetic radiation on humans [1–3].

The research of the effect of electromagnetic waves on biological tissue is not exactly new and has been investigated for a long history since its early discovery [4, 5]. However, earlier research focused on the older generation of the wireless network with a lower operating frequency [6–9]. Some studies relate the mobile phone radiation with physical health symptoms such as fatigue, headache, dizziness, tension, and sleep disturbance [10]. Others looked into the electric field and specific absorption rate (SAR) calculation in the human skin tissues [4, 11]. The cumulative effect of radiation from mobile phones requires research [12]. 4G handphone affects specific vital haematological parameters [13].

Moreover, mobile phones' exposure during the gestation period was associated with increased oxidative stress levels [14]. The Mm-wave is categorised as non-ionising radiation, meaning it does not carry the risk of getting cancer as in ionising radiation carried by an mm-wave photon is 10,000 less time than that of ionising radiation. With the 5G network will implement worldwide soon, the utmost concern about human health risks in future of 5G mm-wave will be the heating of human skin on prolonged exposure. The penetration of electromagnetic radiation in the human body will produce an induced electric field, resulting in power absorption.

Studies that considered 5G electromagnetic radiation were reported in [15]. The authors attempted to investigate the specific absorption rate (SAR) in which the absorption of radiations from the mobile antenna by human head and brain tissues. The mobile antenna used for the simulation is proposed by Jandi et al. [16]. The study of SAR effects on the human head tissue by a monopole printed antenna proposed by Yazdandoost and Laakso [17] is not significant because the numerical simulation was running at a frequency of 10 GHz. Meanwhile, recent studies have also considered different operational frequency exposure from the same antenna, which alters the human tissue [18]. In [19], the calculation of SAR in the human head exposed to the older generation of the mobile network with lower operating frequency has been proposed.

5G mm-wave penetration is superficial at the epidermis layer, and the specific absorption rate (SAR) level decreases as operating frequency increases [18]. However, when a similar simulation is done with operating frequencies of 10.12 and 29.6 GHz near the brain model, the antenna of higher frequencies shows a higher SAR value [15]. However, the author does not mention the distance between head and antenna that might cause such disparity in results. Besides, there is a lack of information about the SAR effect of 5G mobile phones.

In this paper, the effects of the electromagnetic wave of the 5G mobile phone spectrum on humans, particularly human skinhead tissues, are investigated to characterise the SAR. The mm-wave has a shallow depth of penetration. Therefore, the energy carried by mm-wave will mostly be deposited in the first 1–2 mm of human skinhead tissues. It is worth analysing the effect of the heat energy projected by the electromagnetic wave on human skinheads, whether it will have a harmful effect or any potential long-term effect [20].

17.2 Methodology

This paper will investigate the heating effect of energy carried by the mm-wave emitted by mobile phone to the human head, emulating calling through mobile phone. An easy way to determine the effect is to characterise the specific absorption rate (SAR) of the tissue exposed to the electromagnetic wave. SAR is the unit of measurement for the amount of radiofrequency energy absorbed by a body when using a wireless device. SAR can be calculated with the following formula:

$$\text{SAR} = \left(\sigma + E^2\right)/md \tag{17.1}$$

where σ is the material's conductivity, E is the RMS electric field, and md is the mass density.

Calculation of SAR field manually for a non-uniform surface such as a human head is complicated. A simulation will be set up with simulation software capable of simulating high-frequency electromagnetic waves and plotting SAR. For this reason, simulation software is used to simulate the project. An antenna is drawn in the software CST by referring to the schematics of the available antenna. A human head model is placed beside the antenna with a 2 cm distance between head and antenna to emulate making a mobile phone call. Due to limited skills in creating a model and hardware limitations, the model used for this simulation will be just a simple hollow dome. The mm-wave component of 5G radiofrequency has minimal penetrating power. The simulation only utilises skin as the human head model's material will suffice as it is unlikely for the radio wave to penetrate deeper into another tissue type. The material "skin" must be defined as it does not come with the software's default material library. Table 17.1 shows the properties of skin used in the simulation model.

Based on [21, 22], the three inset feed antennas for operating frequencies of 28 GHz, 30 GHz and 33 GHz have been designed. Copper is used for the antenna patch and ground plane, while Substrate Duroid 5880 with a dielectric constant of 2.2 is used as substrate. The antenna's design with a substrate thickness of 0.5 mm and ground thickness of 0.035 mm. The antenna is placed 2 cm from the head. The head model is simply a dome with 2 mm thickness as the average thickness of skin and 8 cm radius. The material properties are made sure to be set to the skin, as shown in Figs. 17.1 and 17.2.

Table 17.1 Dielectric properties of skin

Material	Skin
Epsilon	69.45
Mu	1
Electric conductivity	0.507 S/m
Rho	1100 kg/m^3
Thermal conductivity	0.3 W/K/m
Heat capacity	3.5 kJ/K/kg
Diffusivity	7.79221×10^{-8} m^2/s
Bloodflow	9000 W/K m^3
Metabolic rate	2000 W/m^3

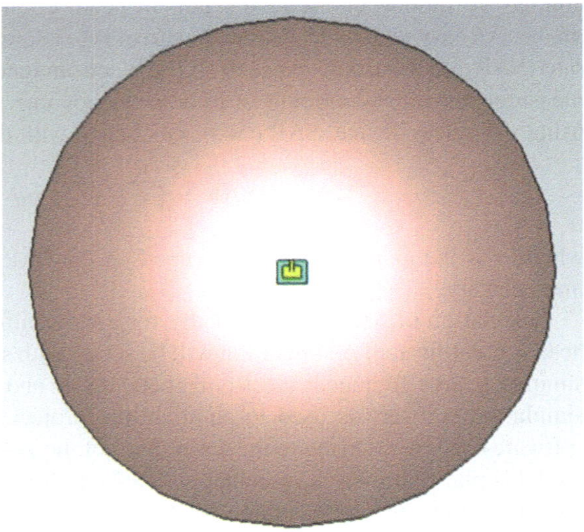

Fig. 17.1 Front view of the simulation model

The antenna will experience frequency shift and return loss change when nearby of biological tissue. For that, optimisation is used to obtain the suitable parameter for the intended parameters. The optimisation is set up to optimise the value of patch length, patch width, feedline insertion length, and feedline gap. The optimisation process is that the system will continuously adjust the program's parameters, then simulate the antenna to calculate the S_{11} parameter until the optimiser's goal is achieved. The goal is set, so the resonance frequency occurs within ±1 GHz at the predetermined frequency with a return loss of at least −10 dB. Three antennas are optimised for operating frequencies of 28 GHz, 30 GHz, and 33 GHz, respectively.

After the optimisation is done, the antennas are examined for radiation pattern and gain, as the excellent antenna should have at least 6 dB gain to be properly functional. By setting up a far-field field monitor for operating frequency, then in

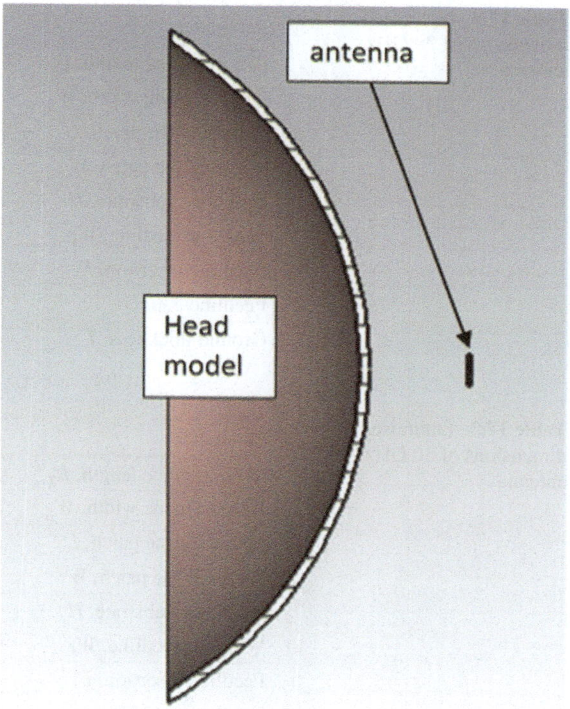

Fig. 17.2 Side view of the simulation model set up with cutting plane

template-based post-processing, the far-field result and SAR results are selected. After the simulation, the 3D plot of the radiation pattern and gain and 3D SAR plot can be obtained in the 2D/3D result section. The VSWR of the antenna can also be checked under the 1D result section.

17.3 Results and Discussion

The antennas' optimised dimensions are as tabulated in Tables 17.2, 17.3, and 17.4. The parameters are optimised with simulation near the head to consider resonance frequency shift and return loss change of antenna in biological tissues' proximity. The antenna will have resonance frequency at the desired frequency to measure the SAR pilot. With the optimisation, the $S11$ the plot of the antenna is as Figs. 17.3, 17.4 and 17.5. From the $S11$ plots shown, all three antennas have return loss lower than −10 dB, which means the antennas accept at least 90% of the power delivered.

The 28 GHz antenna resonates at 27.915 GHz with a return loss of −12.73707 dB. The 30 GHz antennae resonate at 29.95 GHz with a return loss −18.501556 dB, while the 33 GHz antenna resonates at 32.97 GHz with a return loss of −18.473485 dB.

Table 17.2 Optimised dimensions of 28 GHz antenna

Parameter	Dimension (mm)
Ground plane length, L_g	6.271
Ground plane width, W_g	8.126
Length of the patch, L	3.271
Width of the patch, W	5.126
Height of substrate, H	0.5
Width of feedline, W_f	1.2
Feedline insertion, F_i	1.297
Feedline gap, G	0.119
Ground thickness, T	0.035

Table 17.3 Optimised dimensions of 30 GHz antenna

Parameter	Dimension (mm)
Ground plane length, L_g	6.008
Ground plane width, W_g	7.719
Length of the patch, L	3.008
Width of the patch, W	4.719
Height of substrate, H	0.5
Width of feedline, W_f	1.2
Feedline insertion, F_i	1.256
Feedline gap, G	0.115
Ground thickness, T	0.035

Table 17.4 Optimised dimensions of 33 GHz antenna

Parameter	Dimension (mm)
Ground plane length, L_g	5.745
Ground plane width, W_g	7.31
Length of the patch, L	2.745
Width of the patch, W	4.31
Height of substrate, H	0.5
Width of feedline, W_f	1.2
Feedline insertion, F_i	1.01
Feedline gap, G	0.11
Ground thickness, T	0.035

Next, the VSWR of the antenna is examined. A good antenna usually has VSWR below 2 in operating frequencies. Figures 17.6, 17.7 and 17.8 show that all VSWR is less than two, which means they have proper matching. The radiation pattern and gain of the antenna is another antenna characteristic to be examined. The gain measures

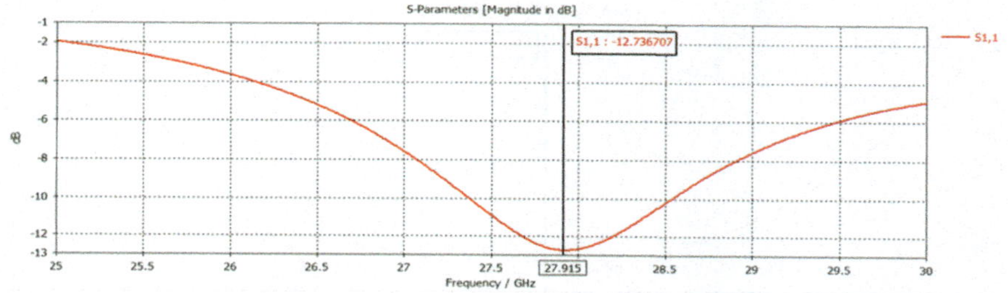

Fig. 17.3 The s_{11} plot of 28 GHz antenna

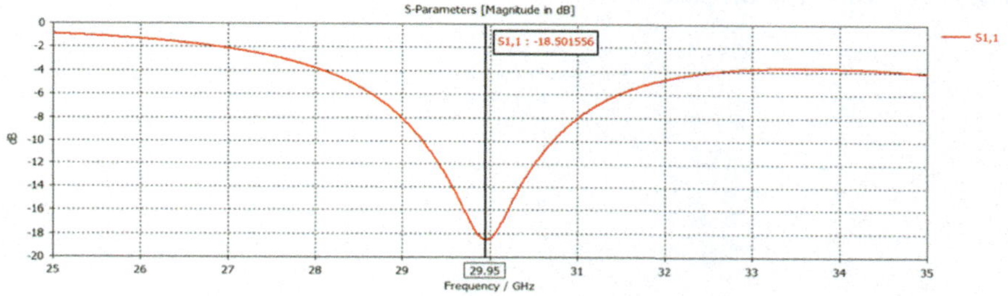

Fig. 17.4 The s_{11} plot of 30 GHz antenna

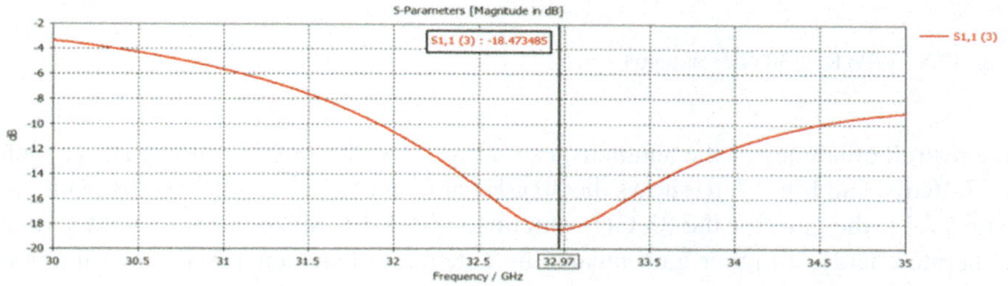

Fig. 17.5 The s_{11} plot of 33 GHz antenna

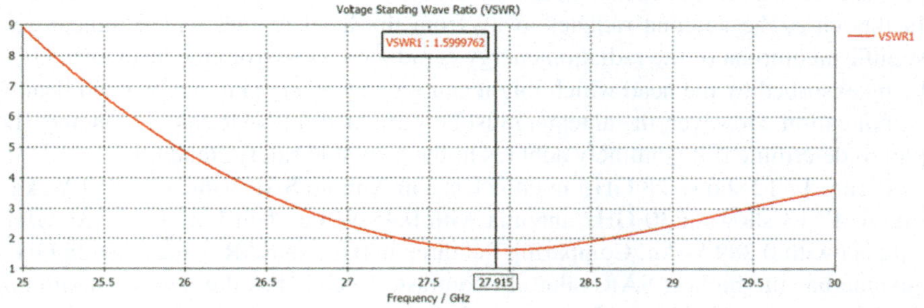

Fig. 17.6 VSWR of 28 GHz antenna

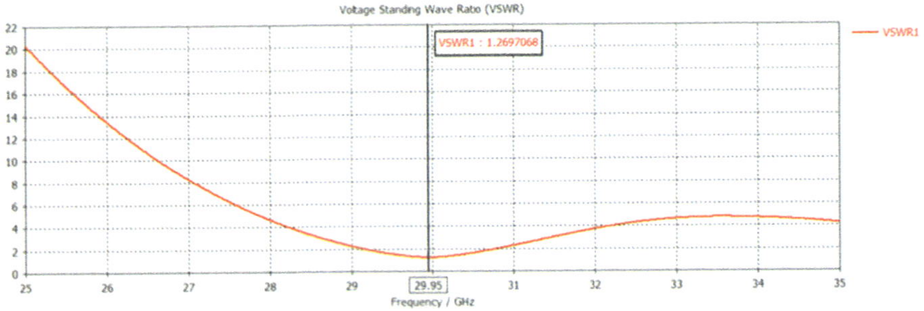

Fig. 17.7 VSWR of 30 GHz antenna

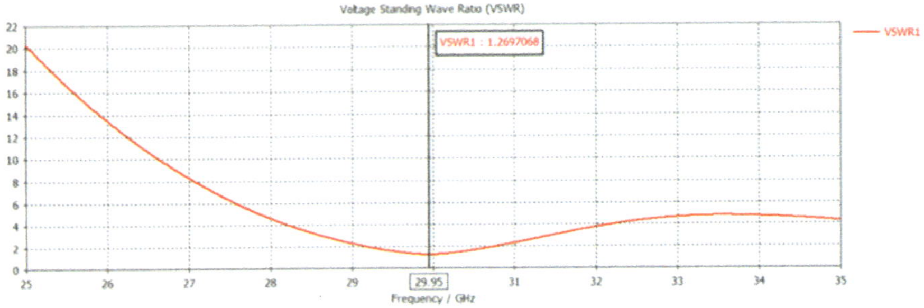

Fig. 17.8 VSWR of 33 GHz antenna

the overall efficiency of the antennas. From Fig. 17.9, the 28 GHz antenna has a gain of 7.46 dB, and Fig. 17.10 shows the 30 GHz antenna has a gain of 6.95 dB, while in Fig. 17.11, the gain for the 33 GHz antenna is 6.93 dB. Since the antenna does not generate energy, a higher gain means the antenna radiates energy in one direction more than the other direction. As observed, compared to lower frequency antenna widely used in older generations which have the more omnidirectional radiation pattern, the high-frequency antenna utilised in 5G tends to radiate in one direction. In this case, the antenna radiates away from the head model, which indicates a significant amount of the radiation energy is radiated away from the head instead of being absorbed by the head which might cause unnecessary skin heating and health complication. However, the antenna must be simulated for SAR calculation and 3D plot to determine if it genuinely adheres to the available safety standard.

Figure 17.12 shows 28 GHz antenna has a maximum SAR value of 0.584 W/kg. Figure 17.13 shows a 30 GHz antenna with 0.45 W/kg, then Fig. 17.14, 33 GHz antenna with 0.289 W/kg. Comparing the three maximum SAR values, the 28 GHz antenna has the highest SAR value. In contrast, 33 GHz has the lowest maximum SAR value, and this shows that the maximum SAR value decreases as frequency increases, and this agrees with the finding by Kaur and Khan [18]. As observed by the SAR plot, all three results show a similar distribution of heat absorbed by the

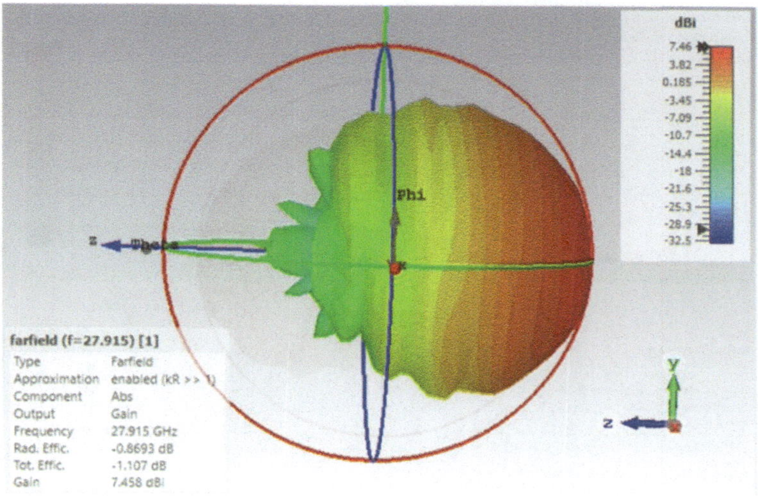

Fig. 17.9 The radiation pattern and gain 3D plot of 28 GHz antenna

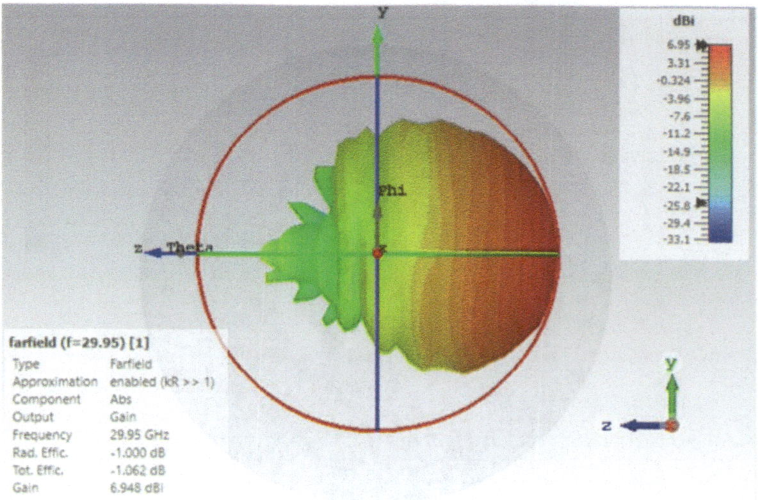

Fig. 17.10 The radiation pattern and gain 3D plot of 30 GHz antenna

head. The head's section near the antenna at the centre absorbs the most energy radiated from the antenna. The energy absorbed decreases quite sharply as moving away from the centre and having close to zero absorption near the model's edge.

In the case of a real human head, the ear positioned the closest to the phone antenna will absorb most of the energy radiated to the head, while the other areas such as the cheek and upper head will have minor effects. FCC's standard from the United

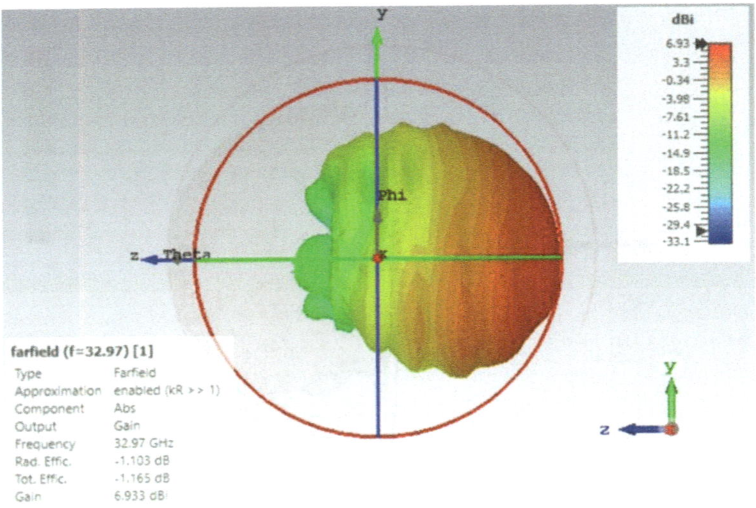

Fig. 17.11 The radiation pattern and gain 3D plot of 33 GHz antenna

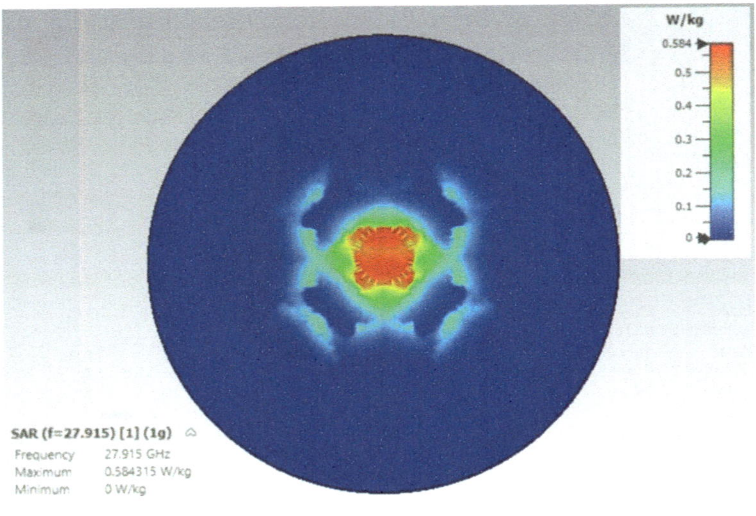

Fig. 17.12 3D SAR plot for 28 GHz antenna

States that agreed-upon globally sets that the SAR level of the mobile antenna must be below 1.6 W/kg per 1 g of tissue, so it does not cause any health complications. According to the stated standard, all three 5G antennas simulated in this project are safe for commercial use in a 5G mobile phone.

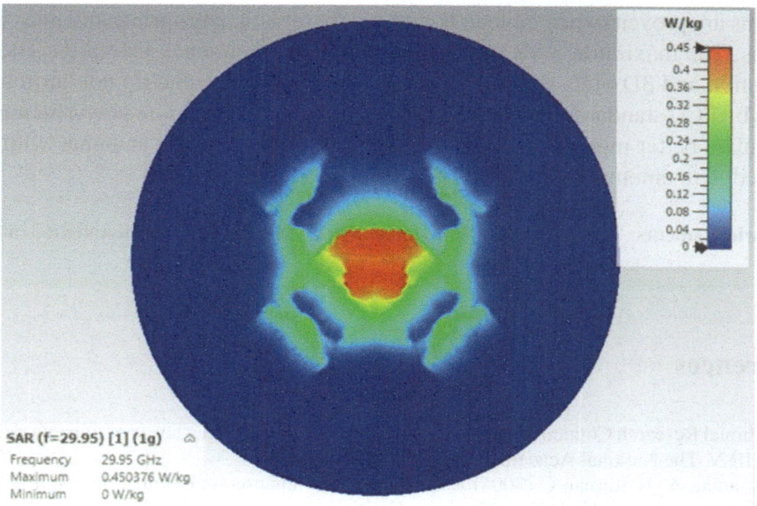

Fig. 17.13 3D SAR plot for 30 GHz antenna

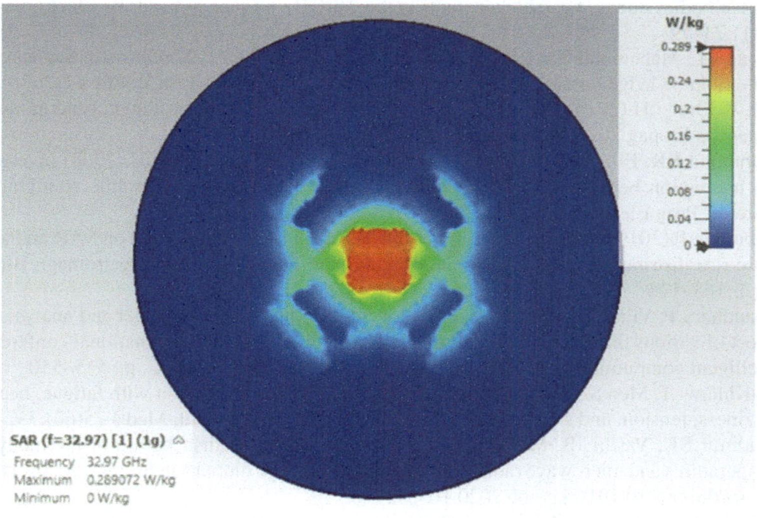

Fig. 17.14 3D SAR plot for 33 GHz antenna

17.4 Conclusion

This project involves designing and optimising antennas for 5G applications with 28 GHz, 30 GHz, and 33 GHz operating frequencies near the human head. The antennas are proven to be viable in commercial use with appropriate antenna characteristics. The maximum SAR value decreases as the frequency increases. The SAR calculation and 3D plots show all the 5G antennas are safe and do not harm human bodies by FCC standard. Future work may include utilising a more powerful machine to simulate better models with higher mesh and more compact antennas with more knowledge in antenna design.

Acknowledgements This research is being funded by USM grant 304/PMEKANIK/6315404.

References

1. National Research Council (1990) Health effects of exposure tolow levels of ionising radiation: BEIR V. The National Academies Press, Washington D.C
2. Zamanian A, Hardiman C (2005) Electromagnetic radiation and human health: a review of sources and effects. High-Freq Electron 4(3):16–26
3. Genuis SJ (2008) Fielding a current idea: exploring the public health impact of electromagnetic radiation. Public Health 122(2):113–124
4. Marta F, Hugo GE, David G, Iván P, David VT, Amaia A (2019) RF energy absorption in human bodies due to wearable antennas in the 2.4 GHz frequency band. Bioelectromagnetics 41(1):73–79
5. Dréan YL, Mahamoud YS, Page YL, Habauzit D, Quément CL, Zhadobov M, Sauleau (2013) State of knowledge on biological effects at 40–60 GHz. C R Phys 14(5):402–411
6. Lak A, Oraizi H (2013) Evaluation of SAR distribution in six-layer human head model. Int J Antennas Propag 2013:1–8
7. Mahmoud KR, El-Adawy M, Ibrahe SMM, Bansal R, Zainud-Deen SH (2008) Investigating the interaction between a human head and a smart handset for 4G mobile communication systems. Prog Electromagn Res C 2:169–188
8. Kaburcuk F (2019) Effects of a brain tumour in a dispersive human head on SAR and temperature rise distributions due to RF sources at 4G and 5G frequencies. Electromagn Biol Med 38(2):168–176
9. Chaudhary P, Vijay R (2019) Calculation of SAR on human brain model and analysis of the effect of various dielectric shielding materials to reduce SAR. In: International conference on intelligent computing and smart communication. Proceedings of ICSC, pp 545–550
10. Al-Khlaiwi T, Meo SA (2004) Association of mobile phone radiation with fatigue, headache, dizziness, tension, and sleep disturbance in Saudi population. Saudi Med J 25(60):732–736
11. Vladimir SK, Vasilii IB, Mikhail VB (2020) Simulation the distribution of thermodynamic temperatures and microwave radiation of the human head. Comput Methods Programs Biomed. https://doi.org/10.1016/j.cmpb.2020.105377
12. Yang H, Zhang Y, Wang Z, Zhong S, Hu G, Zuo W (2020) The effects of mobile phone radiofrequency radiation on cochlear stria marginal cells in Sprague–Dawley rats. Bioelectromagnetics 41:219–229
13. Christophera B, Marya YS, Khandaker MU, Bradley DA, Chew MTPJ, Jojo PJ (2020) Effects of mobile phone radiation on certain hematological parameters. Radiat Phys Chem 166:10844
14. Bektas H, Dasdag S, Bektas MS (2020) Comparison of effects of 2.4 GHz Wi-Fi and mobile phone exposure on human placenta and cord blood. Biotechnol Biotechnol Equip 34(1):154–162
15. Laghari MR, Hussain I, Ali MK, Yaseen KG (2018) Modeling and analysis of 5G antenna radiation effect on human head by calculating specific absorption rate (SAR) using adult brain model. J Inf Commun Technol Robot Appl 9(2):13–18

16. Jandi Y, Gharnati F, Said AO (2017) Design of a compact dual bands patch antenna for 5G applications. In: International conference on wireless technologies, embedded and intelligent systems (WITS). IEEE, Morocco, pp 1–4
17. Yazdandoost KY, Laakso I (2018) Numerical modeling of electromagnetic field exposure from 5G mobile communications at 10 GHz. PIER M 72:61–67
18. Kaur J, Khan SA (2019) Electric field and specific absorption rate (SAR) analysis in multilayered skin tissue exposed to 5G mobile communication frequencies advances in basic science. AIP Conf Proc 2142:110028
19. Chaudhary P, Vijay R (2020) Calculation of SAR on human brain model and analysis of the effect of various dielectric shielding materials to reduce SAR. In: Singh Tomar G, Chaudhari N, Barbosa J, Aghwariya M (eds) International conference on intelligent computing and smart communication 2019. Algorithms for intelligent systems. Springer, Singapore. http://doi.org/10.1007/978-981-15-0633-8_53
20. Ryan KL, D'Andrea JA, Jauchem JR, Mason PA (2000) Radio frequency radiation of millimeter-wavelength: potential occupational safety issues relating to surface heating. Health Phys 78(2):170–181
21. Darboe O, Konditi DBO, Manene F (2019) A 28 GHz rectangular microstrip patch antenna for 5G applications. Int J Eng Res 12(6):854–857
22. Ramesh M, Yip KB (2003) Design formula for inset fed microstrip patch antenna. J Microw Optoelectron 3:5–10

Chapter 18
Design of Green Ship Recycling Yards: A Review

Mohamad Suhaimi Samsudin, Ismail Zainol, Zulzamri Salleh, Siow Chee Loon, and Nuraihan Ismail

Abstract This article presents an overview of green ship recycling yard design. This article presents the basic process and method used in recycling yards. This article also provides an insight to the relevant aspects of yard facilities and layout design. This approach used for designing green ship recycling yards are briefly discussed. The last section of the article explains issues arise in green ship recycling and related works taken to overcome the issues including the prospect of systematic layout planning (SLP) as one of methods for efficient ship recycling yard design. Finally, it can be concluded that there is a gap in green ship recycling yard specifically on layout design that requires further work.

Keywords Green recycling yard · Design layout · Systematic layout planning

18.1 Introduction

Ship recycling is the process of dismantling and reusing the ship when the ship is no longer fit to continue its operation. In practice, when the ship lifespan is around

---

M. S. Samsudin · I. Zainol (✉) · Z. Salleh
Marine Engineering Technology Section,, Universiti Kuala Lumpur Malaysian Institute of Marine Engineering Technology, Dataran Industri Teknologi Kejuruteraan Marin, Bandar Teknologi Maritim, Jalan Pantai Remis, 32200 Lumut, Perak, Malaysia
e-mail: ismailz@unikl.edu.my

M. S. Samsudin
e-mail: suhaimi.samsudin@s.unikl.edu.my

Z. Salleh
e-mail: zulzamri@unikl.edu.my

I. Zainol · S. C. Loon · N. Ismail
School of Mechanical Engineering, Universiti Teknologi Malaysia, Skudai, Johor, Malaysia
e-mail: scheeloon@utm.my

N. Ismail
e-mail: nuraihan@graduate.utm.my

A. Ismail et al. (eds.), *Advanced Materials and Engineering Technologies*,
Advanced Structured Materials 162,
https://doi.org/10.1007/978-3-030-92964-0_18

22 years or older, it should not be operated anymore in which the upkeep cost later becomes uneconomical in accordance with IMO regulation [1–3]. These end-of-life (EOL) vessels are then retired and sold for scrap to the operators of ship breaker and recycling industry around the world. It involves a wide range of activities starting from removing all the equipment and items including hazardous materials from the ship, cutting down the whole structures to the size where it can be transported to recycling centers. Ship recycling is basically operated in growing nations where manpower, and land costs are less expensive, and labor and natural environment are less strict compared to the developing nations. The list of heading ship recycling countries highlights the domination of growing nations in this department, with Bangladesh, China, India, Pakistan, and Turkey leading the list and made up 70–80% of the world's gross tonnage of scrapped vessels [4–6].

Current practice involves labor intensive work and hazardous as most of the ship recycling industry employs manual labor to break ships. Such activity is normally carried out at substandard yards that are not accepted under rule and regulation due to improper facilities which create residue from hazardous materials and waste. This has urged many yard owners to turn their substandard yard to green ship recycling yard in order to comply with safety and health requirements. However, the issue is arising when the competitiveness level is reduced along the process of turning the substandard yard to green ship recycling yard due to extra cost of maintaining the standard as green yard. As the result, current green ship recycling operation is cost ineffective compared to conventional ship recycling yards. Despite the importance of the design aspect for cost effective operation, hitherto, there is a lack of review on design of green ship recycling yards. Thus, this paper investigated the issues and current research on green ship recycling yard design.

18.2 Ship Recycling Process

Sunaryo and Pahalatua [7] describe the processes of ship recycling activities in a ship recycling yard which involves large scale of activities. From extracting all equipment and items such as fittings and engine parts left on ship, to cutting down and recycling the entire ship's infrastructure before handling them to temporarily storage the hazardous materials. Finally, a complete disposal takes place at the safe dumping plant.

18.3 Ship Recycling Method

There are two prominent methods of breaking a ship, which are beaching and dry docking [4]. Others methods include slipway and floating or alongside. The selection of method commonly depends on many factors such as geographical factors, size of ship to be scrapped, facilities and technologies involved during the process [8].

Fig. 18.1 Beaching method. Reprinted with permission from Elsevier [8]

Ship breaking on beaches is the most common method of ship disposal and is also considered as one of the most life-threatening jobs in the world bodies [9–11]. It includes tearing down of the ship on dedicated ship breaking beaches. Thousand's tons of steel are scrapped and disposed by workers using minimum tools and safety precautions. The third world countries such as India, Pakistan, and Bangladesh have been operating this beaching method as shown in Fig. 18.1.

Dry dock is the ship disposal technique that aims at dismantling a vessel in a shipyard dock as illustrated in Fig. 18.2. This is the safest and cleanest method of ship disposals as the process takes place in dry docks. Though such technique provides zero chance of accidental pollution, the method is extremely expensive [8]. The less expensive method is using slipway or ramp as shown in Fig. 18.3. The ship is moved to shore from the water using a winch. On the other hand, the ship dismantling process is also being carried out in floating conditions using air bag method (see Fig. 18.4). EOL ship is berthed in quay side of seaports and shipyards for dismantling. The dismantling is done by cutting and removing the ship parts in vertical direction. Starting from top of the bridge and subsequently reaching the keel. The cutting peripheries do not come in contact with seawater, and ship stability must be carefully maintained. Most ship recycling facilities in China and Europe make use of this method.

18.4 Ship Recycling Yard Facilities and Layout

In general, ship recycling yards are sharing several common facilities with a typical shipyard such as berthing facility, docking equipment, handling machineries such

Fig. 18.2 Dry dock. Reprinted with permission from Elsevier [8]

Fig. 18.3 Slipway

crane and forklift, and cutting equipment. However, ship recycling yards require several prominent facilities for storing dismantle materials and wastes. The facilities that should be available in the recycling yard are such as storage tank, container, and proper drainage system which is most important in ship recycling to prevent any environmental issue. The storage tank is used to store any liquid such as water ballast from ballast tank, oil, and refrigerant need to fully covered and permanently roofed. A container is used to store any useful materials, electricals equipment, and other equipment that can be refurbished and sold. The drainage system is the most

Fig. 18.4 Floating. Reprinted with permission from Elsevier [8]

important in a green ship recycling yard to prevent any leakage during the process of dismantling a vessel.

In case of green ship recycling yards, it should comply with all applicable regulations and requirements such as the Hong Kong Convention (HKC) and Basel Convention (BC) to ensure it is environmentally friendly and safe to the workers. One good example of green ship recycling yards was proposed by Sunaryo and Pahalatua [7]. The author lists the common facilities that need to be considered and presented typical green ship recycling yard layout design as shown in Table 18.1 and Fig. 18.5, respectively. Some of important facilities are discussed in the following paragraph.

Table 18.1 List of common facilities [7]

1. Slipway (primary cutting)	11. Underground storage tank
2. Secondary cutting area	12. Facility garage
3. Structural parts storage open area	13. Internal road
4. Storage tank	14. Office
5. Machinery and outfits storage	15. Canteen
6. Electric and electronics storage	16. Clinic
7. Storage for materials containing asbestos	17. Parking ground
8. Storage for materials containing PCB's	18. Security post
9. Scrapping work facilities warehouse	19. Road
10. Garbage (not harmful) disposal area	20. Future expansion plant

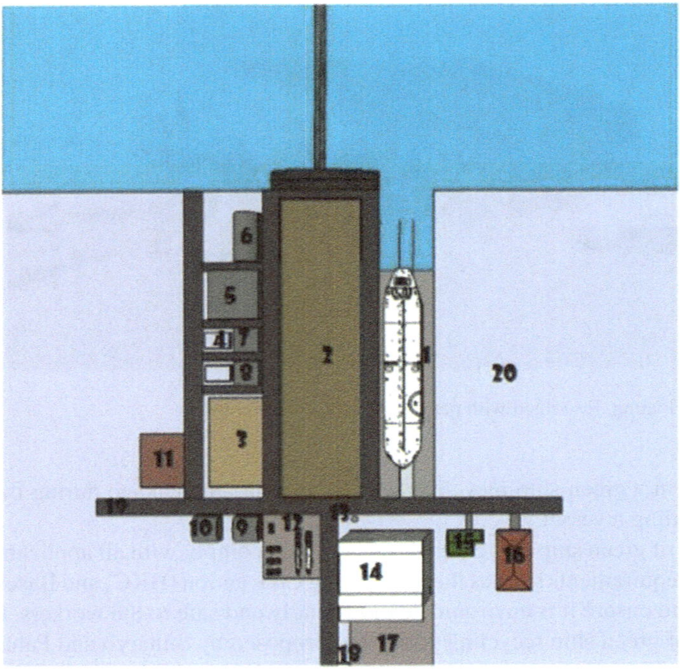

Fig. 18.5 Example of general layout for ship recycling yard. Reprinted with permission from Association of Naval Architects and Marine Engineers [7]

The primary area is the place where the ship is cut by a block assemblies or module. This allowed to reduce the harmful particles that enter the sea water. Secondary cutting area is where parts of the ship will be cut into the size that can be lifted by using forklifts. This area comprises a fully protected cutting area with an effective drainage system that ensures the flow of storm water runoff. The structural area is the place where all recycled ship parts are placed after being cut into small plates and classified according to their types. Meanwhile, chemical and liquids storage tank are provided to ensure that all liquids such as water ballast, oil, and hazardous waste are stored. Machinery and outfits storage are the place where all machineries such as main engines, pumps, generators, and others from the ship will be placed. The electric and electronics warehouse are the place where all these electrical appliances will be stored before the disposal process. Storage facilities containing asbestos will be stored using dry airtight containment and wet containment techniques. Storage containing PCB is where all equipment or appliances such as voltage regulators, switches, electromagnets, and circuit breakers that are categorized containing PCB is stored. Specific storage for PCBs and asbestos is important for safe handling and control as it can cause a serious health impact [2, 12].

18.5 Current Issues and Research in Green Ship Recycling Yards

The main issue in a green ship recycling yard is on its competitiveness. Current green ship recycling operation is cost ineffective compared to conventional ship recycling yards. Jain and Hopman [13] brought to the attention the issue on the price gap between green ship recycling and substandard yards. Several studies emphasize to improve green yard to be more competitive by reducing operational cost and increasing the efficiency of the recycling process. Later, work by Jain [14] focuses on improving the ship recycling process planning by proposing material quantification model and material flow analysis model. Similar effort continued by Sunaryo et al. [15] who carried out financial analysis to identify which business model is suitable for improving the economic situation and they discover that the service provider model has a prospect to provide internal rate of return of 17%.

Sivaprasad [16] suggested a further work on new ship recycling yard layout as to improve the ship recycling process. Welaya et al. [17] proposed a fuzzy model for economical and safe planning of a ship recycling yard. Sunaryo and Pahalatua [7] proposed the concept of green ship recycling layout (see Fig. 18.6) for capacity of recycling 30,000 DWT ship with an estimation of 15–20 ships that can be recycled annually. The author claimed that the proposed yard complies with existing national and international regulations. Mohd and Zainol [18] proposed the first conceptual design layout and green indicator for green ship recycling yard in Malaysia as shown in Fig. 18.7. His work was later continued by Fachry [19] who work on environmental friendly ship recycling yard specifically for recycling a general cargo ship, tug boat, and barge in Indonesia as shown in Fig. 18.8. However, Song and Woo [20] said that so far in many cases, design of a shipyard has been relying on the experienced engineers, resulting in sporadic and poorly organized processes. And thus, economic

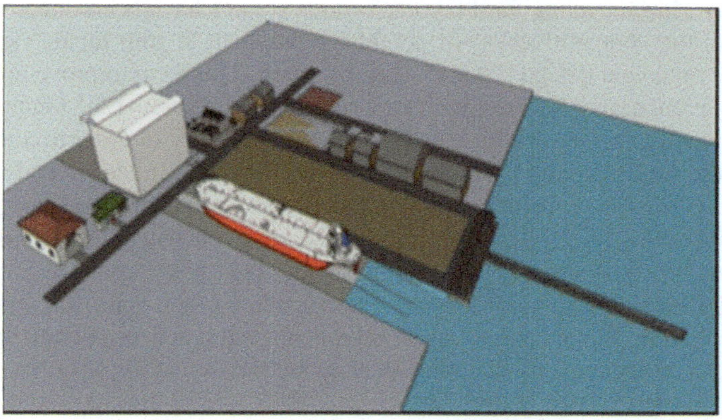

Fig. 18.6 Concept of green ship recycling layout. Reprinted with permission from Association of Naval Architects and Marine Engineers [7]

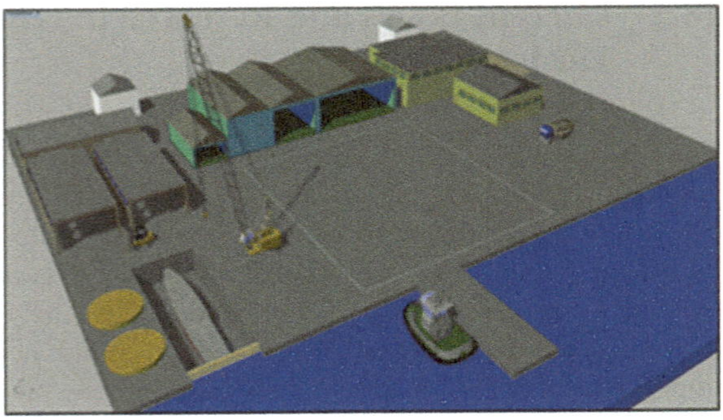

Fig. 18.7 Concept of green ship recycling layout in Malaysia. Reprinted with permission from UNIKL [18]

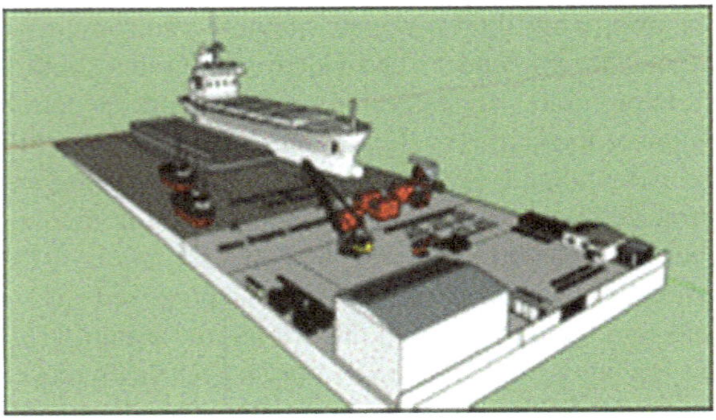

Fig. 18.8 Proposed ship recycling yard for general cargo ship, tugboat, and barge in Indonesia. Reprinted with permission from AIP Publishing [19]

losses as well as trials and errors in that accord have been pointed out as inevitable problems.

According to Janson [21], implementation of the green concept requires the optimization of the yard layout as to balance the related expenditure. In addition to research on ship recycling yard layout, various studies have been done in the past to improve the competitiveness of neighbor yard such in ship repair yard mainly emphasize on layout design [20–24]. Latest reference found in improvement of ship recycling layout design is from Gunbeyaz et al. [25] and Jayaram [26]. Gunbeyaz et al. [25] developed a framework for the ship recycling industry in order to improve and optimize the ship recycling procedures, whereby Jayaram [26] developed a methodology in the form of steps for generation of concept layouts of green ship recycling yards based on the SLP. The author identified a comprehensive list of activity areas required to be considered in the layout design. This is the most recent work in designing a ship recycling yard layout using a step-by-step approach which could be adapted easily for industrial cases. Significant reduction in operational cost to 10–30% can be seen just by planning an effective facility layout [27, 28].

18.6 Summary

This review paper described the current state of green ship recycling yard design and discussed the recent research taken to improve ship recycling yard design. It can be summarized that current works carried out, which are focused on efficient layout design, are still insufficient and application of SLP is promising. Hence, ship recycling yard layout design using SLP could be very interesting subject to be further studied to fill the price gap between substandard recycling yards and green recycling yards.

Acknowledgements The authors would like to thank the Ministry of Higher Education for supporting this work under the Fundamental Research Grant Scheme (FRGS/1/2019/TK08/UNIKL/03/1).

References

1. Kagkarakis ND, Merikas AG, Merika A (2016) Modelling and forecasting the demolition market in shipping. Marit Policy Manage 43(8):1021–1035
2. Du Z, Zhang S, Zhou Q, Yuen KF, Wong YD (2018) Hazardous materials analysis and disposal procedures during ship recycling. Resour Conserv Recycl 13:158–171
3. Toxicol JEA, Hossain KA (2015) Overview of ship recycling industry of Bangladesh. J Environ Anal Toxicol 5(5):312
4. Reddy NGK (2014) Ship recycling: an important mile stone for India. Indian J Sci Technol 7(S6):15–21
5. Karlis T, Polemis D (2016) Ship demolition activity: a monetary flow process approach. Pomorstvo 30(2):128–132
6. Jain KP, Pruyn J (2017) An overview of the global ship recycling industry. In: Hashmi MSJ (ed) Encyclopedia of renewable and sustainable materials, vol 1. Elsevier, Amsterdam
7. Sunaryo S, Pahalatua D (2015) Green ship recycle yard design. J Nav Archit Mar Eng 12(1):15–20
8. Rahman S (2017) Aspects and impacts of ship recycling in Bangladesh. In: Adi MAM, Karim MM (eds) Procedia engineering, vol 194. Dhaka, Bangladesh, pp 268–275
9. Chang YC, Wang N, Durak OS (2010) Ship recycling and marine pollution. Mar Pollut Bull 60:1390–1396
10. Schøyen H, Burki U, Kurian S (2017) Ship-owners' stance to environmental and safety conditions in ship recycling. A case study among Norwegian shipping managers. Case Stud Transp Policy 5:499–508
11. Barua S, Rahman IMM, Hossain MM et al (2018) Environmental hazards associated with open-beach breaking of end-of-life ships: a review. Environ Sci Pollut Res 25(31):30880–30893
12. Hossain MS, Fakhruddin ANM et al (2016) Impact of ship-breaking activities on the coastal environment of Bangladesh and a management system for its sustainability. Environ Sci Policy 60:84–94
13. Jain KP, Pruyn JFJ, Hopman H (2017) Strategic guidance based on the concept of cleaner production to improve the ship recycling industry. Environ Syst Decis 38:250–260
14. Jain KP, Pruyn JFJ et al (2017) Material flow analysis (MFA) as a tool to improve ship recycling. Ocean Eng 130:674–683
15. Sunaryo, Hanura AR, Hussein F (2020) Feasibility study on the financial aspect of green ship recycling yard business model. In: Nahry, Marthanty DR (eds) AIP conference proceeding, vol 2227, no 1, Padang, Indonesia, p 040006

16. Sivaprasad K, Nandakumar CG (2013) Role of naval architects in ship recycling. J Shipping Ocean Eng 3(2013):47–54
17. Welaya YM, Naby MMA, Tadros MY (2012) Technological and economic study of ship recycling in Egypt. Int J Nav Archit 4(4):362–373
18. Mohd S, Zainol I (2017) Development of green ship recycling yard for Malaysian. Undergraduate thesis, Universiti Kuala Lumpur, Lumut
19. Fachry IA (2020) Environmentally friendly ship recycling yard design for general cargo ship up to 30.000 DWT and ship-sets: tug and barge. In: Nahry, Marthanty DR (eds) AIP conference proceeding, vol 2227, no 1, Padang, Indonesia, p 040006
20. Song YJ, Woo JH (2013) New shipyard layout design for the preliminary phase & case study for the green field project. Int J Naval Archit Ocean Eng 5:132–146
21. Janson DI (2016) The development of a green shipyard concept. Master thesis, University of Twente, Enschede
22. Song YJ, Lee DK, Woo JH et al (2008) A shipyard layout design system by simulation. J Nav Archit Mar Eng 45(4):441–454
23. Choi M, Kim SH, Chung H (2017) Optimal shipyard facility layout planning based on a genetic algorithm and stochastic growth algorithm. Ships Offshore Struct 12:486–494
24. Matulja T, Fafandjel N, Zamarin A (2009) Methodology for shipyard production areas optimal layout design. Brodogradnja: Teorija i praksa brodogradnje i pomorske tehnike 60(4):369–377
25. Gunbeyaz SA, Kurt RE, Turan O (2018) Designing efficient contemporary ship recycling yards through discrete event simulation. In: Transport research arena (TRA). Reed Messe Wien, Vienna, Austria
26. Jayaram S, Sivaprasad K, Nandakumar CG (2018) Strategic guidance plan for recycling of ships in India. Int J Adv Res Sci Eng Technol 9(4):91–101
27. Naik SB, Kallurkar S (2016) A literature review on efficient plant layout design. Int J Ind Eng Res Dev 7:43–51
28. Zha S, Guo Y, Huang S et al (2017) Robust facility layout design under uncertain product demands. Procedia Cirp 63:354–359

Chapter 19
Passenger's Behavior Recognition System Using Computer Vision

Nurhayati Hasan and Muhd Khairulzaman Abdul Kadir

Abstract One of the problems that occurs inside public transport is that passengers often overlook or ignore public transport's rules such as in regards to eat and drink. Eating foods and drinks are not even allowed in public transport in avoiding excessive littering and drink spills which could potentially cause unwanted accidents. This system aims to recognize two actions such as eating and drinking through image processing in real-time environment. It also aims to classify and label the behavior of the passengers. This research is on a passengers behavior recognition system using computer vision (PBRSUCV) as a prototyping model. The method consists of image processing with a faster region-based convolutional neural networks (faster R-CNN) classification project that can be implemented in public transport to solve the stated problem. The system consists of a camera and laptop. The camera is used as a sensor to detect the targeted behaviors while the laptop will be the main device where every image processing takes place. The system was tested in real-time and is able to detect and label the eating and drinking behavior correctly with 99% accuracy on a single and two people in the image frame. It is certain that this system is capable to accurately recognize and classify the targeted behaviors in the public transport without any problem with the help of the faster R-CNN deep neural network.

Keywords Computer vision · Image processing · Recognition system

19.1 Introduction

Public transport is very important for Malaysian citizens as it makes us convenience to go anywhere and helps to reduce traffic congestion in the country. To maintain a delicate environment of the public transport, it is important to have rules in place. The rules which are often overlooked or ignored by public transport passengers are in regard to eating and drinking. Consuming food and drinks are not even allowed

N. Hasan · M. K. A. Kadir (✉)
Universiti Kuala Lumpur, British Malaysian Institute, Gombak, Selangor, Malaysia
e-mail: khairulzaman@unikl.edu.my

A. Ismail et al. (eds.), *Advanced Materials and Engineering Technologies*,
Advanced Structured Materials 162,
https://doi.org/10.1007/978-3-030-92964-0_19

in public transport to avoid excessive littering and drink spills that could potentially cause unwanted incidents and dirty environments.

This research intends to examine a high-definition real-time video of passengers in public transport to recognize their wrong behavior. To achieve such a goal, computer vision and image processing with faster R-CNN have been applied in this system. The objective of the system is to recognize two actions such as eating and drinking through image processing in real-time. It also aims to classify and label the wrong behavior on the passenger.

There are a lot of different computer vision techniques used in the past to identify the presence of human and its behavior. A study regarding extraction of human gait is developed using Haar Cascade classifier technique on OpenCV. It applies background subtraction to crop the unnecessary region of interests (ROI) in a video frame which are then transformed into silhouette form [1–3].

In a study to detect human carrying baggage from video sequences, several human body parts are divided such as head, body, leg, and baggage component. Each feature is then extracted and trained using a support vector machine (SVM) classifier [4].

Histogram of oriented gradient (HOG) feature set is used in the study to identify pedestrians. It applies different machine learning algorithms to ensure a successful detection and analyze the speed of the system. HOG performs better with SVM which gives almost linear separable features [5, 6].

19.2 Materials and Methods

The system's need to input data which requires thousands of images. Both images are taken from high-definition (HD) video recordings, which were then splits into thousands of frames to produce samples of the image. Figure 19.1 shows the flow of process to train the images.

Initially, a virtual environment was set up in Anaconda for TensorFlow-GPU using the anaconda command prompt to run as an administrator and install the other required packages by issuing commands.

Once the TensorFlow Object Detection API is all configured, then the images will be used to train a new classifier for detection which will be included.

To train a good classifier for detection, TensorFlow needs thousands of images of an object. The training images should have random entities in the image together

Fig. 19.1 Flow of image training process

with the target objects and should have a various range of backgrounds and lighting environments, to train a versatile classifier. After that, 20% of the images were moved to the test directory, and 80% of them to the training directory. This is to ensure smooth and accurate image processing in real-time mode.

Finally, it is time to generate the TFRecords with the images labeled, that serve as input data for the TensorFlow training model.

First, the.xml image data will be used to create.csv files that contain all the train and test image data. These generates file called 'train.record' and 'test.record'. This will continue to train the current classifier for object detection. If all is set right, TensorFlow will start the preparation. The setup will take 30 s before commencing the actual preparation [1, 7].

The flowchart as in Fig. 19.2 is designed to provide a better understanding of the PBRSUCV process. This program begins by gathering image samples of selected behaviors. After that, the image samples obtained will be trained using the deep neural network. During the system testing, the image obtained from the live stream would be compared to the training images.

If the observed behaviors are not the same as the qualified images, live videos will continue to be collected until a targeted behavior is observed. The cycle stops until the activity has been correctly identified and the public transit manager has been informed of the identification.

It is also important to determine the suitable hardware and software requirements to make sure the objectives of this research are achieved. Table 19.1 shows the hardware requirement used to build PBRSUCV while Table 19.2 describes the software information being applied.

19.3 Results

The system environment has been tested in real-time mode. The video is being analyze for its performance and accuracy of PBRSUCV detection.

For the first test, the process is done for a single person in real-time video and varying the distance of the subject from camera. The distance of the subject from camera was increased by 0.3 m for each step. After that, the percentage of detection accuracy was recorded and analyzed. The results for the detection are shown in Figs. 19.3, 19.4 and 19.5.

The percentage of accuracy is calculated based on the accuracy recorded on the labeled box of each eating person in the image frame as in Eq. (19.1).

$$\text{Accuracy percentage (\%)} = 100\left[\frac{\text{Total number of people eat or drinks in the frame}}{\text{Total number of people in the frame}} \times 100\right] \tag{19.1}$$

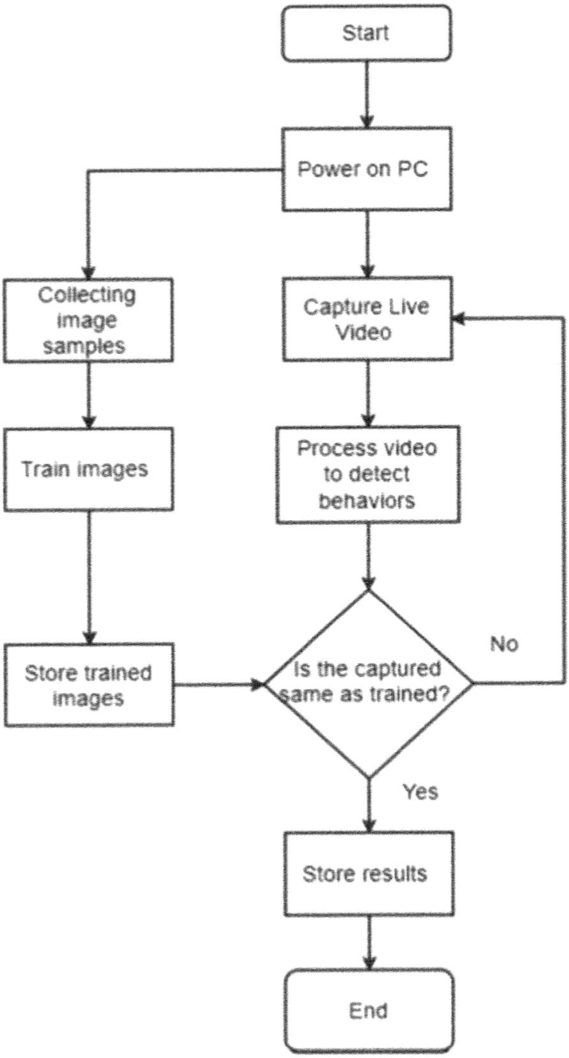

Fig. 19.2 Flowchart of PBRSUCV

The results of percentage accuracy are shown in Tables 19.3, 19.4, and 19.5 for one person, two and three people in a scene of real-time video and then cut into nine frames based on the different distance of the person from the camera.

Table 19.1 Hardware requirements

Hardware	Justification
Laptop	OS: Windows 10 CPU: Processor Intel Core i5 8th Gen GPU: NVIDIA GeForce GTX 1650 RAM: 8 GB DDR4 Storage: 256 GB SSD Display: Full HD (1920 × 1080)
Camera	Dual: 12 MP, f/1.8, 28 mm (wide) Features: Quad-LED dual-tone flash, HDR Video: 4 K@24/30/60 fps, 1080p@30/60/120/240 fps

Table 19.2 Software requirements [8, 9]

Software	Justification
Anaconda	Distribution of the Python language free and open source for scientific computation The release provides software for Windows, Linux and macOS suitable for data-sciences
CUDA	Enable application developers and professional engineers to use a graphics processing unit allowed by CUDA for general purposes
cuDNN	CuDNN provides finely optimized implementations for common protocols such as forward and backward convolution, pooling, normalization, and layers of activation
LabelImg	A graphical tool to annotate files. Annotations are stored in PASCAL VOC format as XML files, the format ImageNet uses
Tensorflow V1.13.1	An end-to-end open source machine learning tool. It has a detailed, dynamic wide range of capabilities, libraries, and community resources that enables researchers to push the state-of-the-art in ML, and developers to easily build and deploy ML powered apps
Python 3.5	A general-purpose, translated programming language

19.4 Discussion

For the results obtained in Tables 19.3, 19.4, and 19.5, the system is showing a high accuracy which is 99% for one and two people. The system also can detect up to 3 people in a time with almost 98% accuracy. However, the accuracy keeps reduced as the number of people and distance from camera is increased.

Currently, in this system the study shows that there is a limitation that can be achieved by using this method. From the experiment that has been conducted in the testing phase, the system recorded quite a low percentage of accuracy for several times with the increase of people in the image frame.

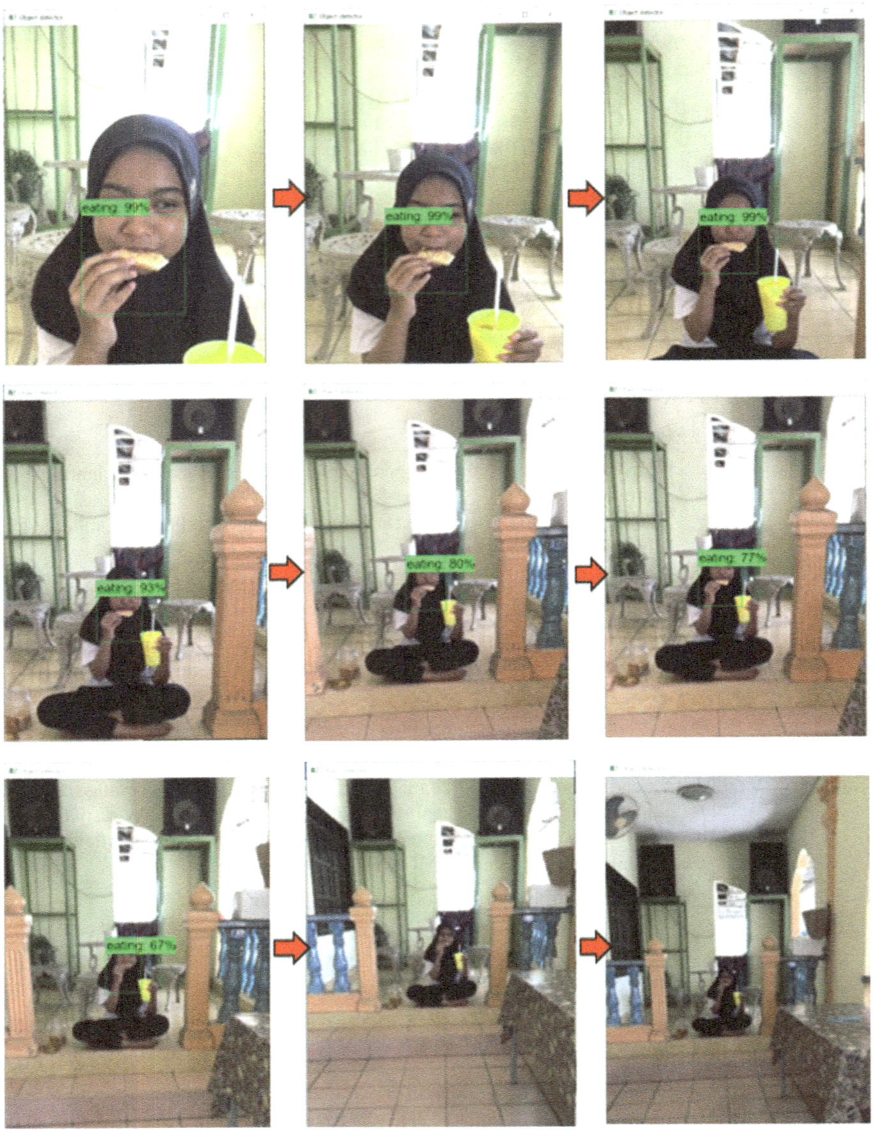

Fig. 19.3 The result of detection percentage for 1 person in real-time video

This limitation of this system can be avoided for future improvements. The accuracy of the detection could reduce due to some factors as follows:

i. The distance of subject from camera being too far away.
ii. Using too low resolution of camera.
iii. Low intensity of light.

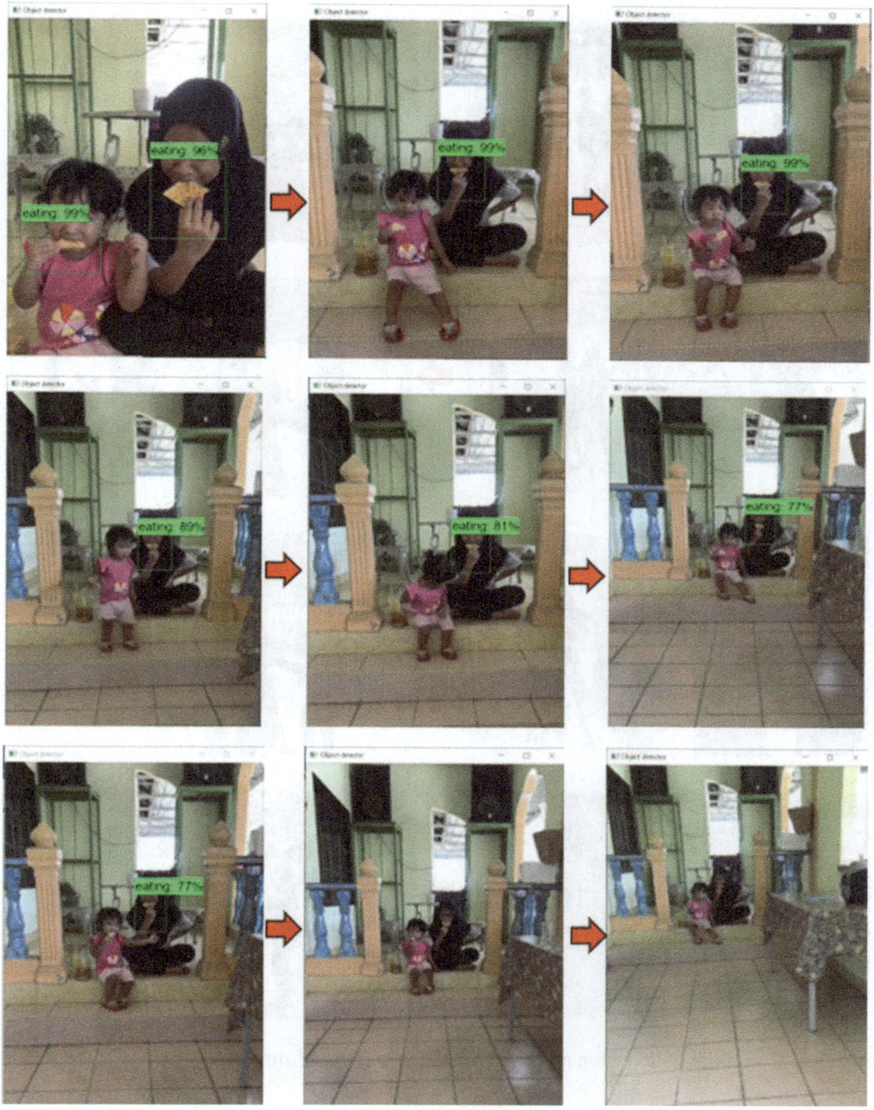

Fig. 19.4 The result of detection percentage for 2 people in real-time video

iv. Too many people (crowded) in a place at one time so the detection can be overlapped.

Therefore, the system's efficiency and accuracy can be improved further by using a camera with higher quality and increase the number of samples for the training data.

Fig. 19.5 The result of detection percentage for 3 people in real-time video

19.5 Conclusion

In conclusion, this PBRSUCV is capable to detect and classify the behavior of eating and drinking through image processing in real-time mode. The result of accuracy percentage for the detection also showing high precision and validity rather than errors.

The system can detect people conducting eating or drinking action up to four people at a time with high accuracy by appointing a green box around the mouth,

Table 19.3 The percentage of accuracy result for one person eating in a scene

Distance of people from camera (m)	Total accuracy percentage recorded in a frame	Total number of people eating in a frame	Accuracy percentage (%)
0.3	99	1	99
0.6	99	1	99
0.9	99	1	99
1.2	93	1	93
1.5	80	1	80
1.8	77	1	77
2.1	67	1	67
2.4	0	1	0
2.7	0	1	0

Table 19.4 The percentage of accuracy result for two people eating in a scene

Distance of people from camera (m)	Total accuracy percentage recorded in a frame	Total number of people eating in a frame	Accuracy percentage (%)
0.3	196	2	97.5
0.6	99	1	99
0.9	99	1	99
1.2	89	1	89
1.5	81	1	81
1.8	77	1	77
2.1	77	2	38.5
2.4	0	2	0
2.7	0	1	0

Table 19.5 The percentage of accuracy result for three people eating in a scene

Distance of people from camera (m)	Total accuracy percentage recorded in a frame	Total number of people eating in a frame	Accuracy percentage (%)
0.3	99	1	98
0.6	99	1	95
0.9	99	1	93
1.2	93	1	90
1.5	80	1	86
1.8	77	1	82
2.1	67	1	0

food and hand gesture with label and percentage of accuracy for each person. While the percentage of accuracy is poor for 5–6 people, it can still be increased to the next point by increasing the number and quality of samples for training images.

Generally, errors consist only of 4–6 people, but for single and two people, the system is capable of being detected and classify with 99% precision. It is also found that the resolution of the camera and intensity of light can affect the result of detection other than the number of people in a frame and the distance of camera from the subjects. Therefore, the type of camera used to construct this system is very important in making sure the reliability and efficiency of the system's function.

The system can be updated to the next level making the system more robust and easier for applications to come. The current system can detect accurately only up to four people at a time.

For future enhancement, the efficiency of the system can be increased by adding more samples of training images with high quality. Currently, about 3000 samples of images were used to create the recognition system. Therefore, the samples can be increased up to 5000 images in the future.

Next, the specification of hardware used is also very important in building this PBRSUCV. A high specification of processing machine can speed up the process of training images which can save a lot of time. A higher specification type of camera also plays an important role in the detection process where the better the quality of the real-time video, the higher the accuracy of recognition can be obtained.

Finally, this PBRSUCV also can be upgraded by including the Internet of Things (IoT) platform and alert system. The current function of the system only can detect the eating and drinking behavior which is displayed on the screen. Therefore, these upgrades can give extra features to the detection system where all the data captured can be recorded and analyzed from time to time regularly.

References

1. Ismail AP, Tahir NM (2017) Human gait silhouettes extraction using Haar cascade classifier on OpenCV. In: UKSim-AMSS 19th international conference on computer modelling and simulation (UKSim). https://doi.org/10.1109/uksim.2017.25
2. Piccardi M (2004) Background subtraction techniques: a review. In: 2004 IEEE international conference on systems, man and cybernetics, vol 4 (IEEE Cat. No. 04CH37583). The Hague, pp 309–3104. https://doi.org/10.1109/ICSMC.2004.1400815
3. Abdul KMK (2018) Tracking and indexing moving object in multitude environment. M.S. thesis. Electrical Engineering, Universiti Teknologi Malaysia
4. Wahyono JKH (2017) Detection of human carrying baggage from video sequences. J Intell Fuzzy Syst 32(2):1601–1613. https://doi.org/10.3233/jifs169153
5. Ardiyanto I, Adji TB, Asmaraman DA (2018) On comprehensive analysis of learning algorithms on pedestrian detection using shape features. J Intell Fuzzy Syst 35(4):4807–4820. https://doi.org/10.3233/jifs-18491
6. Dalal N, Triggs B (2005) Histograms of oriented gradients for human detection. In: 2005 IEEE computer society conference on computer vision and pattern recognition (CVPR'05), vol 1. San Diego, CA, USA, pp 886–893. https://doi.org/10.1109/CVPR.2005.177

7. Abadi M, Agarwal A, Barham P et al (2020) TensorFlow: large-scale machine learning on heterogeneous distributed systems. http://download.tensorflow.org/paper/whitepaper2015.pdf. Accessed 12 June 2020
8. TensorFlow (2020) Anaconda documentation. https://docs.anaconda.com/anaconda/user-guide/tasks/tensorflow/. Accessed 12 June 2020
9. NVIDIA (2020) NVIDIA CUDNN documentation. https://docs.nvidia.com/deeplearning/sdk/cudnn-developer-guide/index.html. Accessed 28 June 2020

Chapter 20
A Review of WSN Clustering Algorithms for Low Powered IoT Protocols

Wan Nur Hurul Asma Wan Hassan, Aizat Faiz Ramli, Hafiz Basarudin, Izanoordina Ahmad, and Aduwati Sali

Abstract In technological advancement, Internet of things or IoT technology is widely used in everyday life such as environmental monitoring, military, smart lifestyle, and transportation. IoT technology produces a smart device that connects sensors or devices to produce and analyses data. Wireless IoT networks consisting of battery-powered devices require highly energy efficient network access. Large communication links between devices will exhaust energy during the transmission process, causing a shift in network topology and degrading the quality of service QoS. Cluster techniques developed for wireless sensor networks can be applied to IoT to increase energy efficiency, spatial reuse, and large IoT networks' scalability. This paper surveys designed clustering techniques philosophy that can be implemented in IoT.

Keywords Internet of things (IoT) · LoRaWAN · Energy efficiency · Design clustering

---

W. N. H. A. W. Hassan · A. F. Ramli (✉) · H. Basarudin · I. Ahmad
Electronics Technology Section, Universiti Kuala Lumpur British Malaysian Institute, Gombak, Selangor, Malaysia
e-mail: aizatfaiz@unikl.edu.my

W. N. H. A. W. Hassan
e-mail: whurul.hassan@s.unikl.edu.my

H. Basarudin
e-mail: hafizb@unikl.edu.my

I. Ahmad
e-mail: izanoordina@unikl.edu.my

A. Sali
Computer and Computer System Section, Universiti Putra Malaysia, Serdang, Selangor, Malaysia
e-mail: aduwati@upm.edu.my

A. Ismail et al. (eds.), *Advanced Materials and Engineering Technologies*,
Advanced Structured Materials 162,
https://doi.org/10.1007/978-3-030-92964-0_20

20.1 Introduction

The Internet of things (IoT) has gained huge worldwide interest in industry or academia over the past few years. Internet of things is a virtual information network technology including radio frequency detection, sensor, and microcontrollers [1]. Today, home appliance, military, and healthcare equipment has been upgraded to automation systems with minimal treatment. IoT technology achievement develops a simple routine for users to access data from IoT devices anywhere and anytime. IoT popularity has generated business growth opportunities.

It is predicted that there will be over 1.5 billion Internet users in the future, generating a massive demand for Internet-enabled smart objects [1]. According to CISCO research each person in the world will have more than one Internet-connected device [1]. By referring data from the MIMOS Malaysia [9, 19], they indicate an expected IoT market growth from 2020 to 2025. IoT technology has developed various business categories such as smart home, smart city, smart environment, and, etc. [1]. In an IoT platform, communication between any Internet-enabled a computer creates several applications in various domains.

IoT is planned to build and deliver a new generation of service and infrastructure using information and communication technology (ICT) [1]. IoT combines technology and system into one application to enable new lifestyle versatility and performance. Depending on traffic load and power consumption requirements and specifications, IoT networks can be generated solely by wired or wireless communication.

In wireless communication, data is transmitted via the gateway and stored in the cloud before data is sent to the application. IoT is a two-way contact, involving two and more devices [3]. IoT networks consisting of battery-powered devices require highly energy-efficient network access. Communication between devices can exhaust energy during transmission. Node with depleted battery will cause node route change in the IoT network resulting in network topology change and can reduce QoS service network quality. This can affect overhead message resulting in network topology change [3]. Therefore, energy consumption is a concern for wireless IoT network.

In WSN topology, clustering generates a node that is self-organizing multi-hoping network that can function independently during data transmission and can be applied to IoT [4]. Clustering is defined as a process of grouping sum nodes into a community that similarly provides improved data transmission efficiency. Network clustering principle is to boost energy efficiency, reliability, and energy consumption during transmission [4]. Clustering can be implemented in one-to-many, many-to-one, one-to any or one-to-all (broadcast) communication [5]. The cluster consists of three main nodes, cluster head, nodes, and gateway nodes [4]. The cluster head acted as local controller in cluster network. The form of the clusters depends on the route for the nodes to transmit the data. The coverage for the large area can create a time delay, scattering, and data loss during the process.

20.2 Clustering Challenge

Clustering creates a promising approach that can help to minimize many IoT challenges such as energy efficiency, power consumption. Energy efficiency is superior to all IoT applications, but it is highly important for the applications such as smart environmental that do not have access to the power supply. The application lifespan would become uncertain thus it caused a drawback to the network such as nodes failure [4].

Each IoT application need a different type of application requirement. Application such as agriculture monitoring and surveillance are homogenous network architecture that require energy efficiency to distribute through the network.

Figure 20.1 demonstrates the sensor forwarding with and without cluster clustering. Data forwarding without clustering showing distance drawback and the data transmission performance to the base station. Clustering establishes a cluster network to capture node data before sent to base station whether in single-hop clustering or multi-hop clustering. Clustering algorithms build for WSN such as low energy adaption clustering hierarchy (LEACH) and hybrid energy efficient distributed clustering (HEED) can therefore be adapted to IoT [4].

20.3 Methodology

Many forms of clustering were implemented for clustering. Clustering design is influenced by many factors, which include network topology and power consumption. Clustering provides various advantages such as energy efficiency, reduces the communication bandwidth requirements and prevents repeated node data exchange [5]. By stabilizing the cluster topology, the overhead for the topology is reduced [5]. Improving energy efficiency, reduce energy consumption or enhancing transmission reliability still consider as a major challenge to clustering [5]. Cluster head (CH) selection is essential to maintain network lifetime and transmit data to endpoint. The clustering objectives will be elaborated as shown in Fig. 20.2.

One of the main objectives of clustering is to reduce nodes energy consumption during the transmission. LEACH introduces a formation of randomized, adaptive and self-configuring network which minimize the energy of CH during transmission [5]. HEED clustering choose the CH by selecting their residual energy and formation of the equal size cluster to reduce the energy in the network [6]. K-means algorithm is known as unsupervised algorithm as the selection of CH is determined by K initial to the center point [7]. Each clustering creates an algorithm to create a good energy efficiency in their clustering. Thus, clustering can be used as one option to create a productive network for energy efficiency.

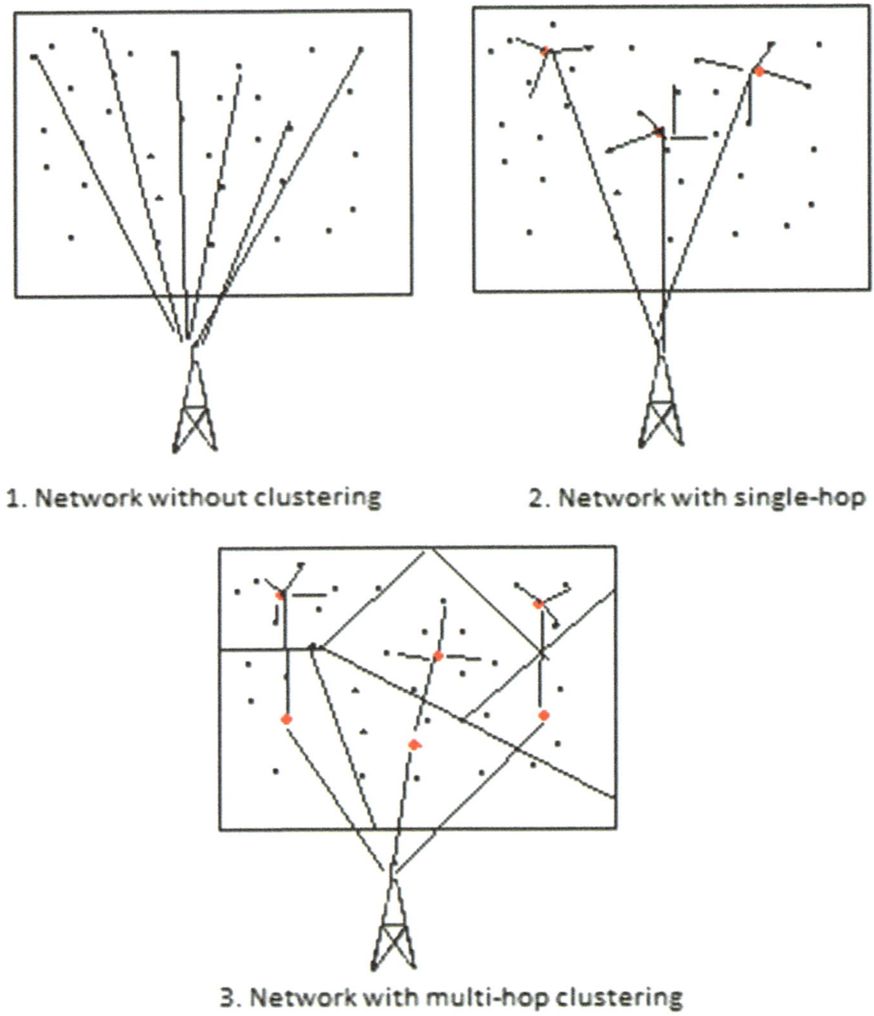

Fig. 20.1 The sensor forwarding data with and without clustering

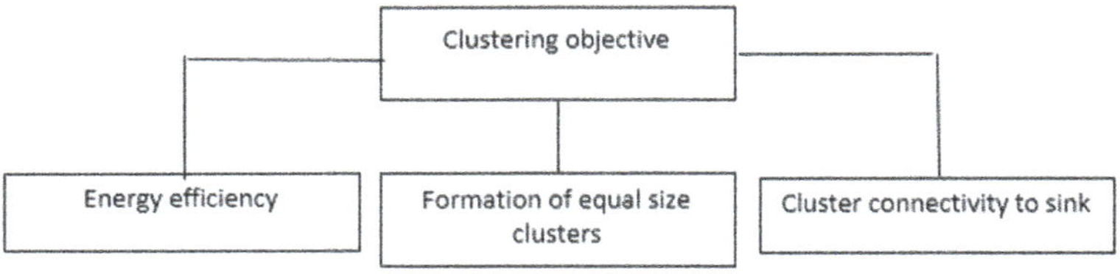

Fig. 20.2 Clustering objective

20.3.1 Low Energy Adaption Clustering Hierarchy (LEACH)

LEACH is a self-organizing adaptive clustering protocol which distributes the energy load equally to the all nodes in the cluster [5]. LEACH is a hierarchical routine which provides a better energy efficiency and scalability in the cluster [8]. There are two phases in LEACH which is the setup phase and the steady-state phase [5]. For this type of protocol, the whole network was divided into cluster and a certain node will be chosen as the cluster head (CH) based on their criteria. The CH will collect, aggregate and compress the data from other nodes in the cluster and transmit to the base station (BS) [8]. The CH acted as lead to the cluster network which consumes more energy compared to other nodes in the cluster. Hence, the energy in the network can be reduced.

The common method used in clustering is the clustering rotation. The rotation is applied in the clustering to balance the energy dissipation within the cluster [8]. The LEACH clustering has two categories which is single-hop and multi-hop communication. The single-hop communication was used in intra-clustering. Intra-clustering and inter-clustering are two type of modes used in LEACH. Intra-clustering, the cluster communicates within the cluster member, but inter-clustering communicates with the neighbor network which outside from their cluster.

The main objective for LEACH is to increase energy efficiency by adapting the rotation CH selection by random number [1]. The CH selection is depending on the value of the threshold $T(n)$ which is given in Eq. (20.1).

$$T(n) = \begin{cases} \frac{1}{1-p\left(r \bmod \frac{1}{p}\right)} & \forall n \in G \\ 0 & \text{otherwise} \end{cases} \tag{20.1}$$

The value of p is the desired percentage for the nodes to become CH, r denoted the current round of cluster and G is the number not participated in the network. The algorithms calculate the highest $T(n)$ to select the CH in the round [1]. The previous CH will not be allowed to join in the next round to make sure the other nodes to get fair and equally participated to be select as CH. During the steady-state phase, the CH transmits the data by TDMA schedule. During the data transmission, the other nodes go to the sleep mode to consume energy. The energy model for the energy efficiency is written as following [1]

$$E_{Tx}(k, d) = E_{Tx\_elec}(k) + E_{Tx\_amp}(k, d) \tag{20.2}$$

The channel for the communication is symmetrical and the energy sending will be k bits/packet to a node in distance, d in meter. The E_{elec} is the energy consumed by transmitter or receiver and E_{amp} is the amplifier parameter of transmission correspond to the multi-path fading model. Each packet data bit transmits the information along with overhead. The overhead contains information related to the packet like coding that contain reliable information for transmission. The CH may communicate with the nodes in the cluster depending on the range of transmission. Therefore, to create a good energy efficiency, the nodes must be within the range of cluster to decide the optimal CH [11].

20.3.2 Hybrid Energy Efficient Distribution Clustering (HEED)

Hybrid energy efficient distributed (HEED) clustering is a clustering process which produces an equal size and the residual energy nodes and intra-clustering communication which plays as significant part in the CH selection [19]. There are three phase of cluster formation in the HEED which are initialization, iteration, and finalization [19]. In the cluster, there will be one node act as CH and in charge to communicate with another CH. The intra-cluster is applied in HEED clustering to increase the energy efficiency and prolong the network lifetime [13]. The energy distribution extends the lifetime of nodes within the cluster, thus stabilizes the neighbor node to operate correctly when nodes are not synchronized.

The selection of CH is determined in the initialization phase. The phase assigns nodes to the probability to become CH. This can be done according to the formula written [19]

$$\mathrm{CH}_{\mathrm{prob}} = \mathrm{CH}_{\mathrm{prob}} * \frac{E_{\mathrm{residual}}}{E_{\mathrm{max}}} \quad (20.3)$$

The C_{prob} is the initial probability (predefined value), E_{residual} is the residual energy and E_{max} is the maximum energy for sensor nodes [7]. For the interactive phase some nodes will become tentative CH. If the other nodes within the range of tentative CH, it will be tentative CH will become another CH [7]. In the finalization phase, the sensor node which does not have CH will form one cluster.

HEED is an equal size balanced cluster. The HEED clustering used both free space and multi-path channel and consume for the error-free communication link. The calculation for the energy module is written [19]

$$E_{Tx} = (E_{\mathrm{elec}} X\, k) + (E_a X\, k\, X d^n) \quad (20.4)$$

The E_{elec} is the energy for the transmitter and receiver, E_a is the energy spent by the amplifier and d is the distance between sender and receiver. The transmission energy E_T is the energy to send the data packet.

20.3.3 K-Means Algorithms

K-means is a statistical, unsupervised, non-deterministic, iterative strategy for grouping the node into the clusters. It is the simplest unsupervised learning algorithms known for its speed, effortlessness, and usability. *K*-Means utilize the *K* as the initial center point from the consummate dataset [10]. This algorithms objective is to minimum the distance between node to the center point. It calculate the Euclidean distance from each data point, to choose the point which is most suitable, then assign it to the suitable cluster. The center point receives the update and will repeat the calculation until it reaches the minimum to remove the error in the cluster. This algorithms can reduce the impact of isolated point and noise to enhance the efficiency in clustering [8].

K-means is a typical clustering in data miming and can be widely used for large set of data cluster. The algorithms consists of two separate phase [10]. The first phase selects the random *K* center, where the value *K* is fixed. The next phase is to find and take the data nearest the center. The Euclidean distance is considered to determine the distance of object to center point. This process will repeated calculate until it reaches its minimum error in the cluster,

$$E = \sum_{i=1}^{k} \cdot \sum_{x \in C} |x - x_i|^2 \tag{20.5}$$

20.3.4 Formation of an Equal Size Cluster

Formation of an equal size cluster attempts to create a cluster which has a similar number of nodes (cluster members). The large variation of cluster sizes will cause a poor load balancing and thus can affect the performance in certain clusters [12]. The process of creating a more equal size cluster can improve the performance of the network in terms of reducing channel contention thereby reducing latency [12].

20.3.4.1 Fuzzy Logic Formation

Fuzzy logic for the formation of uniform size clusters (FUSA) attempts to reduce cluster size variation caused by random placement of nodes in the network. The altruistic decision is used for the nodes to work independently for the selection of CH in the network [14]. The number of neighboring un-cluster nodes in the network is used to determine the selection of CH. The transmission range, *r* between the nodes must be minimized to maintain the network connectivity and energy efficiency. Large cluster size creates low energy efficiency and low transmission coverage which can create data loss during the transaction.

20.3.4.2 Algorithms Cluster Establishment (ACE)

The algorithm for cluster establishment (ACE) plays two parts which are spawning and migrating [15]. The spawn is a node which elects itself to be CH and finding his loyal follower to play the second part. The second part is migrating. The node controls the data migrate to avoid overlap data. During the node iteration, the CH will find his follower to create a path to be synchronized during the data transmission. The iteration reduces the overlap data and avoids repeated path which is used in previous CH leader [14, 15]. The best candidates for CH leaders are the leaders with a loyal follower. This produces a good packing of the repulsion effect between clusters. However, the process to determine the CH leader can create a time delay due to the process of the iteration.

20.3.5 Cluster Connectivity to Sink

Clustering techniques contain two phases which is the cluster head selection and the clustering. The phase remains ongoing until any of the nodes run out of energy. In clustering networks, large wireless sensor are deployed to obtain, monitor the data. Due to this circumstance and limited power limited power consumption of nodes, energy efficiency essentially choose multi-hop communication to create multiple sink [16]. The main modes of communication are multi-hop and single-hop. Single-hop routing is depending on the chain of each other. The chain will be broken when one of the nodes fails to respond. Unlike multi-hop routing which randomly selects the nodes to be the CH in the specific distance to avoid the signal propagation to the BS. The percentage for the chain to be broken is lower than for the single-hop [16].

20.3.5.1 Single-Sink

The cluster was used to cover large area and the single-sink formation is used to transmit the data from nodes to the base station. Single-link minimizes the route of the data and reduces the forward time in the network [11]. Single–hop communication is considerable as straight-forward communication which reduces the time for the data to the base station [17]. However, the main drawback is the lack of redundancy of data.

20.3.5.2 Multi-sink

In the clustering, the network is randomly divided to several clusters. Each cluster is managed by the CH. The sensor node functions as collect data, process data and exchange data with another sensor. A node can be a sink which is generally no energy limitation. Sink collect the information from the sensor nodes [15]. The multi-hop communication is considerable as an energy efficiency for large scale network [18]. The commonly use for multi-hop topology is the aggregation tree rooted at the sink. The multi-hop topology creates several cluster and CH creates the path for the data to transmit to the sink nodes. However, the uneven cluster size can create unbalanced data distributed which causes the data overloading in the certain CH [18]. Multiple sink are deployed in large scale of network to reduce the redundancy, distribution traffic, and network lifetime. The multi-hop localizes the sink to each cluster to minimize the traffic and redundancy of data [17]. Thus, the combination can create an energy consumption for the transmission in the network.

20.4 Conclusion

In this research, IoT network can be configured based on the cluster criteria such as sink point, CH selection, cluster group to improve data routing efficiency, and network energy consumption. LEACH, HEED, and *K*-Means clustering are an independent network that is efficient in terms of stability based on nodes selection in each neighbor independent of CH selection. Network size defines distance between the nodes and the sink point to minimize data transmission scale. In the technology world, as MOSTI notes, IoT has a positive impact on Malaysia lifestyle and economy. There are a lot of ideas for improvement, as an example for the IoT interface mode during the transmission or network equal size cluster.

Acknowledgements This research is supported by Malaysia Education Minister grant (Application ID-295462-313519).

References

1. Behera TM, Samal UC, Mohapatra SK (2018) Energy-efficient modified LEACH protocol for IoT application. IET Wirel Sens Syst 8(5):223–228. https://doi.org/10.1049/iet-wss.2017.0099
2. Bhandari S, Sharma SK, Wang X (2017) Cloud-assisted device clustering for lifetime prolongation in wireless IoT networks. In: 2017 IEEE 30th Canadian conference on electrical and computer engineering (CCECE). https://doi.org/10.1109/ccece.2017.7946815
3. Bouguera T, Diouris J, Chaillout J, Jaouadi R, Andrieux G (2018) Energy consumption model for sensor nodes based on Lora and LoRaWAN. J Sens 18(7):1–23. https://doi.org/10.3390/s18072104
4. Cao L, Xu C, Shao W, Zhang G, Zhou H, Sun Q, Guo Y (2010) Distributed power allocation for sink-centric clusters in multiple sink wireless sensor networks. J Sens 10(3):2003–2026. https://doi.org/10.3390/s100302003
5. Carlos-Mancilla M, López-Mellado E, Siller M (2016) Wireless sensor networks formation: approaches and techniques. J Sens 1–18. https://doi.org/10.1155/2016/2081902
6. Chan H, Perrig A (2004) ACE: an emergent algorithm for highly uniform cluster formation. Lect Notes Comput Sci 154–171. https://doi.org/10.1007/978-3-540-24606-0_11
7. Kumar JS, Zaveri MA (2018) Clustering approaches for pragmatic two-layer IoT architecture. Wirel Commun Mob Comput 1–16. https://doi.org/10.1155/2018/8739203

8. Li Z, Xin P (2017) Evidence-efficient Multihop clustering routing scheme for large-scale wireless sensor networks. Wirel Commun Mob Comput 2017:1–14. https://doi.org/10.1155/2017/1914956
9. MIMOS B (2015) National internet of things (IoT) strategic roadmap: a summary (Online). Available: http://www.mimos.my/iot/National_IoT_Strategic_Roadmap_Summary.pdf. Accessed on 10 Sept 2021
10. Na S, Xumin L, Yong G (2010) Research on *K*-means clustering algorithm: an improved *K*-means clustering algorithm. In: 2010 third international symposium on intelligent information technology and security informatics. https://doi.org/10.1109/iitsi.2010.74
11. Palan NG, Barbadekar BV, Patil S (2017) A retrospective analysis of cluster head selection protocol. In: 2017 international conference on inventive systems and control (ICISC). https://doi.org/10.1109/icisc.2017.8068724
12. Perera C, Liu CH, Jayawardena S, Min C (2014) A survey on Internet of things from industrial market perspective. IEEE Access 2:1660–1679. https://doi.org/10.1109/access.2015.2389854
13. Rajan AA, Swaminathan A, Brundha, Pajila B (2019) A comparative analysis of Leach, teen, Sep and DEEC in hierarchical clustering algorithm for WSN sensors. Lect Notes Comput Sci 395–403. https://doi.org/10.1007/978-3-030-28364-3_39
14. Ramli A, Basarudin H, Abu MA, Yaakop M, Sulaiman MI (2017) FUSA: fuzzy logic based clustering protocol for formation of uniform size clusters. In: 2017 international conference on engineering technology and technopreneurship (ICE2T). https://doi.org/10.1109/ice2t.2017.8215997
15. Sholla S, Kaur S, Begh GR, Mir RN, Chishti MA (2017) Clustering internet of things: a review. JST 3(2):21. https://doi.org/10.31130/jst.2017.61
16. Singh SK, Kumar P, Singh JP (2017) A survey on successors of LEACH protocol. IEEE Access 5:4298–4328. https://doi.org/10.1109/access.2017.2666082
17. Ullah Z, Mostarda L, Gagliardi R, Cacciagrano D, Corradini F (2016) A comparison of HEED based clustering algorithms—introducing ER-HEED. In: 2016 IEEE 30th international conference on advanced information networking and applications (AINA). https://doi.org/10.1109/aina.2016.87.
18. Xu L, Collier R, O'Hare GM (2017) A survey of clustering techniques in WSNs and consideration of the challenges of applying such to 5G IoT scenarios. IEEE Internet Things J 4(5):1229–1249. https://doi.org/10.1109/jiot.2017.2726014
19. Younis O, Fahmy S (2004) HEED: A hybrid, energy-efficient, distributed clustering approach for ad hoc sensor networks. IEEE Tran Mob Comput 3(4):366–379. https://doi.org/10.1109/tmc.2004.41

Chapter 21
Electrochemical Activities of C/GNP Electrode for Ultrasensitive Immunosensors

Fatihatul Zuriati Makmon, Mohd Azraie Mohd Azmi, Suhaili Sabdin, Azman Abd Aziz, and Nur Azura Mohd Said

Abstract Graphene is well known for its electroanalytical activity due to the presence of enormous π-electrons in the structure. For this reason, it is believed that the graphene used for modification of the conventional electrodes have a great potential to become a good electrode. The drop-casting method is the most widely used approach due to simple observation in dispersion composition of the substrate and it is capable to tailor the detection limit and reproducibility of the samples. There are two main factors to ensure a good drop-casting material; the solvent used to disperse and the temperature during drying the substrate. Water has been used as vector to carry graphene in this study. Drying in an oven at 100 °C displayed a good electroanalytical activity compared to unmodified screen-printed carbon electrode. The sample with 100% graphene nanoplatlet (GnP) has been selected as an excellent candidate for producing a good analytical study and excellent current densities.

Keywords Graphene nanoplatelets · Cyclic voltammetry · Electrochemistry · Screen-printed electrode

F. Z. Makmon · M. A. M. Azmi (✉) · S. Sabdin · A. A. Aziz
Multidisciplinary Nanotechnology Centre, Universiti Kuala Lumpur British Malaysian Institute, Batu 8, Jalan Sungai Pusu, 53100 Gombak, Selangor, Malaysia
e-mail: mazraie@unikl.edu.my

F. Z. Makmon
e-mail: fatihatul.makmon@s.unikl.edu.my

S. Sabdin
e-mail: suhaili.sabdin@s.unikl.edu.my

A. A. Aziz
e-mail: azmanaziz@unikl.edu.my

N. A. M. Said
Biotechnology and Nanotechnology Research Centre, Malaysian Agriculture Research Development Institute, 43400 Serdang, Selangor, Malaysia
e-mail: nazurams@mardi.gov.my

A. Ismail et al. (eds.), *Advanced Materials and Engineering Technologies*, Advanced Structured Materials 162,
https://doi.org/10.1007/978-3-030-92964-0_21

21.1 Introduction

A Graphene sheet is defined as a single layer of graphite with high conductivity, rapid heterogenous electron transfer and intrinsic flexibility [1–4]. One of the most promising applications of graphene is in electrochemical sensing. Since every atom in a graphene sheet is a surface atom, the molecular interaction causes rapid electron transfer and eventually becomes highly sensitive to adsorb molecules [5, 6]. For this reason, it is believed that graphene used for modification of the conventional electrodes has equal great potential for screen-printed electrode in producing good electrodes for sensing purposes. Several methods of SPE modification have been done such as spin coating, aerosol, and drop-casting.

Drop-casting is the most widely used method in modification of SPE where graphene is the underlying substrate. Besides, this method enables to observe the effect of dispersion, composition, and reproducibility of the sample that tailors the limit of detection compared to others. Several organic solvents such as ethanol, acetone, hexane, and dimethylformide have been used to disperse graphene molecules [7, 8] whilst some researchers used di-water as medium to disperse graphene. Both ways work through wonders. However, once dried, the formation of a coffee ring is observed. This effect, which also known as coffee ring effect, occurs during the drying process, where drop edges became pinned to the substrate and capillary flow outward from the centre of the drop brought suspended chemical to the edge as evaporation occurs. Therefore, two main factors that may contributed to the coffee ring effect were substrate temperature and the solvent system [9]. Most of the studies described the coffee ring effect is due to organic solvents [4, 9, 10] and some because of water [9]. Temperature also played an important role in producing the effect of coffee ring where several studies have been conducted on drying properties of the coffee ring effect by inkjet printing [4, 9], but not much studies reported drying by oven.

Here in this study, di-water was used as medium to disperse graphene nanoplatelets (GNP). Two methods of drying were conducted using both oven and hot plate. Hence, two objectives were carried out in these studies; (1) to observe the morphology of SPE and GNP in water and (2) to determine the electrochemical activities via cyclic voltammetry (CV) on different drying methods.

21.2 Methodology

Three-electrode system screen-printed the carbon electrode (SPCE), which acts as medium for the electrochemical sensor (Metrohm DropSens). A carbon working electrode is surrounded by another carbon as counter electrode and silver/silver chloride as reference electrode (Fig. 1a). GNP was obtained from GO Advanced Sdn Bhd. The GNP produced is stable in aqueous solution and ready to be used as monolayer

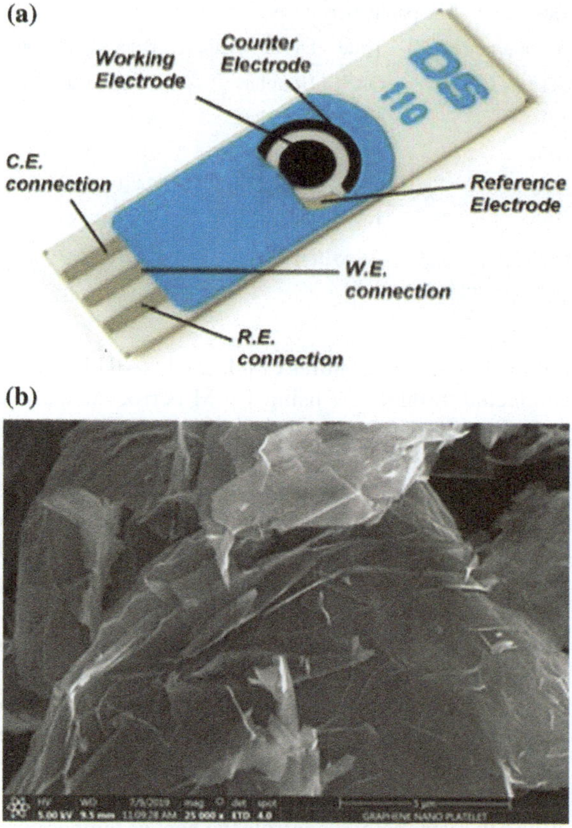

Fig. 21.1 **a** Screen-printed carbon electrode before surface modification. *Reprinted from* [11]. **b** Scanning electron microscopy image of GNP before drop-casting

or multilayer graphene sheets. Figure 1b shows the scanning electron microscope (SEM) image of GNP at 5 μm, respectively.

All chemicals were purchased from Sigma -Aldrich (Gillingham, UK) of analytical grade and were used as received without any further purification. All solutions were prepared with deionized water of resistivity not less than 18.2 Ω cm.

Voltammetric measurements were carried out using PC portable BiPotentiostat/Galvanostat (Spain) with a connector for screen-printed electrode purchased DropSens. Experiments were performed using 20 μL (amount after drop-casted) of GNP onto 1 mm diameter carbon working SPE (ref 110).

21.3 Results and Discussion

To understand the principal of this work, 1 mg/mL of GNP was dispersed in water and drop-casted onto SPCE. Later, drying took place by means of hot plate or oven at

Table 21.1 The parameter of C/GNP at different drying method and time

Temperature (°C)	Time	Drying method	Peak separation (mV)	Current density (mA/cm^2)
RT[a]	ON[b]	Hot plate	86 ± 0.00	0.012 ± 0.000
60	1 h	Hot plate	84 ± 0.00	0.006 ± 0.000
100	1 h	Oven	70 ± 0.00	0.005 ± 0.000
100	ON[b]	Oven	74 ± 0.02	0.010 ± 0.001
120	1 h	Hot plate	92 ± 0.03	0.006 ± 0.000

[a]*RT* room temperature (27 °C)
[b]*ON* overnight (16–18 h)

four different temperatures (Table 21.1). All electrochemical activities of GNP were conducted through CV using 1 mM Ferrocene carboxylic acid (FCA) in phosphate buffers solution (pH 7.4). FCA was used as redox solution due to excellent electrochemical properties and stability in both oxidation and reduction state [5, 9]. Raman was conducted for determination of different GNP percentage fingerprints.

The substrate temperature plays a role on the behaviour of the droplets which influenced the result of the end products. Table 21.1 shows the experimental conditions of GNP drop-cast. The temperature conditions reported are higher as compared to other studies using the inkjet printing method [9, 10]. Drying using hot plate showed a uniform dome shape after completion. These profiles are suggestive to be a coffee ring effect as reported by Deegan et al. [10]. He described that when higher temperature is employed, the evaporation rates over entire feature increased, and the amount of evaporation time has effects on the substrate [4]. Similar results were seen in Soltman and co-worker using inkjet printing [9]. In contrast, drying by oven which acts as contained environment did not show any coffee ring effect. This suggests that drying in oven method controls the heat as temperature rises, hence reduce the effect of coffee ring.

A rather popular tool to study electrochemical reaction is CV, qualitatively. The experiment was performed at the stationary phase by applying the voltage to the working electrode and scanned linearly within the unstirred solution. Diffusion happens by means of transporting the reactant to the working electrode surface. Hence, estimation of the reduction potentials, formation constants and number of electron transfer per reactant molecule could be made [12, 13]. Figure 21.2 depicts the CV signatures of unmodified SPCE (bare carbon) and GNP-modified SPCE of drying using hot plate and oven. Both displayed oxidation and reduction characteristics at a scan rate of 100 mV s^{-1}. All modified SPCE showed higher peak separation compared to unmodified SPCE in all drying methods. This suggested that graphene nanomaterials provided excellent electrical conductivity and large surface area which facilitated electron transfer activity [8].

The integration area under the anodic and cathodic peaks indicated the optimum electrochemical activity and peak separation was calculated based on the potential differences between cathodic and anodic peaks. The ideal peak separation was selected based on the lowest value of peak separation. Hence, in this study, drying

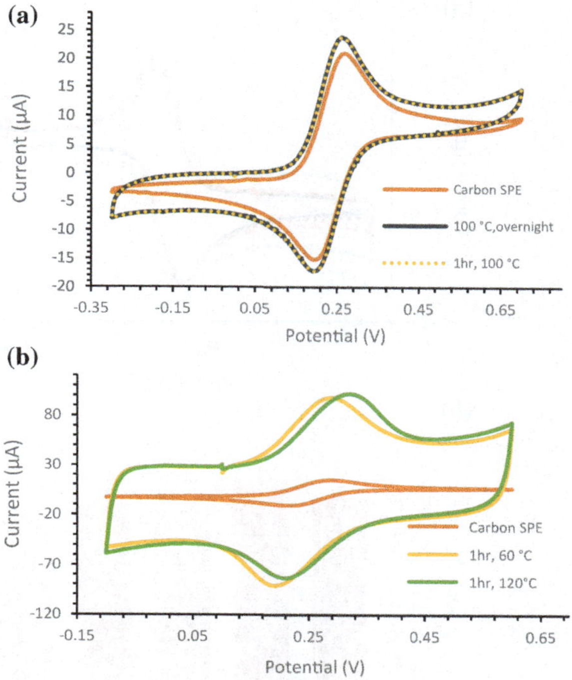

Fig. 21.2 **a** CV peaks obtained from drying using oven (potential range −0.35 to 0.7 V). **b** CV peaks obtained from drying using hot plate (potential range −0.15 to 0.7 V). Scan rate: 100 mV s^{-1}

by oven at 100 °C for 1 h (Table 21.1) with peak separation of 70 ± 0.00 mV were ideal drying condition for GNP. A huge peak separation was observed in drying by hot plate compared to oven. This may be due to unstable condition such as moisture from the environment which fluctuated the temperature during incubation time [10].

As described earlier, 1 mg/mL of GNP displayed an excellent electrochemical activity in all drying methods, thus showing a good electroanalytical activity. However, not many studies on lower concentration of GNP were conducted for evaluating electrochemical and electroanalytical activities. CV was conducted on a series of GNP percentages (4, 8, 12, 16, 20, and 100%). Figure 3a exhibits the CV of different GNP percentages drop-casted on single printed working carbon electrode in a potential range between −0.20 and +0.70 V at a scan rate of 100 mV s^{-1}. Each percentage displayed excellent electrochemical and electroanalytical activities with typical Nernstian electrochemical reaction. Figure 3b shows the peak separation values between anodic and cathodic peaks based on CV from Fig. 3a. The values were between range of 65–77 mV. The range of peak separation was significantly higher than 59 mV (theoretical value) and the separation of these two signals increased with faster scan rates. Hence, this reaction is quasi-reversible.

Current densities and diffusion rate occurred during the whole CV process was calculated using the Randles Sevcik equation. In this equation, current peak (Ip) is predicted to be proportional to the square root of the scan rate (Eq. (21.1)).

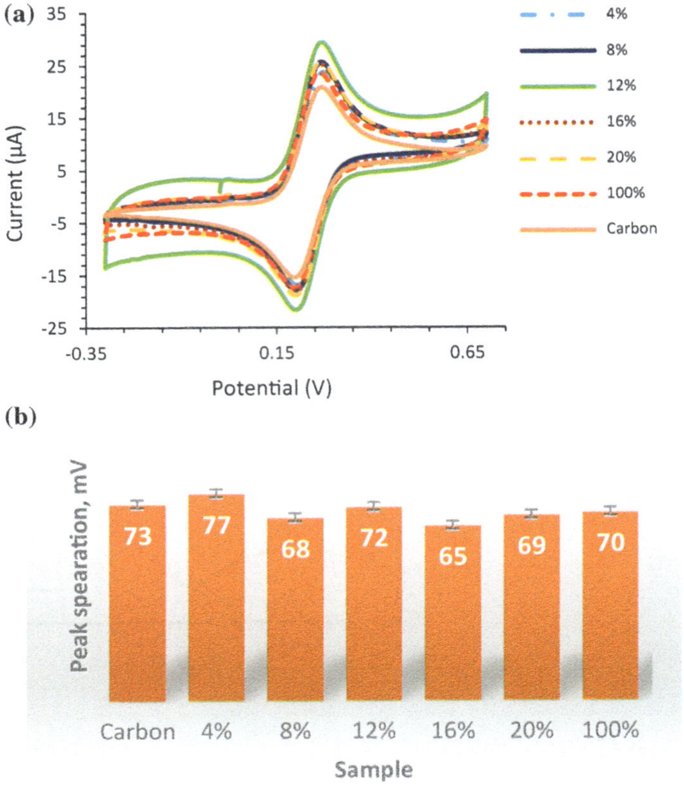

Fig. 21.3 **a** CV peaks presented as voltammograms. **b** Peak separation of GNP-modified SPCE at various percentage and unmodified (carbon) in 1 mM FCA phosphate buffer (pH 7.4) at scan rate: 100 mV s^{-1} presented as peak separation

$$\mathrm{Ip} = 2.69 \times 10^5 \times n^{3/2} \times \mathrm{ACD}^{1/2} \upsilon^{1/2} \tag{21.1}$$

where n is the number of electrons involved in the reaction, A is the electrode surface area (cm^2), C, the bulk concentration (mol cm^{-3}), D, the diffusion coefficient (cm^2 s^{-1}) and v, the sweep rate (V s^{-1}). Taking Eq. (21.1) is in the resemblance with $y = mx + c$ linear graph equation, the m value or the slope replaced the Ip to produce Eq. (21.2) as follows.

$$\mathrm{Slope} = 2.69 \times 10^5 \times n^{3/2} \times \mathrm{ACD}^{1/2} \tag{21.2}$$

For this experiment, the number of electrons involved in this reaction, n, was 1, the electrode area of carbon working SPCE, A, was 0.04 cm^2 with the concentrations of 10^{-6} mol ferrocyanide. The slope and current were given in Table 21.2 for each

Table 21.2 The slope, diffusion coefficient and current densities of unmodified and modified GNP

Sample	Slope	Diffusion coefficient ($cm^2\ s^{-1}$), D	Current densities ($mA\ cm^{-2}$)
Unmodified (carbon)	$Y = 67.886x - 2.5876$	1.126×10^{-10}	0.33 ± 0.064
Modified GNP			
4%	$Y = 75.701x - 3.3445$	1.28×10^{-10}	0.33 ± 0.002
8%	$Y = 87.756x - 5.2654$	1.26×10^{-10}	0.53 ± 0.002
12%	$Y = 84.982x - 4.9892$	1.20×10^{-10}	0.54 ± 0.02
16%	$Y = 80.616x - 4.8453$	2.92×10^{-9}	0.54 ± 0.02
20%	$Y = 78.516x - 3.4782$	1.17×10^{-10}	0.55 ± 0.02
100% (1 mg/ml)	$Y = 758.526x - 3.478$	2.52×10^{-9}	0.55 ± 0.00

percentage at 100 mV s^{-1} scan rates. After the substitution of the parameter and unit, the diffusion coefficient, D values for each percentage were as displayed in Table 21.2. Table 21.2 indicated the slope, diffusion coefficient and current densities of unmodified and modified SPCE at different GNP percentages. Every percentage of GNP displayed higher current densities than unmodified SPCE, suggesting an increase in electroanalytic activity of GNP on carbon working electrode surface. Diffusion coefficient was the highest in 16% GNP compared to others, followed by 100% GNP. In contrast, 100% GNP (1 mg/mL) displayed the highest current densities which proposed that high concentration provided high current densities. Raman spectroscopy in combination with electrochemistry was used in the following analysis to better understand the characteristic of graphene when drop-casted onto SPCE.

21.4 Conclusion

In summary, we have demonstrated proof-of-concept that the method of drying GNP heavily influenced the electrochemical and electroanalytical activities of modified SPCE. Drying in oven at 100 °C for an hour was selected based on better and low peak separation. The highest electrochemical and electroanalytical activities with highest current densities were observed using 1 mg/mL (100%) GNP concentration, thus, reflecting the graphene characteristic of having high electroconductivity further studies on functionalization are yet to be established to produce a sensitive electrochemical biosensor.

Acknowledgements Financial supports received under Short Term Research Grant UniKL/CoRI/str16028 and Fundamental Research Grant scheme FRGS/1/2016/STG04/UNIKL/02/1 are highly acknowledged. We are thankful to MARDI for their assistance throughout the duration of this work.

References

1. Dimiev AM, Ceriotti G, Metzger A, Kim ND, Tour JM (2016) Chemical mass production of graphene nanoplatelets in ~100% yield. ACS Nano 10:274
2. Pumera M (2010) Graphene-based nanomaterials and their electrochemistry. Chem Soc Rev 39:4146
3. Wang Z et al (2012) A nanoscale graphene oxide–peptide biosensor for real-time specific biomarker detection on the cell surface. ChemComm 48:9768
4. Secor EB, Prabhumirashi PL, Puntambekar K, Geier ML, Hersam MC (2013) Inkjet printing of high conductivity, flexible graphene patterns. J Phys Chem Lett 4:1347
5. Chen L, Tang Y, Wang K, Liu C, Luo S (2011) Direct electrodeposition of reduced graphene oxide on glassy carbon electrode and its electrochemical application. Electrochem Commun 13:133
6. Serodre T et al (2019) Surface silanization of graphene oxide under mild reaction conditions. J Braz Chem Soc 30:2488
7. Cinti S, Arduini F (2017) Graphene-based screen-printed electrochemical (bio) sensors and their applications: efforts and criticisms. Biosens Bioelectron 89:107
8. Badruzaman NA, Mohd AMA, Mohd SNA (2020) Electrochemical immunosensor based on highly sensitive amino functionalized graphene nanoplatelets-modified screen printed carbon electrode. Key Eng Mater 833:171–175
9. Soltman D, Subramanian V (2008) Inkjet-printed line morphologies and temperature control of the coffee ring effect. Langmuir 24:2224
10. Van DBAM, de-Laat AW, Smith PJ, Perelaer J, Schubert US (2007) Geometric control of inkjet printed features using a gelating polymer. J Mater Chem 17:677
11. Cheng W, Stuart E, Tschulik K, Cullen J, Compton R (2013) A disposable sticky electrode for the detection of commercial silver NPs in seawater. Nanotechnology 24:505501
12. Mabbott GA (1983) An introduction to cyclic voltammetry. J Chem Educ 60:697
13. Seger B, Kamat PV (2009) Electrocatalytically active graphene-platinum nanocomposites. Role of 2-D carbon support in PEM fuel cells. J Phys Chem C 113:7990

Chapter 22
Development of an Atmospheric Drinking Water Generator

Suhaimi Bahisham Yusoff and Muhammad Amirul Rafizat

Abstract Some communities have difficulties accessing clean water sources because of a lack of water supply facilities especially in rural areas and during disasters such as droughts and floods. Thus, the atmospheric drinking water generator is proposed to overcome this problem by providing a system that can convert air into safe drinking water. This atmospheric drinking water generator uses the principle of condensation of air, cooled to its dew point, and becomes so saturated with water vapor that it can no longer hold the liquid. Peltier or thermoelectric cooler (TEC) is used as a condensation process medium to cool the air to the dew point. Water that has been converted from the air is collected in a collector tank. When the collection tank is full of water, water will be pumped out through a filter set to remove possible bacteria and add minerals to the water. The proposed system with a condenser size of 4.5 cm $\times$ 4 cm $\times$ 2 cm can produce 6–9 ml of water per hour depending on surrounding conditions. To increase the amount of water converted from the air, the size of the condenser used needs to be enlarged to increase the surface area of the condensation process.

Keywords Thermoelectric cooler · Atmospheric water generator · Peltier effect

22.1 Introduction

Water is very important for human life, water is the source of life. With the water resources available on earth, human beings can live in a prosperous way to enable us to carry out activities well. But not all communities in the world have access to clean water. Some communities find it difficult to get clean water supply due to the lack of water supply facilities such as the rural population [1]. The government has spent

S. B. Yusoff (✉) · M. A. Rafizat
Electronics Technology Section, Universiti Kuala Lumpur British Malaysian Institute, 53100 Gombak, Selangor, Malaysia
e-mail: suhaimib@unikl.edu.my

M. A. Rafizat
e-mail: amirul.abu05@s.unikl.edu.my

A. Ismail et al. (eds.), *Advanced Materials and Engineering Technologies*,
Advanced Structured Materials 162,
https://doi.org/10.1007/978-3-030-92964-0_22

millions of moneys to implement infrastructure projects to supply clean water to the rural population. However, there are still communities that still do not have access to clean water. In addition, natural disasters can also affect water supply problems such as drought and prolonged floods in the country which can make water supply difficult for the community.

To overcome this problem, a system that can convert water from the air has been introduced. Several projects have been implemented to produce systems that can convert air into water [2–6]. To convert air into water, the temperature needs to be lowered to reach the dew point and there are several steps and ways to do this [2]. Several projects have used a microcontroller as a controller circuit, the fan serves as air to let air enter the condenser and through a coil that has been cooled using TEC [2–6]. Once the water is formed, the water will be collected in a storage tank after going through the filtration process. Suryaningsih and Nurhilal [2] had proposed a system consists of a PWM controller circuit, air fan, an array of Peltier modules, the hot side channel as radiator, blade valve, and condenser. The TEC module will remove heat from the air flowing along the cold side channel and push this heat in the hot side channel of the module. The humid air source is required to supply the cold side channel for condensation process. Tripathi et al. [3] had proposed to convert air into water by using a compressor, and then the air is passed through a condenser pipe which lowers the temperature to the dew point. The water is then filtered through several filter such as carbon, reverse osmosis, and UV sterilization lights. Nitheesh et al. [4] had developed a water condensation system based on thermoelectric cooler. The system consists of cooling elements, heat exchange unit, and air circulation unit. A solar cell panel unit with a relevant high current output drives the cooling elements through a controlling circuit. It uses the principle of latent heat to convert water vapor into water droplets.

Nandy et al. [5] had developed a system based on a thermoelectric cooler. The developed system consists of cooling elements, heat exchange, air circulation, TEC couple, conventional compressor, and evaporator system be used to condense water by simply exchanging the latent heat of coolant inside the evaporator. Ramya et al. [6] had proposed a system consisting of cooling elements, heat exchange unit, air circulation unit, pH sensor, conductor with fan, and Arduino Uno. They have used a Peltier which is portable and eco-friendly to create the atmospheric water generator. Using dehydration tactics will be more expensive and not portable, so they generate water from the air using Peltier tools at a lower cost and efficient in terms of the amount of water collected and easy to carry.

In this paper, we have developed the atmospheric drinking water generator which will convert the air into water by using the principle of condensation. The idea of this project is to help the community to get clean water supply for those living in the rural area and during natural disasters that cause the clean water supply is disrupted. The system to convert the air into water is proposed by using a microprocessor-based system which consists of Peltier or TEC that has been used as a condensation process medium, cooler, etc. The rest of the paper is organized as follows. Section 22.2 gives a detailed explanation of the proposed system. Section 22.3 presents the experiment results for the proposed system with discussion. Section 22.4 concludes with a summary of the proposed system and discusses the future works.

22.2 Methodology

The main purpose of this project is to convert air into water using a condensation process. Theoretically, there will be water droplets through the condensation process but it has some problems. For example, if the temperature is too cold, water droplets will freeze and if the temperature is not cold enough, water droplets will not form. So, a suitable temperature is required to produce water droplets from the air. This project uses the principle of air condensation, cooled to its dew point, and becoming so saturated with water vapor that it can no longer hold liquids. Peltier or TEC is used as a condensation process medium to cool the air to the dew point. The water that has been converted from air accumulates in the collector tank.

Figure 22.1 shows the design of the proposed atmospheric drinking water generator. The condenser functions as an air to water conversion medium where the condenser requires the cooler to allow temperatures in the condenser to be lower than the incoming air temperature. The incoming and outgoing air are controlled using fans that act as an air sucker. Once the incoming air is converted into water droplets, the water formed will be collected in the first water storage tank until full. Once the first water storage tank is full of water, it will be transferred and filtered to the second tank. Minerals have also been added for health benefits. Upon completion

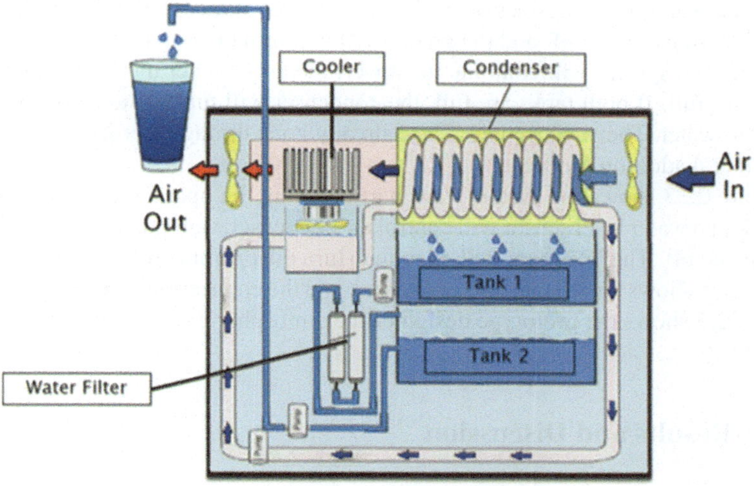

Fig. 22.1 Design of the atmospheric drinking water generator

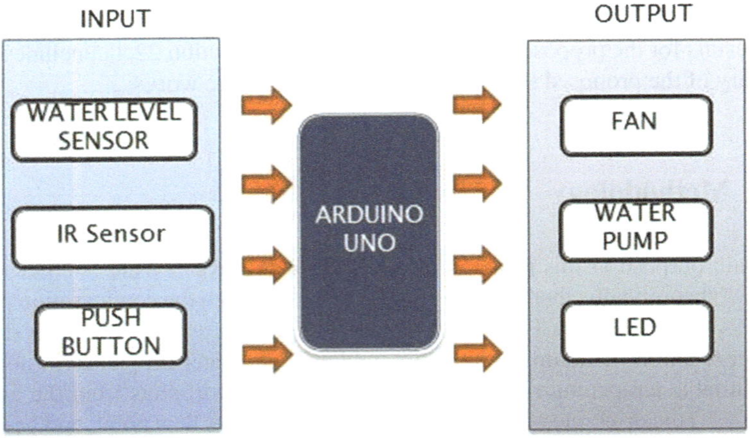

Fig. 22.2 Block diagram

of the filtration process, the water will be stored in a second tank to allow the water to be used safely.

Figure 22.2 shows the block diagram of the proposed system. The proposed atmospheric water generator system is operated by a microcontroller. The project requires sensors to act as inputs that detect system changes and functional outputs relying on input detection. Water converted from the atmosphere will be collected in tank 1. When the water in the first tank is full, the water pump will turn on. The water pump in tank 1 will operate until the water level sensor in tank 2 reached the desired level. The water pumped to tank 2 will pass through a filter before going to tank 1. When the water level sensor in tank 2 detects water at the desired level, the water pump in will be turned off and the green LED will turn on to indicate the tank is full of water. The system will continue to convert the air to water until both tank 1 and tank 2 are full. If both tanks are full, the condenser will turn off to stop converting the air to water. The water in tank 2 has already gone through a process of filtration and mineral addition for health.

Figure 22.3 shows the system flowchart detailing the operation of the proposed system. The water level sensor will control the condenser, water pump 1, and LEDs to turn on and off. The IR sensor will operate to turn the operation of the water pump 2. Figure 22.4 shows the system circuit diagram and the connection of each component. Figure 22.5 shows the prototype design of the atmospheric water generator.

22.3 Results and Discussion

A prototype has been developed for this project. The prototype designs are produced in sizes that are not too large to allow the project to be moved easily without having

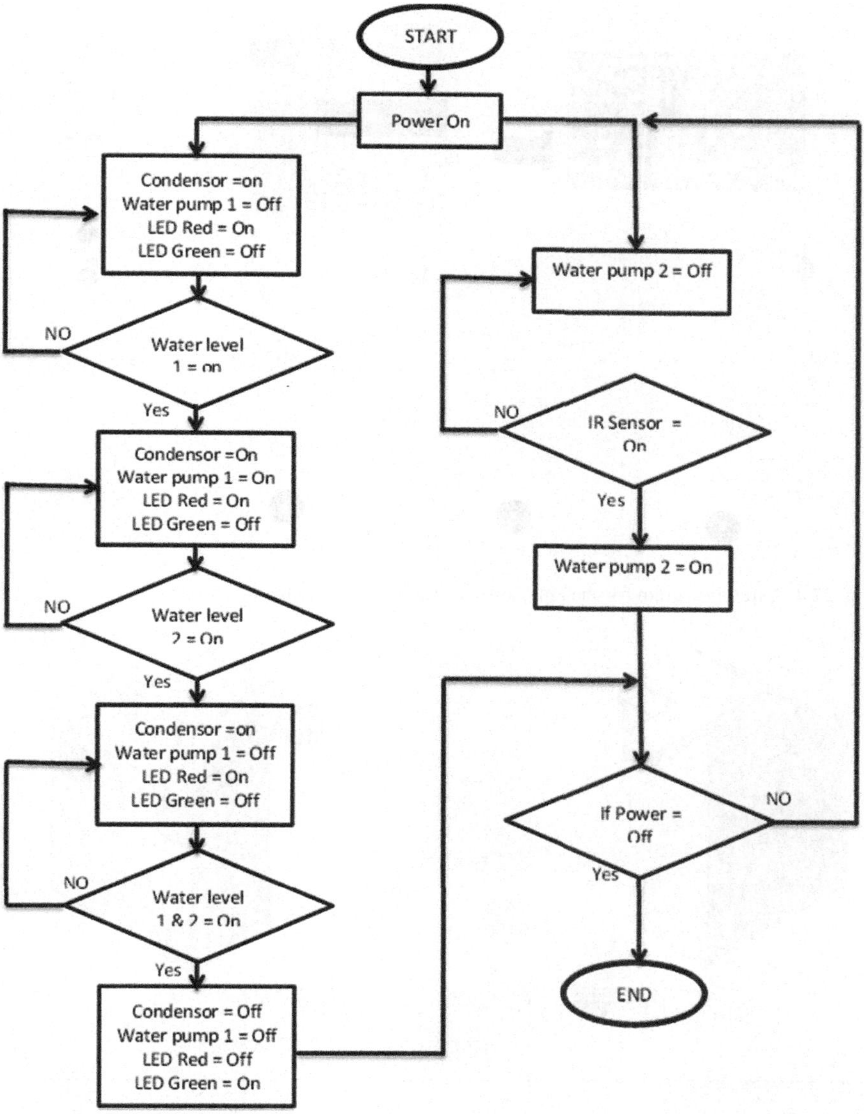

Fig. 22.3 Flowchart of system

to use a lot of energy. In addition, it is also designed without the need to add additional features to the tap water source. Figure 22.6 shows the prototype atmospheric drinking water generator developed in this project, while Fig. 22.7 shows the arrangement of the important parts found in the project, namely condenser, storage tank, coolant, fan, filter, and water pump.

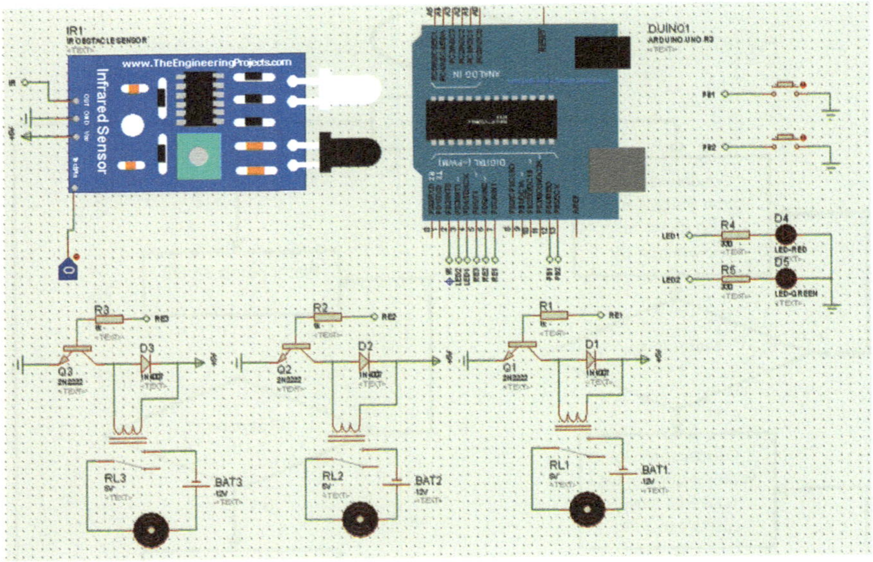

Fig. 22.4 Circuit diagram for the atmospheric drinking water generator

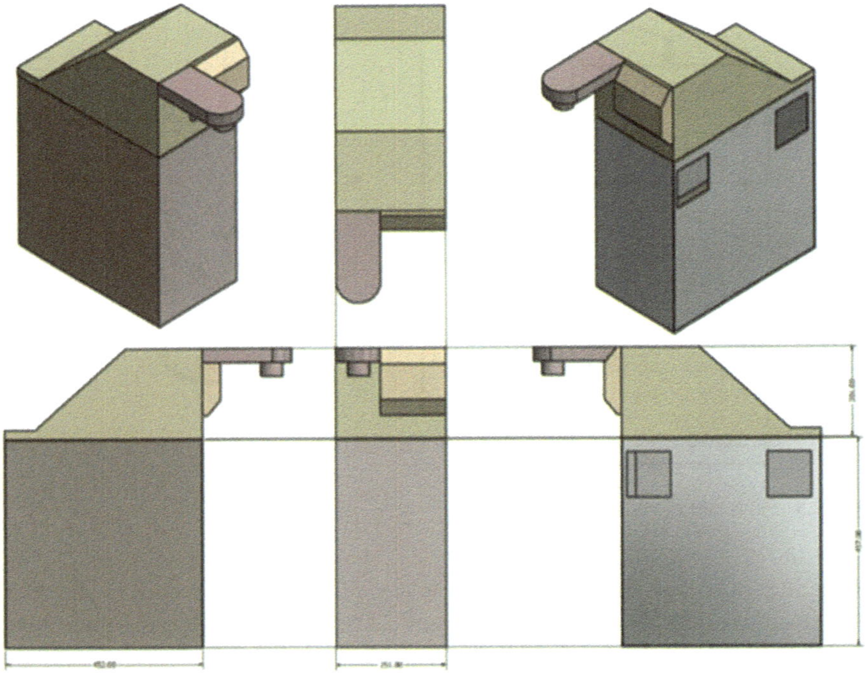

Fig. 22.5 Prototype design of the atmospheric water generator

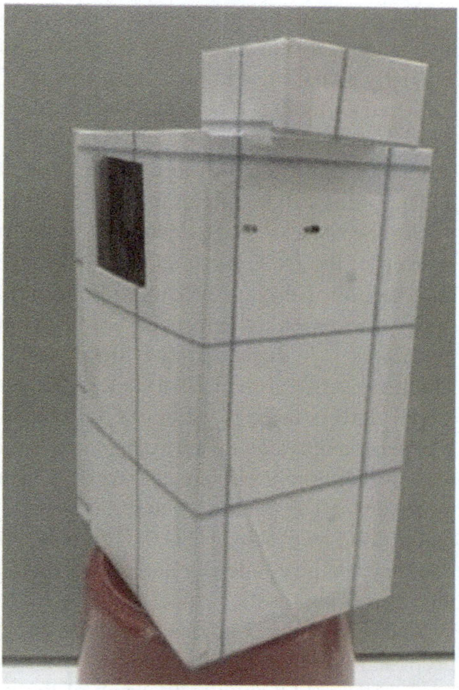

Fig. 22.6 Prototype of atmospheric drinking water generator

Fig. 22.7 Internal view of the prototype

Figure 22.7 shows the three main parts of the system, namely tank 1, tank 2, and condenser. The condenser is the part that does the process of converting air into water by removing the heat from the air. Once water is formed, the water will be collected in tank 1. The water will be collected until the sensor detects water at a certain level. When the water reaches the setting level and the sensor detects the water, the controller will turn on the pump and pump the water out of tank 1 to tank 2. The water will be stored in tank 2 for drinking purpose.

For the process of dispensing the water for drinking, the user needs to place a glass in front of the IR sensor under the dispenser host and the water will be poured into the glass for drinking. When the IR sensor detects the glass in the dispenser, it will send information to the processor to activate the water pump in tank 2. The purpose of pump 2 is to dispense water from tank 2 that has been purified and safe to drink. The level sensor functions to detect the water level. It works when 0 indicates, the sensor detects there is water, while 1 means the level sensor does not detect water. The sensor operates that way because the sensor is connected to ground. The level sensor 1 and the level sensor 2 work to turn on the pump 1 and the condenser if the level sensor 1 is equal to 0 and the pump 1 is turned on (1). Pump 1 will be turned on until the level sensor 2 is 0. Otherwise, pump 1 continues to operate. If all sensor levels are 0, the condenser and pump 1 will be turned off (0). Table 22.1 summarizes the operations for the project.

The size of the condenser used for this prototype is 4 cm × 4.5 cm × 2 cm. The proposed system with the condenser size can produce 6 ml to 9 ml of water per hour. Figure 22.8 shows the graph of the water produced from the system. The test period is from 9:00 a.m. to 8:00 p.m. For each hour, the produced water is measured. The average amount of water produced per hour is around 7 ml. Water output is determined by the surface size of the condenser, the temperature produced by the condenser and the humidity level of the air at that time. To increase the amount of water converted from the air, the size of the condenser used needs to be enlarged to increase the surface area of the condensation process.

Table 22.1 System operation

Input			Output		
IR sensor	Level sensor 2	Level sensor 1	Pump 2	Condenser	Pump 1
0	0	0	0	0	0
0	0	1	0	1	0
0	1	0	0	1	1
0	1	1	0	1	0
1	0	0	1	0	0
1	0	1	1	1	0
1	1	0	1	1	1
1	1	1	1	0	0

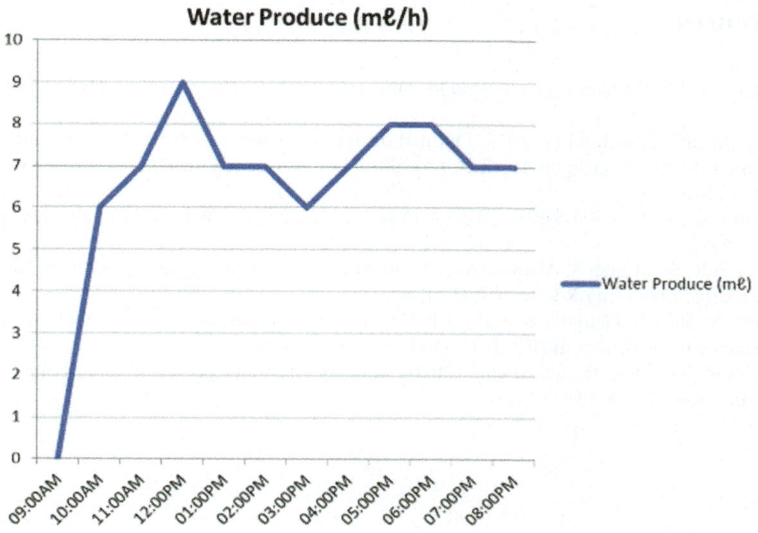

Fig. 22.8 Hourly water production results

22.4 Conclusion

The development of the atmospheric drinking water generator has been proposed. This project provides a system that can convert air into safe drinking water. This project can provide many benefits to communities who face problems to get clean drinking water, especially those in rural areas and also during disasters such as major floods. This project uses the principle of condensation of air, cooled to its dew point, and becomes so saturated with water vapor that it can no longer hold the liquid. Peltier or TEC is used as a condensation process medium to cool the air to the dew point. The condenser size of 4.5 cm × 4 cm × 2 cm can produce 6–9 ml of water per hour. In future, to increase the amount of water converted from the air, the size of the condenser used needs to be enlarged to increase the surface area of the condensation process.

Acknowledgements The authors wish to deepest thanks to all people who have been very helpful and provided guidance for the completion of this journal.

References

1. Gleick PH (1993) Water in crisis: a guide to the world's freshwater resources. Oxford University Press
2. Suryaningsih S, Nurhilal O (2015) Optimal design of an atmospheric water generator (AWG) based on thermo-electric cooler (TEC) for drought in rural area. In: 2nd international physics symposium, pp 1–6
3. Tripathi A, Tushar S, Pal S et al (2016) Atmospheric water generator. Int J Enhanc Res Sci Tech Eng 5(4):69–72
4. Nitheesh K, Saravanan S, Ahamed AA (2019) Atmospheric water generator with the concept of Peltier effect. Int J Eng Res Tech 7(11):1–4
5. Nandy A, Saha S, Ganguly S et al (2014) A project on atmospheric water generator with the concept of Peltier effect. Int J Adv Comp Res 4(2):481–486
6. Ramya M, Roja KR, Roopa M et al (2020) Atmospheric water generator using Peltier device. Int J Eng Res Tech 8(13):182–185

Chapter 23
Development of Mobile Applications for Monitoring and Managing Hajj and Umrah Pilgrimage

Nur Shaliyana Ilias, Mohd Azraie Mohd Azmi, and Khairul Nizam Mat Nor

Abstract Performing Hajj or Umrah is an Islamic pilgrimage ritual and the fifth essential pillar of Islam. Preparation to perform Hajj or Umrah is very vital before and in the middle of traveling to Mecca including learning all the rituals, obligatory, and rules [1]. The existing mobile applications in the market that are related to Hajj or Umrah are less interactive, too simple, have limitations and do not offer smart features to assist the user, and do not meet the target users' demand. Hence, the proposed project, Umrah and Hajj application are used to provide pilgrims and "*mutawwif*" with guidance and information about the Hajj or Umrah pilgrimage. The "*mutawwif*" can utilize the features to manage his group while pilgrims can learn and see all the mandatory and obligatory rituals to be performed at various locations. The other features are the global positioning system (GPS) tracking system and the Lilypad Tawaf counter that can help the pilgrims to track the number of completed rounds when the Tawaf is complete. The development of this mobile application used the Android Studio software and JAVA programming language and is exclusively for the Android platform only. Moreover, the Android platform for the mobile application is cost-effective and the open-source idea is very popular nowadays for all generations [2]. This application also can help the pilgrims of Android device users to learn, understand, and perform the Hajj and Umrah ritual perfectly. The methodology used in this application is the ADDIE method that contains 5 phases, namely analysis, design, development, implementation, and testing.

Keywords Mobile applications · Global positioning system · Lilypad Tawaf counter · Tracking system · Mapping system

N. S. Ilias · M. A. M. Azmi (✉)
Multidisciplinary Nanotechnology Centre, Universiti Kuala Lumpur British Malaysian Institute, Batu 8, Jalan Sungai Pusu, 53100 Gombak, Selangor, Malaysia
e-mail: mazraie@unikl.edu.my

K. N. M. Nor
Pegawai Tadbir Agama, Pejabat Agama Islam Daerah Manjung, 32040 Seri Manjung, Perak, Malaysia

A. Ismail et al. (eds.), *Advanced Materials and Engineering Technologies*, Advanced Structured Materials 162,
https://doi.org/10.1007/978-3-030-92964-0_23

23.1 Introduction

In today's era of globalization, mobile applications referred to an application software designed to run on a mobile device, such as a smartphone or laptop. A mobile application is known as an app, web app, online app, or smartphone app. The number of pilgrims that traveled to Mecca from around the world keeps increasing every year. In the year 2018, the quota for Malaysian pilgrims was 30,200 [3] from 2.4 million pilgrims visiting Mecca [4]. The pilgrims do not have to memorize all the prayers for all the rituals, and they can refer to the reference books or "*mutawwif*" throughout performing the duty.

This application aims to create an android platform to build an application for the mobile application. So, besides asking people or referring to the books in a very crowded place, the pilgrims can utilize their mobile devices to review and refresh the upcoming rituals. There are two main features in this application, which are the global positioning system (GPS) tracking system and a pedometer that can help the pilgrims to track the number of completed rounds when the Tawaf is completed. Besides, there is interactive information provided for the user's reference before or during performing the ritual. By having an application that can be accessed from mobile devices, the preparation for the pilgrims and performing it will be effective and enjoyable.

23.2 Methodology

This section will discuss the methodology used to develop the application. The ADDIE methodology contains 5 phases such as analysis, design, development, implementation, and evaluation as shown in Fig. 23.1. More information about each phase of this model, hardware, and software will be explained in this chapter.

23.2.1 Analysis Phase

At this stage, it involves several determining processes and identifying problems in the process of developing the mobile Umrah and Hajj application. Once the problem is identified, an analytical process will be undertaken to find out what factors that cause the problem. This process involves various aspects including analysis of Umrah or Hajj pilgrims, learning environment, and identify the aims of creating this application. The requirements to complete the project are collected from various sources which include the online sources, document and report analysis, or other similar applications from previous Android application developers.

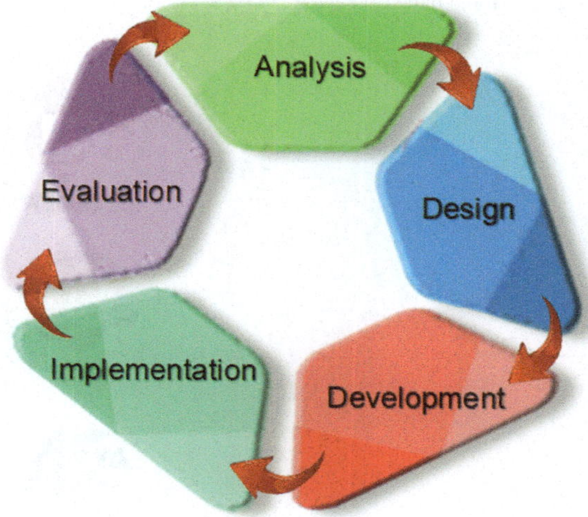

Fig. 23.1 5 phases of the ADDIE model

23.2.2 Design Phase

During this phase, identify the design of the system and develop a storyboard based on the functionalities that will be built. The data or requirement obtained during the analysis phase is transformed into the design. The storyboard is designed using the Photopea software to show the look and feel of the application that will be developed [5]. Figure 23.2 shows the design of the home page of Umrah and Hajj mobile application.

23.2.3 Development Phase

Building a real Umrah and Hajj mobile application was done by using the Android Studio software by the program with JAVA language with all the elements of multimedia selected based on the requirements of this application. The output in the design phase will be input to the development phase and each development will be tested to ensure that it is consistent and effective. Figure 23.3 shows the development of the Islamic video page. VideoView has been implemented on this page to display or load a video file from various sources. Besides, VideoView allows the user to play, pause, rewind, fast-forward, or back to specific paces in the video. Moreover, listview is a vertically scrollable list below the VideoView that displays the title of the video. When the user clicks the video title in the listview, the selected video will play at VideoView. This allows the apps to minimize the content on one page and make it simple and interactive to use.

Fig. 23.2 Home page

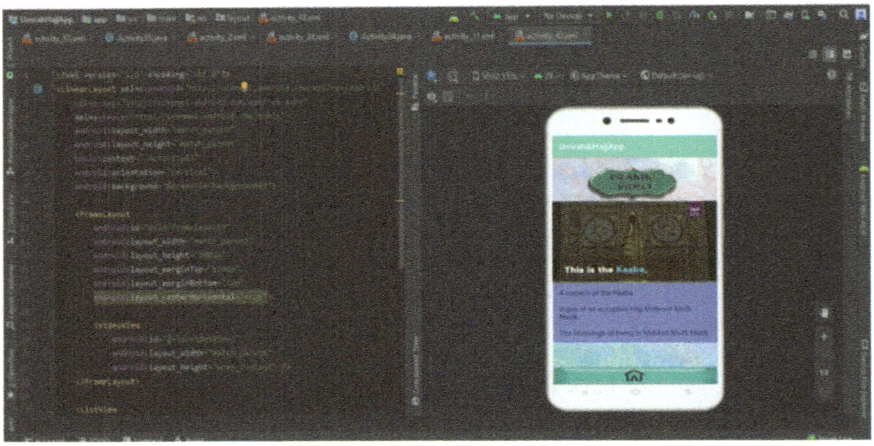

Fig. 23.3 Development of Islamic video page

Fig. 23.4 Implementation of Tawaf page

23.2.4 Implementation Phase

There are two parts in this phase which are the test implementation phase and the final implementation phase. The implementation phase is a testing phase using the Nox emulator. Besides, the final implementation phase will be tested by the supervisor. This phase is important to detect the weaknesses and unobserved errors during the authoring process. The implementation phase aims to demonstrate that the mobile application meets its requirements. Figure 23.4 shows the implementation of the Tawaf page using the Nox emulator. This is important to test the audio and video play smoothly and consistent with on-time frame playback [6].

23.2.5 Evaluation

This phase aims to detect the effectiveness and efficiency of the instruction for the Umrah and Hajj mobile application. This assessment involves the process of obtaining feedback from users on the content, image, audio, video, animation, and interfaces contained in the application.

23.3 Results and Discussion

This chapter discusses the results by program the Java programming language in the Android Studio software. The result of the application is installed in the smartphone and the application is tested in the real environment using the target device which is the Android device. The device used is a VIVO Y55s smartphone with a 5.2-inch screen size and resolution of 720 × 1280 pixels.

Figure 23.5 shows the opening page and Fig. 23.6 shows the main page of Umrah and Hajj mobile application. The user can see the objective of this project on the overview page. There is also information about Umrah and Hajj pilgrimage with an interactive device like video and sound for their guidance. Besides, the two main features are GPS tracker and Lilypad Tawaf counter that can help the pilgrims to track the number of completed rounds when the Tawaf is completed. The nametag page contains a pilgrim's information like name, operating company name, country, and emergency number. This information is stored in the Google Firebase Storage

Fig. 23.5 Opening page

Fig. 23.6 Main page

[7]. Lastly, the services page contains information about the medical center, visiting Madinah, post office, Islamic video, Friday prayer, mustajab place, and Al-Quran.

Figure 23.7 shows the Tawaf of the Umrah page. On this page, the user can see the video by clicking the play button. The video can be started and stopped at a specific time. The user can hear the audio by clicking the sound icon. The audio plays the Arabic language and the English language. This interactive device can help the user recite the intention and Talbiyah accurately. Lastly, the home button is connected to the Umrah and Hajj mobile application main page.

Figure 23.8 shows the GPS tracker and mapping system to track missing pilgrims and this is a system that can track the position or location of the object at the specific place that had been decided on the earth space.

Figure 23.9 shows the Lilypad Tawaf counter that helps the pilgrims to track the number of completed rounds when the Tawaf is complete by using a wearable Arduino or known as Lilypad Arduino. The pilgrims should "ON" the Bluetooth connection at the wearable Lilypad Arduino and connect with the Umrah and Hajj mobile application by clicking the scan Bluetooth button. Then, when the pilgrims click the selected Bluetooth it will link to Fig. 23.10 Lilypad Tawaf counter information page that can notify pilgrims on the Tawaf cycle counting by showing the steps and distance.

Fig. 23.7 Tawaf page

23.4 Conclusion

The objective of developing the Umrah and Hajj mobile application is achieved by developing and implementing a basic platform of a mobile application using Android Studio for Muslims on pilgrimage. This project was also successful to integrate the GPS tracking system and Lilypad Tawaf counter. The recommendations for this mobile application are to add more information and features like prayer times, tasbih, hadith, and others. Besides, Umrah and Hajj mobile application should link the emergency contact number in the nametag page to a clickable link that opens a dial dialog in the smartphone in case of an emergency. In the future, an enhancement of this application can be developed to make it compatible with iOS or Apple phones.

Fig. 23.8 GPS tracker and mapping system page

Fig. 23.9 Lilypad Tawaf counter page

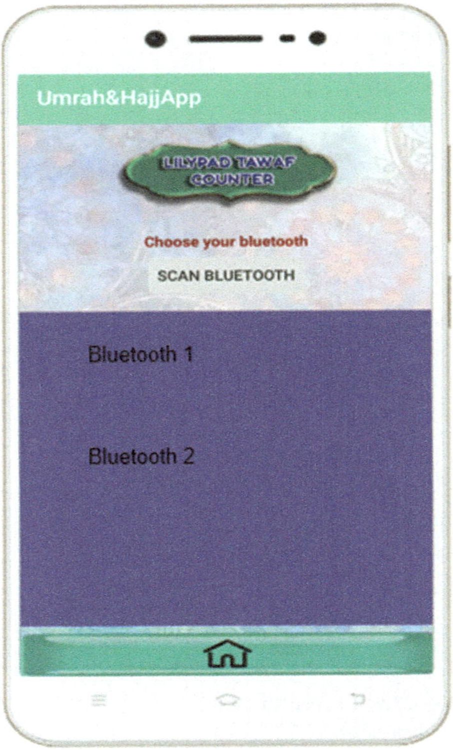

Fig. 23.10 Lilypad Tawaf counter information page

Acknowledgements I would like to express my deep appreciation and gratitude to my parents and family members for their priceless support and encouragement upon my project completion. I offer my sincerest gratitude to Dr. Mohd Azraie Mohd Azmi who has supported me throughout the process of completing this project.

References

1. El-Shahabi S (2002) The Pilgrim's guide to the Ka'bah (Allah's Sacred House). Dalil-alhaj.com. https://www.dalil-alhaj.com/en/index.htm. Accessed 5 Jan 2020
2. En.wikipedia.org (2019) Android Studio—Wikipedia. https://en.wikipedia.org/wiki/Android_Studio. Accessed 5 Jan 2020
3. Ling SCS (2018) More than half of total pilgrims already in Holy Land, one death reported. https://www.nst.com.my/news/nation/2018/08/396865/more-half-total-pilgrims-already-holy-land-one-death-reported. Accessed 5 Jan 2020
4. Puri-Mirza A (2018) Saudi Arabia: Hajj population | Statista. https://www.statista.com/statistics/617696/saudi-arabia-total-hajj-pilgrims/. Accessed 5 Jan 2020
5. Endicott S (2019) Photopea is basically Photoshop. But in your browser. And free. https://www.windowscentral.com/photopea-review-free-photoshop-alternative-works-through-your-browser. Accessed 5 Jan 2020

6. Bignox.com (2021) NoxPlayer—the perfect android emulator to play mobile games on PC. https://www.bignox.com/. Accessed 5 Jan 2020
7. En.wikipedia.org (2019) Firebase—Wikipedia. https://en.wikipedia.org/wiki/Firebase. Accessed 5 Jan 2020

Chapter 24
IoT-Based Indoor Air Quality Monitoring System Using SAMD21 ARM Cortex Processor

Zaki Khaslan, Noor Hidayah Mohd Yunus, Mohd Shahrul Mohd Nadzir, Jahariah Sampe, Noorazlina Mohamad Salih, and Kemal Maulana Alhasa

Abstract In this paper, an IoT-based technology is demonstrated to efficiently monitor indoor air quality in real-time monitored online through Wi-Fi. The idea was derived from the issues of toxic gas poisoning that cause environmental pollution, general health as well as death due to excessive inhalation of toxic gases in a closed area. This paper proposed a device of air quality monitoring system which should be able to provide an alert for the presence of toxic gases effectively via any mobile applications. The device is composed of a SAMD21 ARM cortex processor as microcontroller, pollutant detection sensors and data transmit to a Web server. The device was designed to measure a concentration of carbon monoxide (CO) and temperature-humidity for monitoring the air quality. SD card has been integrated into a Seeeduino Xiao microcontroller for analyzing and visualizing the air quality data from the device. A Web-based mobile application was developed to monitor the air quality at anytime and anywhere. To demonstrate the feasibility, the device has been tested in three closed areas which are inside a room, inside a static car with

Z. Khaslan · N. H. M. Yunus (✉)
Communication-Electronics Technology Section, Universiti Kuala Lumpur British Malaysian Institute, Batu 8, Jalan Sungai Pusu, 53100 Gombak, Selangor, Malaysia
e-mail: noorhidayahm@unikl.edu.my

Z. Khaslan
e-mail: zaki.khaslan@s.unikl.edu.my

M. S. M. Nadzir · K. M. Alhasa
Department of Earth Sciences and Environment, Faculty of Science and Technology, Universiti Kebangsaan Malaysia, 43600 Bangi, Selangor, Malaysia
e-mail: shahrulnadzir@ukm.edu.my

J. Sampe
Institute of Microengineering and Nanoelectronics, Universiti Kebangsaan Malaysia, 43600 Bangi, Selangor, Malaysia
e-mail: jahariah@ukm.edu.my

N. M. Salih
Marine Electrical Engineering Technology Section, Universiti Kuala Lumpur Malaysian Institute of Marine Engineering Technology, Lumut, Malaysia
e-mail: noorazlinams@unikl.edu.my

A. Ismail et al. (eds.), *Advanced Materials and Engineering Technologies*, Advanced Structured Materials 162,
https://doi.org/10.1007/978-3-030-92964-0_24

burning fuel energy to power the car and a kitchen area that has cooking facility. The result indicates that CO gas produced from the kitchen area when the gas stove being used shows more than 63.75% and 82.5% compared to condition from inside a car and inside a room, respectively. Additionally, the device has been successfully implemented to provide notification for community to prevent from the areas that contaminated with toxic gases.

Keywords IoT technology · Indoor air quality monitoring · Seeeduino Xiao microcontroller · Sensor · Web-based

24.1 Introduction

Air quality is a major concern in developed countries and has been associated with ecosystem health effects. The scientific community reveals that pollutants affecting air quality are of major concern in urban areas, including with presence of carbon monoxide (CO), nitrogen dioxide (NO_2), unpredictable natural mixes (VOC), secondary gas pollutants ozone (O_3), smoke and some uncommon components [1, 2]. Numerous modern housing by close arrangements in urban areas have been the challenging concern for the indoor air quality issue. Closed or indoor spaces can provide a higher potential for pollution compared to open or outdoor space conditions. The effects of indoor air pollution can reach levels 100 times higher compared to outdoor air pollution [3]. Furthermore, polluted air in indoor space has brought direct impact on human health in which the air contaminants transmitted to the lungs, causing internal diseases, dizziness and multiple sensitivities. Thus, the situation is getting worse when indoor air pollution issue is not solved because most of the modern populations now spend about 80–90% of their time indoors [4–6].

Cooking and heating foodstuffs are common routine prepared in a household. Most households rely on solid fuels energy such as liquid petroleum gas, coal and biomass for the routine cooking and heating desires [7]. In poorly ventilated households, higher levels of various toxic compounds such as CO, NO_2 and particulate matter (PM) are indeed produced from the incomplete combustion of solid fuels energy in traditional cooking stoves. In addition, too much and continuous consumption of cooking fuel by using solid fuels can cause to a decreasing fuels energy resources and lead to a chronic health problems [8].

People need a situation of the surrounding air that is free from pollutants. This is regularly significant for substantial health and environmental benefits. Air pollution can causing climate change and severe diseases. Many people are unconcerned to the severity of air pollution or in recent times recognized the issue. Majority of the population now breathes polluted air. Based on the World Health Organization (WHO) report, air pollution is the cause of death for 7 million population every year [9].

The occurrence of individuals who died due to inhalation of excess CO gas at fireplaces and during fire incidents in enclosed buildings have been recorded several times in Malaysia. The impact of indoor air pollution is not limited to the incidents in enclosed buildings. In addition, there are also some cases of individuals drowned in the car while sleeping and has caused death due to inhalation of the CO gas that has entered the car from the burning fuel energy to power the car [10].

CO consists of one atom carbon and one atom oxygen which is a colorless and odorless less dense than air. CO is an omnipresent element of fire atmospheres produced when organic material is burned in an insufficient source of oxygen. In high concentration levels, CO gas is toxic to humans and animals as it induces disorders in the blood from the obstructing of the oxygen supply which potentially causing sudden death [11, 12].

To overcome the issues of toxic gas poisoning in a closed area, the development of indoor air quality monitoring system which is capable in providing notification updates on the air quality levels is very important. Therefore, an Internet of things (IoT)-based indoor air quality monitoring with low-cost energy by using SAMD21 ARM cortex processor based on integration of Web server is presented in this paper. IoT system is associated with the embedded air quality sensors, microcontroller unit (MCU) and electronically software frameworks to give correspondence data and information exchange. Furthermore, the data outcomes can be utilized by enhancing communication technology system to provide instant updates besides to monitor the impact of indoor air quality levels in the particular areas.

24.2 Methodology

The most important factor for an indoor air quality monitoring system is the accuracy to detect air quality platform. The primitive design of the smart air concept was motivated through the development by [1] that is a low-cost air quality sensor (LAQS) system known as DiracSense for surface O_3, NO_2 and CO measurement. Raspberry Pi-3 on the single board computer was the programmable system for the required air quality levels monitoring operation. The air quality levels were sent using IoT setup to the Web server such as Google drive or Dropbox for detailed monitoring.

The designs, communication technologies and hardware development must be determined in depth to handle data updates associated with air quality levels. Here, an IoT-based application was implemented remotely to monitor and get the notification updates for the air quality detection. Figure 24.1 describes the block diagram of the IoT-based indoor air quality monitoring system of this paper.

In this paper, the monitoring system includes several sensors, communication connectivity and MCU programmable by Seeeduino Xiao as the main part of the indoor air quality monitoring system. The data collected is further stored into a data storage system. Furthermore, the data is analyze and employed on data visualization for the end users to receive information updates of the indoor air quality levels. Figure 24.2 describes the flowchart functionality insights into the system.

Figure 24.3 shows the circuit diagram of the indoor air quality monitoring system. Arduino Uno libraries could be used in the diagram instead of Seeeduino Xiao

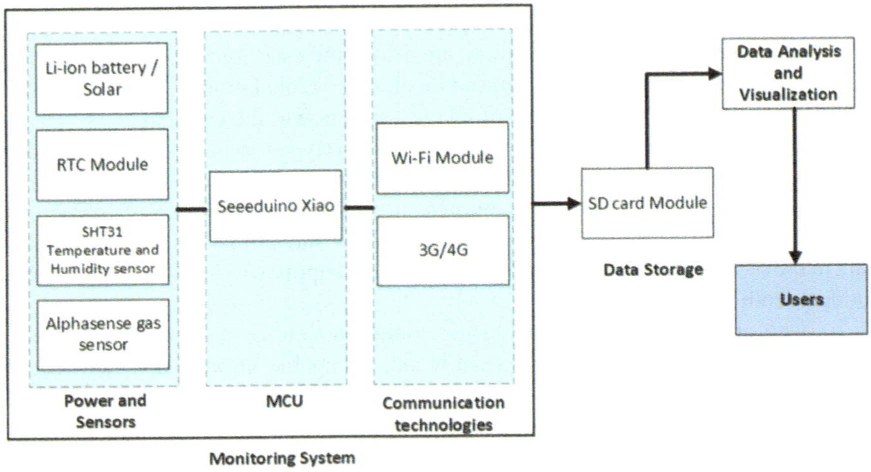

Fig. 24.1 Block diagram of the indoor air quality monitoring system

libraries for the unavailable module pin. In this paper, a low power microcontroller embedded with SAMD 21 ARM processor was selected. The core of the microcontroller is the high performance in processing with less operating power.

The reliability data measurements for temperature, humidity and CO concentration from the sensors during the monitoring application were observed. The data such as date, day, time, temperature in degree Celsius (°C), percentage humidity and CO concentration in parts per billion (ppb) were extracted from the SD card before display in Web server.

24.3 Results and Discussion

The paper presented a development and implementation of a monitoring indoor air quality system based on an Arduino compatible board in the Seeeduino Platform. Figure 24.4 shows the indoor air quality monitoring system prototype which is placed inside a covering box. The covering box needs to have a bit of holes to allow the surrounding air to enter inside the box. To test the sensors capabilities which are the SHT31 temperature-humidity sensor and Alphasense CO-A4 gas sensor, the air quality monitoring box has been placed in three different places which are; (1) inside a room, (2) inside a static car while the car's engine is turned on and (3) at the kitchen area that has cooking stove fume. The experiment was conducted in the three different particular areas during the day for around 2 h.

The data collected from the sensors used in the air quality monitoring system are shown in Figs. 24.5, 24.6, and 24.7. The average data results for CO concentration from inside a room is 1409.4254 ppb for the average temperature and humidity of

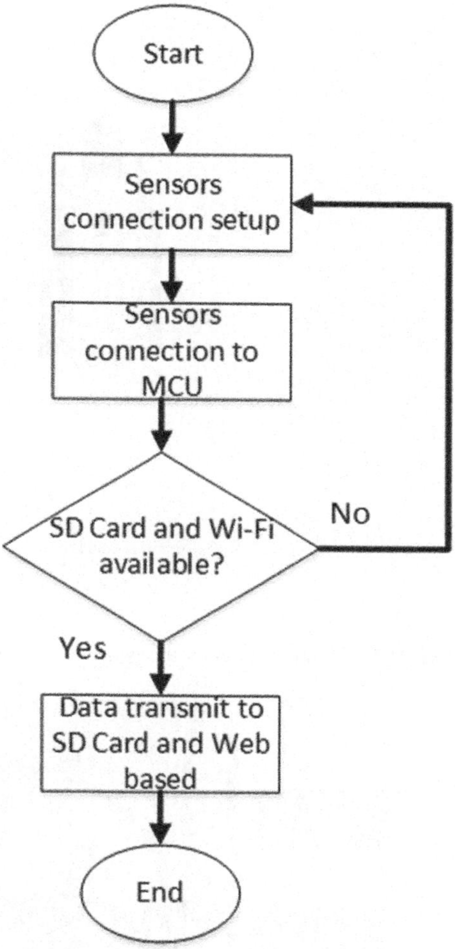

Fig. 24.2 Flow chart of system

30.96 °C and 84.0315%, respectively. The results show for the normal data of the surrounding conditions.

The average data results for CO concentration from inside a static car while the car's engine was turned on indicates 2901.03 ppb for the average temperature and humidity of 28.01 °C and 44.4038%, respectively. The average temperature is low because from the air conditioner that has been turned on inside car besides causing a slight more reduction of humidity percentage compared to the condition inside a room. However, the CO concentration from inside the static car indicates 51.42% higher CO concentration compared to the condition from inside a room.

The average data results for CO concentration from the kitchen area that has cooking stove fume is 8000.44 ppb for the average temperature and humidity of 29.44 °C and 82.0285%, respectively. The average temperature and humidity

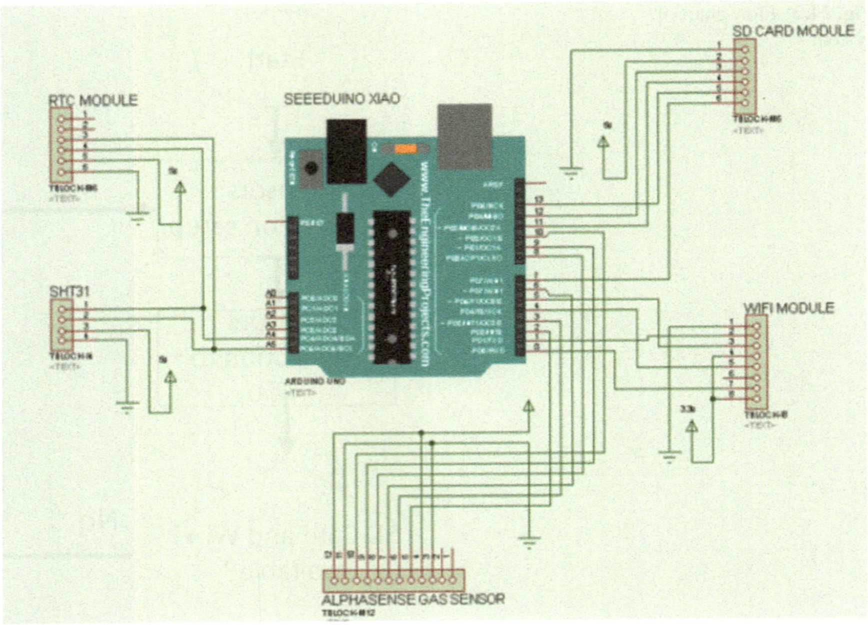

Fig. 24.3 Circuit diagram for the system

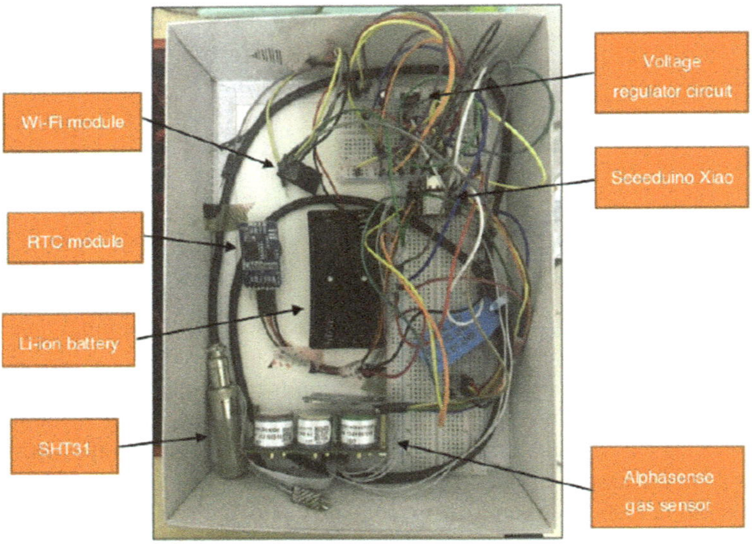

Fig. 24.4 Indoor air quality monitoring system prototype

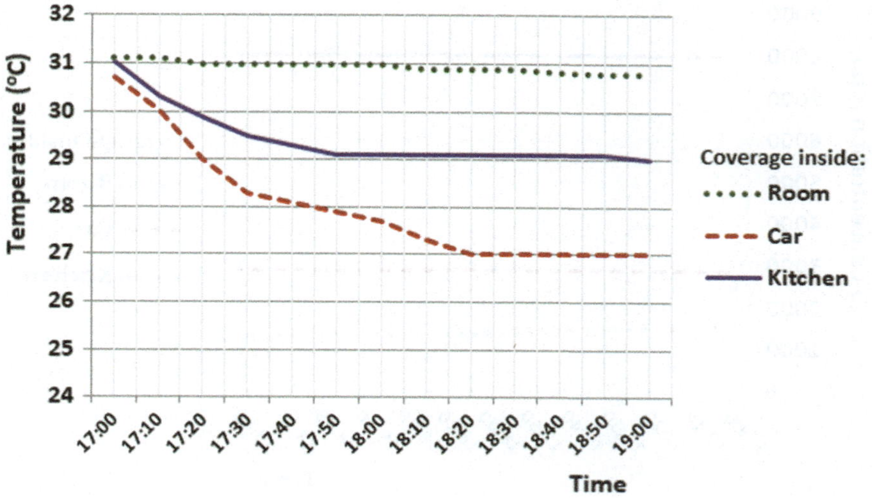

Fig. 24.5 Test results for temperature sensor

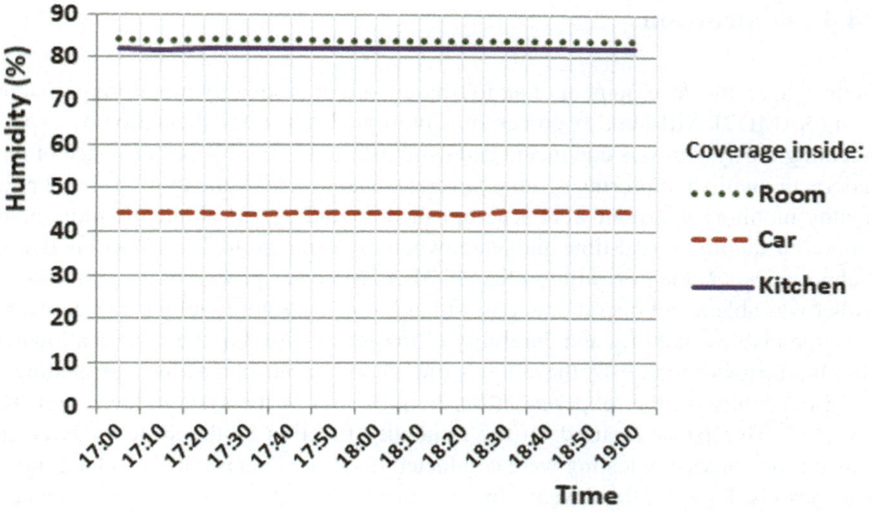

Fig. 24.6 Test results for humidity sensor

percentage show a value approximately similar to the condition inside a room. However, the CO concentration from the kitchen area indicates the highest CO concentration compared to the other conditions from inside the car and inside a room.

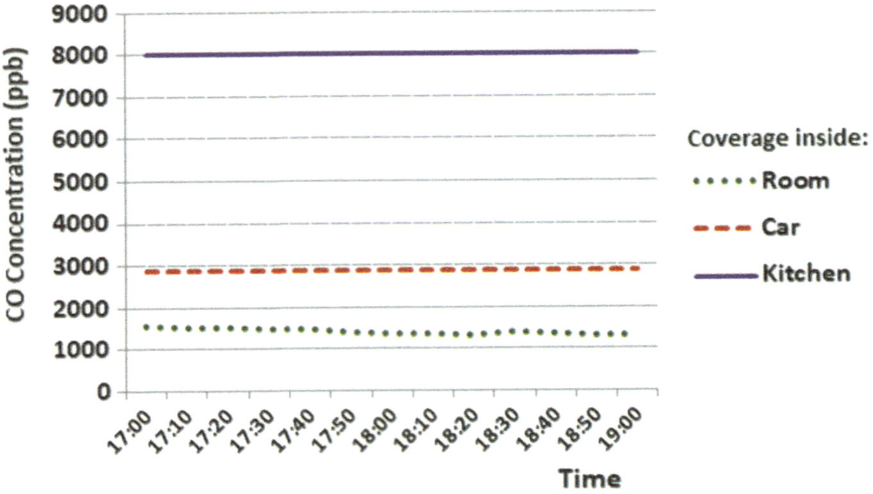

Fig. 24.7 Test results for CO concentration sensor

24.4 Conclusion

In this paper, the development of an IoT-based indoor air quality monitoring system using SAMD21 ARM cortex processor is presented. The implementation by experimenting the system was conducted and demonstrated a suitable performance of the indoor air quality monitoring system. Several valuable achievements of the indoor air quality monitoring system were accomplished, including; (1) efficiently monitored indoor air quality in real-time and at anywhere by using an IoT and Wi-Fi network technologies, (2) the system used SAMD21 ARM cortex processor as a microcontroller was able to cut operational cost of the power consumption by the system and (3) expandable reliability and durability of the sensors used in the system allowing easy platform installation by the user to various appropriate monitoring surroundings.

In the future, further experiment on the precision of the system device will be involved. This paper focused on the reliability function of the system device in which more necessary testing were conducted to ensure data accuracy in long duration periods. Besides, the device can be future improved by connecting automated ventilation system in which the system can operates when detecting the presence of polluted air in surroundings.

Acknowledgements The authors wish to thank Earth Sciences and Environment research team of UKM for granting permission to conduct this project. We also like to thank UniKL and UKM for the support given in facilitating and contributing to the success of this project.

References

1. Alhasa KM, Mohd NMS, Olalekan P et al (2018) Calibration model of a low-cost air quality sensor using an adaptive neuro-fuzzy inference system. Sensors 18(12):4380–4400
2. Gulia S, Nagendra SS, Khare M et al (2015) Urban air quality management—a review. Atmos Pollut Res 6(2):286–304
3. Ma N, Aviv D, Guo H et al (2021) Measuring the right factors: a review of variables and models for thermal comfort and indoor air quality. Renew Sustain Energy Rev 135:110436–110453
4. Phongphetkul P, Mangkang S, Praditsmanont A et al (2021) Evaluation of indoor air quality in high-rise residential buildings in Bangkok and factor analysis. Environ Monit Assess 193(1):1–11
5. Saini J, Dutta M, Marques G (2021) Sensors for indoor air quality monitoring and assessment through Internet of Things: a systematic review. Environ Monit Assess 193(2):1–32
6. Canha N, Teixeira C, Figueira M et al (2021) How is indoor air quality during sleep? A review of field studies. Atmosphere 12(1):110–131
7. Gould CF, Schlesinger SB, Molina E et al (2020) Household fuel mixes in peri-urban and rural Ecuador: explaining the context of LPG, patterns of continued firewood use, and the challenges of induction cooking. Energy Policy 136:111053
8. Yunus NHM, Sampe J, Yunas J et al (2018) Comparative study of Si based micromachined patch antenna operating at 5 GHz for RF energy harvester. In: 2018 IEEE international conference on semiconductor electronics (ICSE), pp 238–241
9. World Health Organization (2018) Air pollution and child health: prescribing clean air. https://www.who.int/publications/i/item/air-pollution-and-child-health. Accessed 9 Sept 2021
10. Valente-Aguiar MS, Magalhães T, Dinis-Oliveira RJ (2019) Suicide by inhalation of carbon monoxide of car exhausts fumes. Curr Drug Res Rev 11(2):145–147
11. Manisalidis I, Stavropoulou E, Stavropoulos A et al (2020) Environmental and health impacts of air pollution: a review. Front Public Health 8:14
12. Jo J, Jo B, Kim J et al (2020) Development of an IoT-based indoor air quality monitoring platform. J Sens 1–14

Chapter 25
The Development of a Fish Counting Monitoring System Using Image Processing

Abdul Muaz Abdul Aziz, Izanoordina Ahmad, Siti Marwangi Mohamad Maharum, and Zuhanis Mansor

Abstract The ability to accurately quantify the survival of fish in culture systems has been a challenge for the fish-farming aquaculture industry for many years. The fish hatcheries in many developing countries still rely on manual and volumetric counting to enumerate the post-larvae (PL) for sale to customers and used a lot of manpower whilst performing the task. Thus, this research reviews on the image processing technology in aquaculture that could be promoted in fish farming and hatcheries for the application of monitoring purposes. The fish counting monitoring system used the Arduino module to detect the post-larvae and capture the image through the IR sensor module. Arduino's information will be displayed on the LCD screen in which the farmer could monitor the fish counting via the system. MATLAB processes the captured image by implementing the image processing system. It is expected that the system will be able to count the post-larvae (PLs) with small mean absolute error, for large and small PLs, and the system has the potential for faster counting as compared to the conventional methods. Furthermore, it will open up possibilities for counting other small aquatic animals such as shrimp, crab, oysters in their nursery process. As for the recommendation, a colour sensor could be applied in the system to improve the readability on the LCD display screen.

Keywords Artificial intelligence · Image processing · Aquaculture · Fish hatcheries · Post-larvae

---

A. M. A. Aziz · I. Ahmad (✉) · S. M. M. Maharum
R4R Research Cluster, Electronics Technology Section, Universiti Kuala Lumpur British Malaysian Institute, Batu 8, Jalan Sg Pusu, 53100 Gombak, Selangor, Malaysia
e-mail: izanoordina@unikl.edu.my

A. M. A. Aziz
e-mail: a.muaz.aziz@s.unikl.edu.my

S. M. M. Maharum
e-mail: sitimarwangi@unikl.edu.my

Z. Mansor
Communication Technology Section, Universiti Kuala Lumpur British Malaysian Institute, Batu 8, Jalan Sg Pusu, 53100 Gombak, Selangor, Malaysia
e-mail: zuhanis@unikl.edu.my

A. Ismail et al. (eds.), *Advanced Materials and Engineering Technologies*, Advanced Structured Materials 162,
https://doi.org/10.1007/978-3-030-92964-0_25

25.1 Introduction

Malaysia has 4055 km of coastline, 1640 km of which are in Peninsular Malaysia, and 2415 km of which are in Sabah and Sarawak. With the creation of a 200-mile exclusive economic zone (EEZ), Malaysia has spent a total fishing area of 160,000 square nautical miles. In spite of this large fishing sector, fishing is an important sector in the Malaysian economy. In 2003, 1.5 million mt of fish priced at approximately RM5 B was produced by the industry.

The production of marine fisheries at RM4 B was estimated at 1.3 million mt, representing 1.4% of gross domestic product (GDP). Aquaculture production was estimated at 196,874 mt above RM1.2 B, representing just 13 per cent of the overall production of fisheries. Though cockles (*Anadara granosa*) are the dominant crop in aquaculture, 37% of total aquaculture production is accounted for. Approximately, 89,400 fishermen and 21,100 aquaculture employees are working in the fishery sector, with 110,500 people in total. Traditionally, both in terms of growth and socio-economic considerations, Malaysian fisheries' principal or backbone is the inshore subsector. Nonetheless the inshore sub-sector has entered a saturation point, as seen by the increasing capture rate in recent years. This is coupled with the significant fishery opportunities in Malaysia's EEZ waters and the country's large potential for aquaculture production [1].

Fish farming is one of the important industry that anticipates tremendous growth in the coming years. New technology or method to increase fish production has been integrated commercially into the world's best practise on-farm management production systems to overcome most of the regulatory hurdles that have limited local development. The farmers, however, use many ways to count the number of fish, but it obviously took some time, and some methods are unreliable.

A traditional method is used to count the number of fishes, catch and weigh multiple fishes, estimating the total number of fishes in the pond. Thus, a new mechanism is needed to help the working environment and ease the farmer's day-to-day work.

Monitoring the production of fish population using image processing is one of the latest approaches proposed to reduce the inaccuracy of counting the number of fishes in a field. The system will reduce the time for the customer to count the number of fishes which in turn, will have a positive impact on the farmer, such as lowering their stress levels whilst raising productivity.

The research for fingerling counting algorithm also utilising the digital picture technique. To achieve this aim, a robust segmentation method, feature extraction approach and machine learning algorithm for fingerling classification and counting are formulated. At the completion of this research, the proposed method is predicted to count different sizes of fingerlings with great accuracy [2].

A catfish recognition and counting method using detection before counting consist of six steps which are picture capture, pre-processing, segmentation, feature extraction, recognition and counting. The images were captured and pre-processed, whilst the segmentation was conducted using picture binarization using Otsu thresholding, morphological operations utilising fill hole, dilation and opening operations and boundary segmentation using edge detection. The boundary features are used to train an artificial neural network (ANN) to conduct the recognition. The new counting approach was employed to determine the number of fish of any size. The proposed counting algorithm achieved 100% accuracy [3].

The combination of image processing techniques consists of image enhancement, edge detection and thresholding process are implemented to count the Nile Tilapia (*Oreochromis niloticus*). The samples data set is divided into three parts which are small data set, medium data set and large data set. The performance of the proposed method and the manual approach method is compared based on the number of fish. The results of the experiment show that the proposed method is outperformed compared to the manual counting due to the number of fish larvae measurement is similar [4].

The techniques based on machine learning to measure the number of shrimp larvae and juvenile on increasing population are proposed in [5]. Numerous image processing techniques were applied such as thresholding process, edge detection and morphological operator.

Due to the widely used image processing in developing the fish larvae counting system that gives high potential in fisheries area, therefore, this research aims to create the image processing system for counting the number of fish to reduce the time consumption and manpower utilisation in counting the number of fish. Therefore, the key scope is to create a prototype of a fish counting method that can be applied in many fish-farming areas to obtain the data and the results. This device is capable of detecting the number of fish in a given area using two infrared sensors mounted on the tube where the sensors can detect the presence of fish flowing through it and then use an image processing method to compare the number of fish. Therefore, this project could reduce the problem of time consumption caused by the abundance of fish that traditionally need to be counted, which consumes a lot of energy and time.

25.2 Methodology

The entire system consists of two IR sensors as the system input. The camera and seven-segment display are the system output. All of the input and output are interfaced to the Arduino as shown in Fig. 25.1.

The IR sensor module is mounted at the centre of the prototype is used to detect the presence of the fish through the transparent tube. The camera is used to capture the image of the object. Thus, the captured image could be further analysed to identify the correct image of the fish using a MATLAB software. The LCD display is used to display the quantity of the captured image of the fish that passes through the tube.

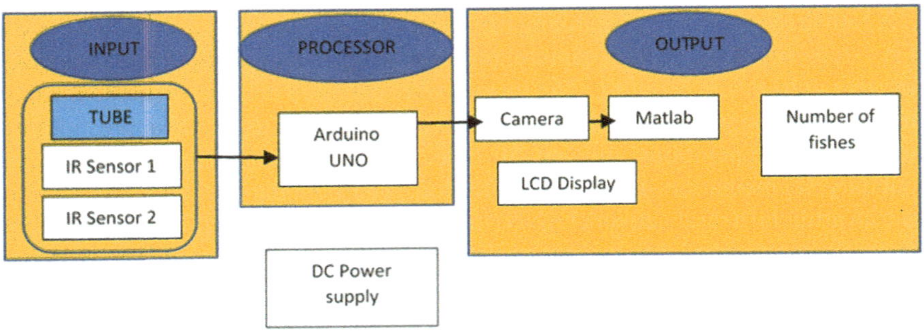

Fig. 25.1 Block diagram of the development of a fish counting monitoring system using image processing

The overall project block diagram as shown in Fig. 25.1 consists of several parts such as the sensor for the IR module, Arduino Uno, ring light, LCD display, tube, container and the camera.

The flowchart in Fig. 25.2 shows the whole process of the system starting from initialisation of the system until the process of image processing using MATLAB.

The IR sensor module used in this project is shown in Fig. 25.3. This sensor is able to detect the object's presence accurately. The time gap in coding that is relevant to 1 s in real time to detect the presence of the object is delayed by 1000. These sensors work well during testing, and the presence of fish moving through the tube can be detected without causing any errors on the LCD display. Three types of experiments have been conducted to assess this sensor's capabilities and to detect the presence of the object that passes through it. The following are the list of experiments on this sensor that have been carried out.

These are the three experiments that have been carried out.

1. To test the ability to detect the present of the fish that are passing through the tube when the time of the sensor detection is delayed for 0.5 s.
2. To test the ability to detect the present of the fish that are passing through the tube when the time of the sensor detection is delayed for 1 s.
3. To test the ability to detect the present of the fish that are passing through the tube when the time of the sensor detection is delayed for 2 s.

25.3 Results and Discussion

Table 25.1 shows the sensor's working state. The sensor time delay is set at 0.5 s, 1 s and 2 s, respectively. During the test, the number of fish is measured, and the sensitivity is being identified. The result shows that when the delayed period is set to 0.5 s, 1 s and 2 s, the exact number of fish detected on the LCD display is 10, 5 and 3, respectively. With a delay of 1000, which is equal to 1 s, this IR sensor module can detect the presence of any object that passes through the tube. Then, the LCD

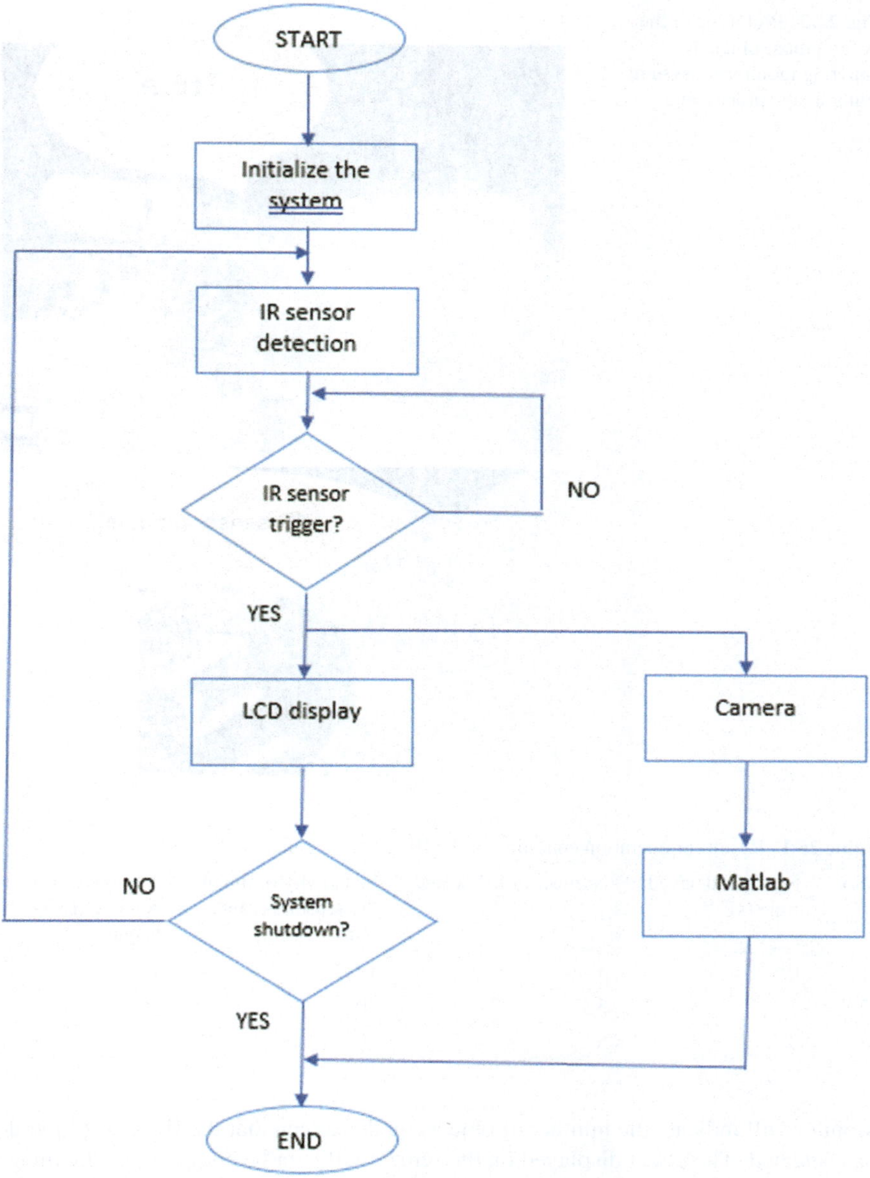

Fig. 25.2 Flowchart of the development of a fish counting monitoring system using image processing

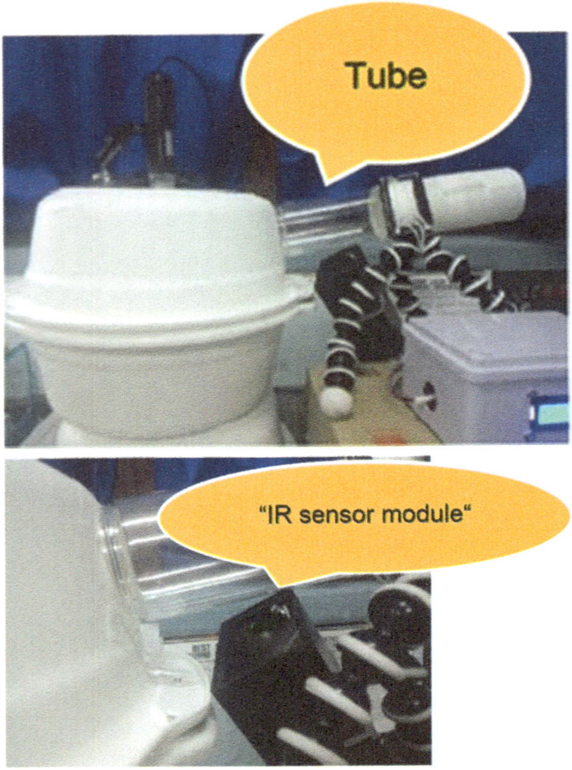

Fig. 25.3 Prototype of the development of a fish counting monitoring system using image processing

Table 25.1 Result for operation condition of the IR sensor module

No.	Sensor's delayed time (s)	Number of fish tested	Sensitivity of the of the sensor's range (cm)	Number of fish counted (LCD display)
1	0.5	5	10	10
2	1	5	10	5
3	2	5	10	3

monitor will indicate the number of objects on the screen that the IR sensor module has detected. The effect displayed on the screen will later be compared to the image processing outcome.

The image processing method can be seen in Fig. 25.4. The camera, via the laptop, transfers the image to MATLAB. The original image that is in RGB image will be received by MATLAB and transformed into greyscale image. The last step is to transform the greyscale frame to binary image. The background becomes black and the fish in the foreground becomes white as the number is equal to the number of white images in the final step of image processing, which is the binary image.

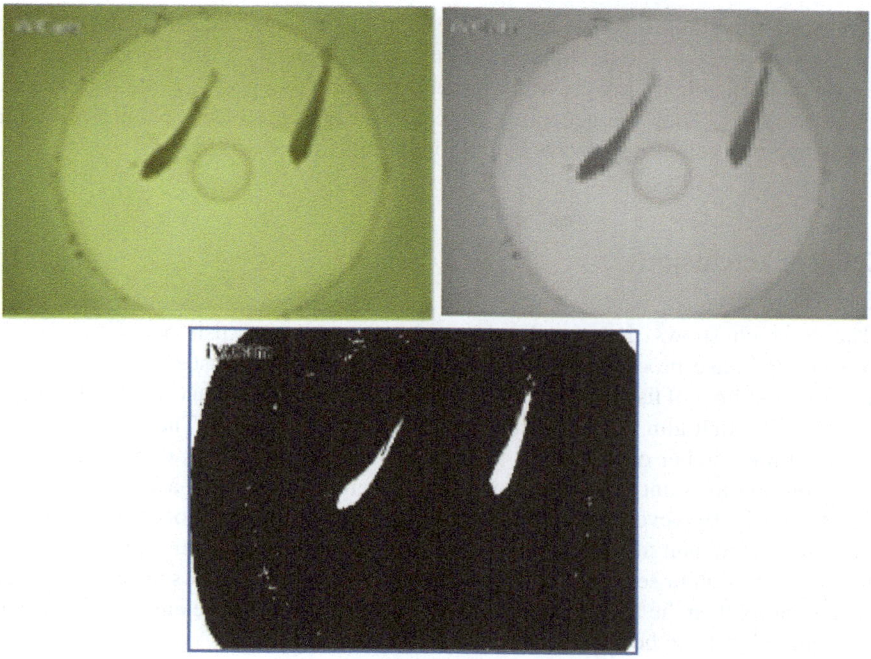

Fig. 25.4 Result in greyscale and binary image

The camera, via the laptop, transferred the image to the MATLAB. The original image that is RGB image will be received by MATLAB and transformed into the greyscale image. Greyscale is a monochromatic array of colours ranging from black to white. With multiple picture editing applications, it could convert a colour image to black and white or greyscale. This method extracts all colour details and leaves only the luminance of each pixel. The optical pictures show a mixture of red, green and blue colours in each pixel, which has three distinct values. Consequently, these three values should be merged into a single value when extracting the colour from an image. In order to differentiate between a picture object and the background, binary photos are often created by keeping down a greyscale or colour snapshot. The last step is to transform the greyscale frame to binary image. The colour of the entity which is usually white is considered as the first colour. The rest (normally black) is referred as the base colour. However, depending on the image to be thresholded, this polarity may be reversed, in which case the object appears with 0, and the background has a value non zero.

Figure 25.4 shows the images on the MATLAB that appeared after pressing the 'Run' button. Clearly, the greyscale image on the left is one in which the only shades of grey are the paints. The reason for distinguishing those images from some other kind of colour image is that fewer data has to be generated for each pixel. In fact, a grey colour is one in which all the red, green and blue components have equal RGB space intensity; thus, it is only appropriate to specify a single intensity value for each pixel, as opposed to the three intensities required to specify a full colour image for each pixel. The binary image, whose pixels only have two possible intensity values, is on the right. They are generally depicted as white and black. Numerically, for black, the two values are always 0, and for white, either 1 or 255.

25.4 Conclusion

The research shows that the development of monitoring fish population using MATLAB image processing is successfully achieved. This system can effectively count the number of fishes. There are three primary objectives that drive this project's growth. The first aim is to build the image processing system. The second goal is to count the number of fishes in a given area, and the last goal is to minimise time consumption to count the number of fishes. Overall, this project accomplished the desired goals. However, there is a constraint with the counting process using the IR module sensor, as it also detects the number of foreign items. This can be done by replacing the colour sensor with the IR module sensor. The colour sensor has greater functionality than the IR sensor. Therefore, in future, all issues related to identifying a foreign object can be solved using a colour sensor.

Acknowledgements The author would like to thank University Kuala Lumpur British Malaysian Institute and R4R Research Cluster for the provision of laboratory facilities and financial support.

References

1. Hashim M, Kathamuthu S (2005) Shrimp farming in Malaysia. Shrimp News Int 21:6–8
2. Aliyu I, Gana KJ, Musa AA et al (2017) A proposed fish counting algorithm using digital image processing technique. J Sci Educ Technol 5:9–15
3. Aliyu I, Gana KJ, Musa AA et al (2020) Incorporating recognition in catfish counting algorithm using artificial neural network and geometry. KSII T Internet Info 14:4866–4888
4. Awalludin EA, Wan MWNA, Arsad TNT et al (2020) Fish larvae counting system using image processing techniques. J Phys Conf Ser. https://doi.org/10.1088/1742-6596/1529/5/052040
5. Raman V, Perumal S, Navaratnam S et al (2016) Computer assisted counter system for larvae and juvenile fish in Malaysian fishing hatcheries by machine learning approach. J Comput 11:423–431

Chapter 26
NodeMCU Based Monitoring System for Individuals with Covid-19 Self-quarantined Order

Siti Marwangi Mohamad Maharum, Mohammad Zharif Mohd Adenan, Izanoordina Ahmad, and Zuhanis Mansor

Abstract One of the factors that contributes to the increased number of Covid-19 cases in Malaysia is according to the authorities that Covid-19 patients, persons under investigation (PUI) and persons under surveillance (PUS) had deliberately left their home quarantine without valid reason and permission. There were irresponsible people who act selfishly by cutting their quarantine wrist tags provided by the Ministry of Health (MOH) and casually left their quarantine zone. The authorities are having a hard time to track down these irresponsible people especially among those who are ordered for self-quarantined at homes. These challenges had inspired for this work to propose a monitoring system to notify and remind respective individuals if they left the permissible area. The developed system could also update the authorities when these individuals continuously defy orders to return to their quarantine area. Through prototype testing, the proposed system offers an acceptable battery consumption since its tracking device consumed an average 150 mAh for a whole day operation. It is anticipated that this system could be an effective technology tool to ensure the quarantine session is undergone perfectly by these individuals and thus, combating the spread of deadly Covid-19 virus in the country.

Keywords Covid-19 · Self-quarantined · Contagious disease control · Internet of things (IoT)

S. M. M. Maharum (✉) · M. Z. M. Adenan · I. Ahmad
Electronics Technology Section, Universiti Kuala Lumpur British Malaysian Institute, Selangor, Malaysia
e-mail: sitimarwangi@unikl.edu.my

M. Z. M. Adenan
e-mail: mzharif.adenan@s.unikl.edu.my

I. Ahmad
e-mail: izanoordina@unikl.edu.my

Z. Mansor
Communication Technology Section, Universiti Kuala Lumpur British Malaysian Institute, Selangor, Malaysia
e-mail: zuhanis@unikl.edu.my

A. Ismail et al. (eds.), *Advanced Materials and Engineering Technologies*,
Advanced Structured Materials 162,
https://doi.org/10.1007/978-3-030-92964-0_26

26.1 Introduction

An initiative called a movement control order (MCO) had been implemented by the Malaysian government on all 14 states starting on 18th March 2020 to restrain the contagious and deadly Covid-19 virus from spreading. During the early MCO implementation, Malaysia is seen to successfully control the active cases reported daily. Thus, the government started to allow several economics sectors and social activities to be operated as usual in several stages through implementations of conditional movement control order (CMCO) and recovery movement control order (RMCO). Unfortunately, the active Covid-19 cases reported daily started to increase again starting from September 2020 with reported cases entering 4-digit values. Since then, Malaysia is struggling to flatten the reported statistics and sadly, the highest statistics were reported on 29th May 2021 with 9020 cases [1]. Many factors contribute to the increasing number of Covid-19 cases in Malaysia, as discussed in [2]. Due to a shortage of medical facilities at the available hospitals and quarantine centers, Ministry of Health (MOH) Malaysia could no longer isolate Covid-19 patients with no symptoms shown at hospitals nor quarantine centers for meticulous quarantine observation.

Alternatively, MOH will perform a thorough assessment to all infected individuals and decide who are eligible for self-quarantined at their homes only. Once a person is diagnosed with Covid-19, their traveling history will be investigated for close contact tracing. According to MOH, there are two types of close contacts namely persons under investigation (PUI) and persons under surveillance (PUS). PUI is referring to symptomatic close contact who will be screened for Covid-19. Conversely, PUS refers to asymptomatic close contact who will not be screened for Covid-19 but is required for self-quarantined until the PUI Covid-19 test result is released. Note that PUI or PUS who are tested positive with Covid-19 will be brought to hospitals or quarantine center for observations if their health condition deteriorates. However, those who obtained negative results during the first screening are still compulsory to undergo 14 days home-quarantined as they will need to perform the second screening on the 13th day. PUI or PUS will be released from self-quarantined order if both screening tests are confirmed to be negative.

Undergoing home-quarantined order may sound easy. Yet, it requires individual's strong will, upright attitude and accountability to perform several guidelines set by MOH without being monitored by them. Sadly, polices and MOH had testified several cases involving people disobeying the home quarantine order. Offenders were traveling in public with quarantine tags still on their wrists. There were also offenders who cut and removed their quarantine tags before traveling elsewhere. For example, MOH had confirmed that 63 quarantine wrist tags were found without proper disposal and thus, police need to track the wrist tag owner as they might be infecting others with this contagious virus [3]. Based on the highlighted issues, there is a need for an electronic tracking device to be fitted on Covid-19 patients, PUI and PUS to monitor their movements if they break the quarantine rules [4, 5].

In [6], a telemedicine system was developed based on a WeChat application to monitor the health progress of home-quarantined patients. The patients will need to self-assess their conditions and update them to a multidisciplinary team stored on a cloud service. The self-assess update is very useful for medical staff identifying the disease progression for appropriate and timely treatment decisions. On the contrary, a geofencing technology-based was developed in [7] with an integration of smartphone application in the iPhone Operating Software (iOS). This system is able to notify the authorities if the quarantine wristband is cut or removed by the wearer. Users are given 15 s to validate their current location if they leave their quarantine zone. Aside from that, a wearable bracelet prototype was developed in [8] to monitor a patient's blood oxygenation and body temperature. The recorded data is connected to geo-localization in which the sanitary authorities will easily identify close contact of the wearer.

Although the telemedicine system in [6] is hassle-free as it requires only the smartphone application, it may not effectively trace the locality of a person with Covid-19 self-quarantined order mainly when the patients left their smartphone and travel to somewhere else. Nevertheless, the system proposed in [7] will drain the user's smartphone battery consumption for its operation since it requires the integration of WiFi, Bluetooth and global positioning system (GPS) assistance to identify the wearer's locality. Moreover, only iOS smartphones can be supported in which this approach may not cover other smartphone users. Having said that, the close contact tracing method proposed in [8] may not be effective at all if the person with Covid-19 light symptoms purposely left their wearable bracelet when outing. As of now, Malaysian authorities had made it compulsory for all its citizens and those staying in the country to use *MySejahtera* application or manual details writing upon entering any premises. Thus, close contact tracing for newly infected people could be easily identified by the authorities.

Based on the above discussion, there is still room for improvement in having an effective and efficient monitoring system. In this work, a prototype of monitoring system is proposed to offer an acceptable battery consumption throughout the quarantine period and straightforward system deployment for individuals with Covid-19 self-quarantined order at their homes.

26.2 System Description

In this section, the proposed monitoring system is described in detail. It consists of two devices called parental device (refer Fig. 26.1) and tracker device (refer Fig. 26.2a). The parental device is used to transmit and receive data to and from the tracker device to ensure the tracker device is always connected to WiFi and in online mode. Basically, the parental device will be accountable to alert the users if they leave the permissible self-quarantined area. It will also notify the authorities if they fail to return to the quarantine area or purposely remove the tracker device from their wrist. Meanwhile, the tracker device will update the latest user's locality

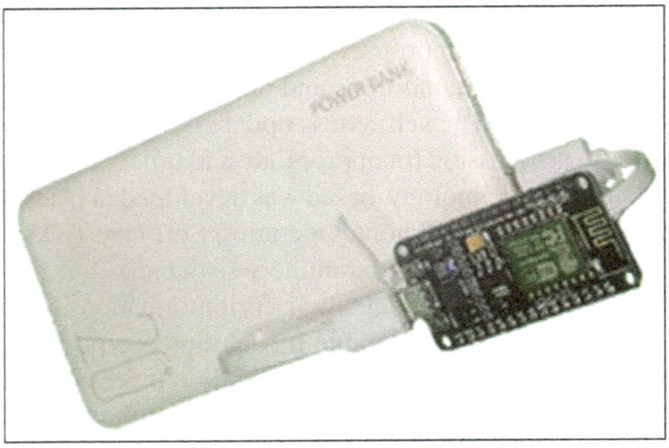

Fig. 26.1 Parental device of the proposed monitoring system

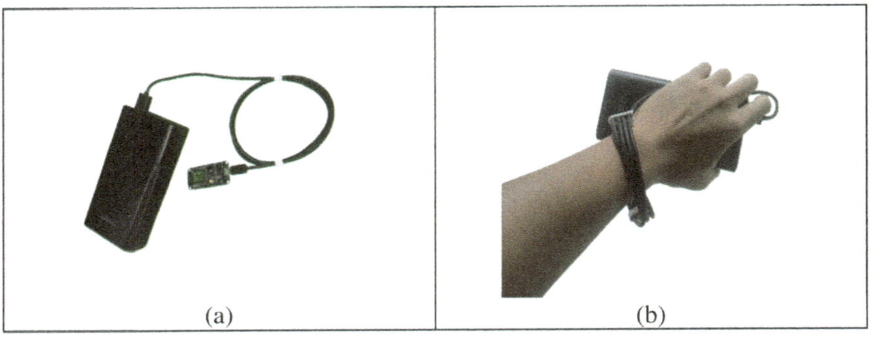

Fig. 26.2 **a** Tracker device of the proposed monitoring system **b** when it is placed on the user's wrist

information throughout the self-quarantined order (i.e., 14 days). The tracker device must be worn on the user's wrist, as portrayed in Fig. 26.2b.

This proposed monitoring system requires two units of NodeMCU ESP8266 (each for the respective parental and tracker devices), user's existing smartphone and Google Firebase. The NodeMCU at the tracker device is programmed to retrieve the user's latest location from the built-in GPS module in the user's smartphone. An open-source platform called Blynk mobile application in the user's smartphone is used to send the acquired location information from the built-in GPS module to the NodeMCU microcontroller in the tracker device. The latest locality information provided by the Blynk application helps the microcontroller in the tracker device to perform (i) the distance calculation from the quarantine zone and (ii) make a declaration on the user's locality status. These two pieces of information will be updated in Google Firebase.

The Google Firebase serves as the real-time database to keep the user's records reported from the tracker device. Thus, the NodeMCU microcontroller in the parental device will track the reported information in Google Firebase to (i) send an alert/reminder messages to the users if they are about to leave the permissible self-quarantined area and (ii) to notify the authorities if the users fail to return to the quarantine zone. Note that, the quarantine zone (i.e., user's home location) can only be set/registered by the authorities in the Google Firebase to avoid the user from changing their quarantine zone details. The system operational flow is described as follows.

Upon system initialization, the parental device transmits a signal to the tracker device until the tracker device is confirmed to receive the signal. When the tracker device receives the transmitted signal from the parental device, it updates the user's current distance from the quarantine zone and locality status onto the Google Firebase. The parental device will continuously monitor such information updated in the Google Firebase to identify if any person with Covid-19 self-quarantined order has breached the quarantine procedures. In this proposed system, Telegram application is used as a medium for users to receive automatic reminder messages on their locality status.

There are three locality statuses in this monitoring system: green zone, yellow zone and red zone. A user is declared to be in a green zone when they are located within 10 m from the quarantine point set by the authorities in the Google Firebase. Nevertheless, a user will be declared to be in a yellow zone if they are located within 20 m but exceeding 10 m from the quarantine area. Red zone will be updated as user's locality status if they travel beyond 20 m from the quarantine area. Reminder messages will only be sent to the users if they are declared in yellow and red zones.

Besides monitoring the user's locality status, the parental device could also track if the tracker device experiences network disconnection or system damage. For actual system implementation, the tracker device will be designed as an electronic wristband and will be fitted according to the user's wrist size. Hence, the wristband will not be easily removed by adding a system detection on trials made to remove the electronic wristband. However, if the wristband is cut, it is impossible to reconnect back due to wiring damage. Therefore, the users could not stop the parental device from submitting several reminder messages to put on the wristband back. Consequently, the authorities will know that there is an event where the user totally removed the wristband.

The system operation will continuously perform the reporting of user's locality status in real-time, although no quarantine violation is detected. Therefore, the system shall be designed with an acceptable battery consumption to ensure the tracker device withstands 14 days of operation without recharging it. Recall that Figs. 26.1 and 26.2 show the proposed system prototypes in which both parental and tracker devices seem bulky due to the powerbanks capacities are 20,000 mAh and 30,000 mAh, respectively. These huge powerbank capacities are purposely chosen for prototype functionality testing to evaluate the battery discharging rate when operating the devices. It is observed that the powerbank battery used on the tracker device depleted only 1% from a fully charged 30,000 mAh in two days. Hence, Eqs. (26.1) till (26.3) show that the tracker device requires at least 2100 mAh supply to withstand 14 days of operation.

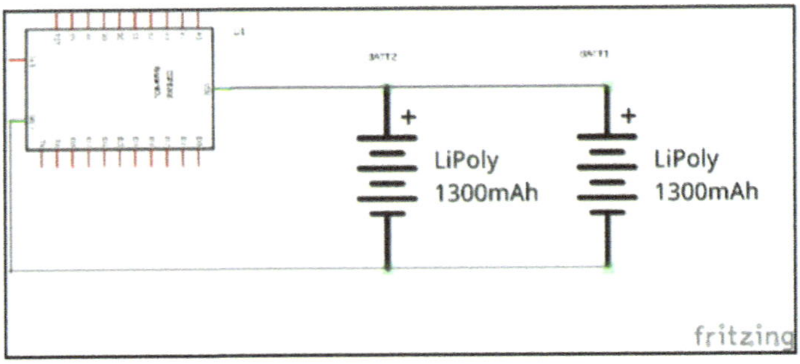

Fig. 26.3 Circuitry diagram for battery supply on the proposed system

$$\frac{1}{100} \times 30{,}000\,\text{mAh} = 300\,\text{mAh for 2 days} \tag{26.1}$$

$$\text{Average 1 day battery consumption} = \frac{300\,\text{mAh}}{2\text{ days}} = 150\,\text{mAh} \tag{26.2}$$

$$\text{Average 14 days battery consumption} = 150\,\text{mAh} \times 14\text{ days} = 2100\,\text{mAh} \tag{26.3}$$

Based on the above calculations, the power source for the tracker device will be replaced with a Li-Poly battery with a capacity of at least 2100 mAh. With the replacement of the Li-Poly battery of specific capacity, the battery supply size could be reduced to its minimal size. Thus, it ensures the users are comfortable when applying the tracker device on their wrist throughout the self-quarantined period.

Figure 26.3 shows the circuitry diagram for battery supply on the proposed tracker device, constructed in Fritzing software. Since Fritzing software has a 1300 mAh Li-Poly battery unit only, the tracker device is powered up with 2600 mAh Li-Poly batteries. However, in the actual system deployment, it is targeted to use 2800 mAh Li-Poly battery to withstand 14 days of system operation.

26.3 Results and Discussion

Experimental measurements have been carried out to assess the capabilities of the proposed system prototype to monitor the user's locality status during self-quarantined order. A house location of a test subject with a coordinate 3.102188, 101.551506 was chosen and updated in the Google Firebase. The test subject distance from the quarantine coordinate can be calculated using Eq. (26.4). Note that in the theoretical calculation, 10 m in the Google Map equals a value of 0.000100 in the Google Firebase.

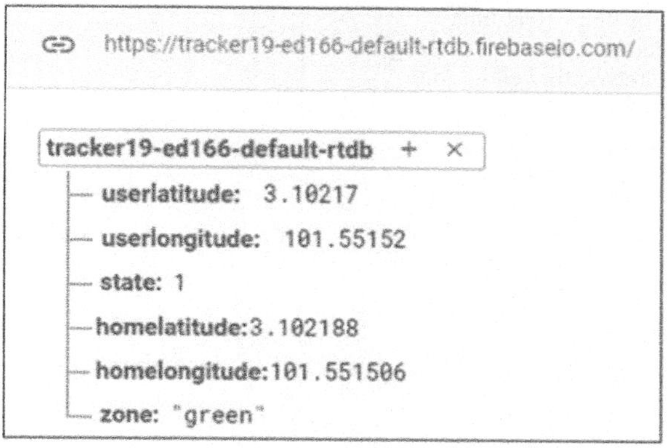

Fig. 26.4 Real-time database recorded in Google Firebase for green zone status

Distance from quarantine coordinate

$$= \sqrt{(\text{homelatitude} - \text{userlatitude})^2 + (\text{homelongitude} - \text{userlongitude})^2} \quad (26.4)$$

Figure 26.4 displays the real-time database recorded in the Google Firebase when the tracker device detected the test subject within the green zone (i.e., within 10 m from the quarantine coordinate). From Eqs. (26.5) and (26.6), it is confirmed that the distance calculation performed by the tracker device prior to locality status reporting on the Google Firebase is correct. Other than that, the state value "1" shown in Fig. 26.4 indicates that the active communication between parental device and tracking device via internet connectivity.

Distance from quarantine coordinate

$$= \sqrt{(3.102188 - 3.102170)^2 + (101.551506 - 101.55152)^2} = 0.000023 \quad (26.5)$$

$$\therefore 0.000023 \equiv 2.3\,\text{m} \quad (26.6)$$

The test subject was then performed an active movement from the quarantine coordinate. As soon as the tracker device detects the test subject distance from the quarantine coordinate to be in the range of 10–20 m, the Google Firebase will get a real-time update stating that the test subject has entered the Yellow Zone. Hence, the user needs to be reminded by the parental device to return to its quarantined station. Figure 26.5 displays the database update when the test subject has entered the yellow zone and Fig. 26.7a shows the notification message sent to the user smartphone via Telegram application when yellow zone is detected. The calculations shown in Eqs. (26.7) and (26.8) confirm the proposed system correctly measured the distance calculation.

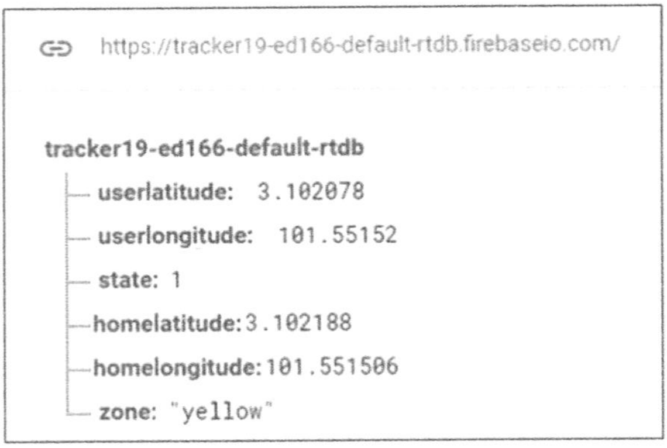

Fig. 26.5 Real-time database recorded in Google Firebase for yellow zone status

Distance from quarantine coordinate

$$= \sqrt{(3.102188 - 3.102078)^2 + (101.551506 - 101.55152)^2} = 0.00011 \quad (26.7)$$

$$\therefore 0.00011 \equiv 11\,\text{m} \quad (26.8)$$

Last but not least, Fig. 26.6 displays the real-time database recorded in Google Firebase in an event when the test subject left the quarantine coordinate for more than 20 m. From Eqs. (26.9) and (26.10), it is confirmed that the test subject has traveled for 21 m from the quarantine coordinate and red zone was declared. Therefore, a notification message in Fig. 26.7b was sent to the user's smartphone. The parental device will keep sending the notification messages to the user for every 10–12 s if the Google Firebase still receives red zone status from the tracker device. If the red zone status remains in a specific duration set by the authorities, the police will be able to track down the user's latest location based on the Google Map link attached herewith the notification message if the person defies orders to return to his/her quarantine

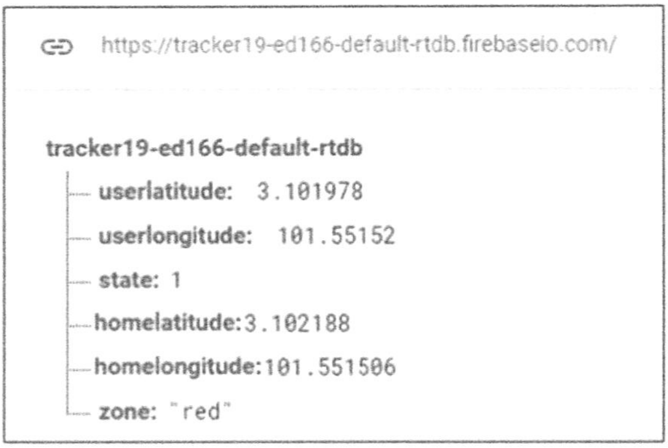

Fig. 26.6 Real-time database recorded in Google Firebase for red zone status

Fig. 26.7 Notification message sent by parental device via Telegram application when **a** yellow zone and **b** red zone are detected

area. Besides that, the police may also make visits to the home-quarantined premises when there is no response reported from the tracker device.

Distance from quarantine coordinate

$$= \sqrt{(3.102188 - 3.101978)^2 + (101.551506 - 101.55152)^2} = 0.00021 \quad (26.9)$$

$$\therefore 0.00021 \equiv 21\,\text{m} \quad (26.10)$$

26.4 Conclusion

In this work, a prototype of monitoring system for individuals with Covid-19 self-quarantined order is proposed. The obtained results run by the prototype measurements demonstrated that the proposed system could be a solid starting point for an actual system implementation of a monitoring system for individuals with Covid-19 self-quarantined order in Malaysia. The system deployment is realistic and straightforward because it requires only a programmed microcontroller unit at both parental and user end, integrated with sophisticated and accurate GPS reading from the user's existing smartphone. Besides, the proposed system has realized the industrial revolution 4.0 (IR4.0) in combating the spread of the deadly Covid-19 virus in a country. This is achieved by using the internet of things (IoT) concept for a device-to-device communication and to manage the real-time data in a cloud database.

Acknowledgements This project was conducted by student and staffs of Universiti Kuala Lumpur British Malaysian Institute (UniKL BMI) and its publication is financially supported by the university. Therefore, the authors would like to thank UniKL BMI and Advanced Telecommunication Technology (ATT) Research Cluster for the provision of laboratory facilities and financial support.

References

1. Hirschmann R (2021) COVID-19 number of daily cases Malaysia 2021. Statista. https://www.statista.com/aboutus/our-research-commitment/1956/r-hirschmann. Accessed 1 July 2021
2. Rampal L, Liew BS (2021) Malaysia's third COVID-19 wave—a paradigm shift required. Med J Malaysia 76(1):1–4
3. Saiful AM, Amir AH, Hafidzul HMN (2020) Covid-19: Polis buru 63 pemakai 'wristband' keluar rumah [METROTV]. Harian Metro. https://www.hmetro.com.my/utama/2020/07/602475/covid-19-polis-buru-63-pemakai-wristband-keluar-rumah-metrotv. Accessed 1 July 2021
4. Sharifah MA, Takiyuddin (2021) Electronic tracking devices only for selected PUI, PUS patients. New Straits Times. https://www.nst.com.my/news/nation/2021/03/671709/takiyuddin-electronic-tracking-devices-only-selected-pui-pus-patients. Accessed 4 July 2021
5. Veena B (2021) Tracking devices could turn the tide. New Straits Time. https://www.nst.com.my/news/nation/2021/03/671744/tracking-devices-could-turn-tide. Accessed 4 July 2021
6. Xu H, Huang S, Qiu C et al (2020) Monitoring and management of home-quarantined patients with COVID-19 using a WeChat-based telemedicine system: retrospective cohort study. J Med Internet Res. https://doi.org/10.2196/19514
7. Hui M (2021) Hong Kong is using tracker wristbands to geofence people under coronavirus quarantine. Quartz. https://qz.com/1822215/hong-kong-uses-tracking-wristbands-for-coronavirus-quarantine. Accessed 4 July 2021
8. Manekiya M, Donelli M (2021) Monitoring the Covid-19 diffusion by combining wearable biosensors and smartphones. Prog Electromagn Res M 100:13–21

Chapter 27
Women Safety Device with Real-Time Monitoring

Siti Ramlah 'Aqilah Arshad, Zuhanis Mansor, Siti Marwangi Mohamad Maharum, and Izanoordina Ahmad

Abstract In every way, the world is becoming less safe for women. Because of the rise in crime, women feel unprotected. According to WHO estimates, one in every three women in the world has experienced physical or non-physical violence at some point in their lives. This paper proposes a quick responding mechanism that helps women during trouble, specifically in a public toilet and house or public parking. This work aims to develop a wearable device for the safety monitoring system for women. The device consumes less power and is created on an open-source single-chip ATmega328 microcontroller with 8-bit reduced instruction set computer RISC processor core based on Harvard architecture. The coding software is written in Java and can be run on Windows, Mac OS X, and Linux. A smart device for women's safety is proposed in this paper, which automates the emergency alert system by using a motion-detectors sensor and a body temperature sensor to detect a potential incident instantly using a detection algorithm. This device detects and sends alerts to close friends and families with the women's location coordinates without requesting communication in crucial circumstances.

Keywords Woman security · Monitoring device · IoT · GSM · GPS

S. R. 'Aqilah Arshad · Z. Mansor (✉)
Advanced Telecommunication Technology Research Cluster, Communication Technology Section, Universiti Kuala Lumpur British Malaysian Institute, Batu 8, Jalan Sungai Pusu, 53100 Gombak, Selangor, Malaysia
e-mail: zuhanis@unikl.edu.my

S. M. M. Maharum · I. Ahmad
Electronics Technology Section, Universiti Kuala Lumpur British Malaysian Institute, Batu 8, Jalan Sungai Pusu, 53100 Gombak, Selangor, Malaysia
e-mail: sitimarwangi@unikl.edu.my

I. Ahmad
e-mail: izanoordina@unikl.edu.my

A. Ismail et al. (eds.), *Advanced Materials and Engineering Technologies*,
Advanced Structured Materials 162,
https://doi.org/10.1007/978-3-030-92964-0_27

27.1 Introduction

Women's safety can be a major concern in countries such as Malaysia, where women play a striking position in each sector. Malaysia-Asia may be a country caring for peace and one among other healthy tourist destinations across the world. But a few recent history incidents have proven that there is a need for security for ladies. Nowadays, in this technology and modern world, women are exposed to danger. The safety of women is in trouble. Danger might happen anywhere, even at public toilets and house parking. These days, violence against women remains a widespread problem. It is very sad and scary to read and hear news about violence against women. The rate of crimes against women is not decreasing but increasing at an alarming rate, especially harassment, molestation, eve teasing, rape, and kidnapping. Sexual violence against women greatly increases.

This becomes a serious issue as our country lacks adequate laws and enforcement of existing laws. The situation will affect a women's life until the end of her life. Women will be traumatized. In addition, no safety mechanisms are available to protect girls from misbehavior activities. No alert mechanisms for girl's safety; it should be done manually only. For example, before going out, we call husband or parent to inform them but they do not know what will happen to her on her way to the place. The women working feel insecure because of rising crimes. The government has taken several preventive steps to avoid these misbehaving practices but has still not affected the through rate of these crimes and barely changed.

This project's goal was to generate a security system that can systematically monitor women. The coding software is written in Java and can be run on Windows, Mac OS X, and Linux. The location will be sent to the preset contacts in a single click. If the device detects any movement, the global system for mobile (GSM) communications also will be sent message links to preset contacts in terms of latitude and longitude coordinates in google maps, which will be sent from a device to the preset contacts in a single click, whereas in PIR sensor can detect the presence of human by detecting movement.

Our main objective was to improvise a safety monitoring system for women's security. Other than that, it is to detect using current locations by using the global positioning system (GPS) tracker. The second objective is to share the current location of the user with preset contacts using GSM. Finally, our last objective is to detect human presence by detecting body temperature using a passive infrared sensor (PIR). The main scope of this project is to monitor for women's security. It also includes sharing current locations and detects the presence of a human. The objective must be fulfilled so, and there are several scopes outlined.

i. For women, 18 years and above.
ii. Target for students.
iii. Independent women working until late at night.
iv. Alone women.

This project is divided into two main parts: hardware part and software. The outline of this paper is organized as follows. The analysis of the previous related studies that motivated this work is discussed in Sect. 27.2. Section 27.3 explains the methodology used in the development of this device. Results are described and analyzed in Sect. 27.4. Finally, conclusions are drawn in Sect. 27.5.

27.2 Analysis of the Previous Works

According to the previous studies, numerous functions and characteristics evolve in response to technological advancements in women's safety systems [1–5]. In [6], the designed Firebase is a mobile phone app that integrates all the features of existing apps. The app requires initial registration with a login id and password along with emergency contacts. Each contact can be manually added or added from the inbuilt contacts application. Users are registered with the Firebase database to keep the app secure. The app allows users to track their emergency contact via dynamic GPS tracking. The user can also call the toll-free helpline numbers with the inbuilt phone application. When in distress, the user needs to press the SOS button then an alert message is sent to the emergency contacts.

The research in [7] looked at the relationship between stress and skin resistance, stress, and body temperature. Findings suggest that sitting, standing, sleeping, and struggling could be linked to stress levels in the human body. Punjabi et al. in [8] have designed a portable device that has a pressure switch. A pressure sensor can detect when an assailant is about to attack a person. When the person senses any insecurity from a stranger, they can put pressure on the device by squeezing or compressing it. It then senses if the attacker is going to attack them.

However, the device and system that deal with detecting emotional stress levels to predict danger and body position exist today but they do so separately. In this work, a women safety device with real-time monitoring is designed and does not need manual effort where it can detect the user using current locations by using GPS and share the current location user with preset contacts using GSM in a single click.

27.3 Methodology

This proposed device is a smart device for women and children's safety that automates the emergency alert system by using a motion sensor and a body temperature sensor to detect a potential incident instantly using a detection algorithm. Without requesting communication in critical circumstances, this device detects and sends alerts to friends and families with the women's location coordinates. The hardware and software are the two main components of this system development. The GPS (GYGPS6MV2), PIR sensors, GSM (SIM900), and Arduino Nano are the six main components of the hardware system.

Other features such as a push button and a phone will be used as the controller's inputs and outputs. The hardware system can be installed in any location where users are present.

GSM and sensors are the most important hardware components in this design for software implementation. This device uses less power because it is based on an open-source single-chip ATmega328 microcontroller with an 8-bit reduced instruction set computer RISC processor core based on Harvard architecture. The coding software is written in Java and can be run on Windows, Mac OS X, and Linux. The GPS keeps track of the user's location and sends it to the GSM network. The data are sent via GSM to pre-determined contacts using the C programming language. The schematic diagram and layout design for the project are created with Proteus. The project's layout is created by manually routing the connections of each electronic component involved in the system's development. Users can be used to replace the entire system. In general, both can act independently as well as in a pre-determined order.

The complete block diagram of a safety monitoring system for women's security is shown in Fig. 27.1. GPS (GYGPS6MV2) keeps track of the current location and sends data to the GSM phone (SIM900). The GSM (SIM900) sends a message to pre-set contacts with the current location. After the data have been received, it will be displayed on the user's phone screen. The PIR sensor is used to detect changes in body temperature, even in dark areas. Within 10 m of a PIR sensor, a human can be detected moving. The precise detection limit is 5–12 m. PIRs are mainly made up of a pyroelectric sensor that can detect infrared radiation levels. Figure 27.2 shows the complete schematic diagram of this project where Fig. 27.3 illustrates the complete flowchart of the safety monitoring system for women's security. Proteus is used to draw the schematic diagram and design the layout for the project. The working layout

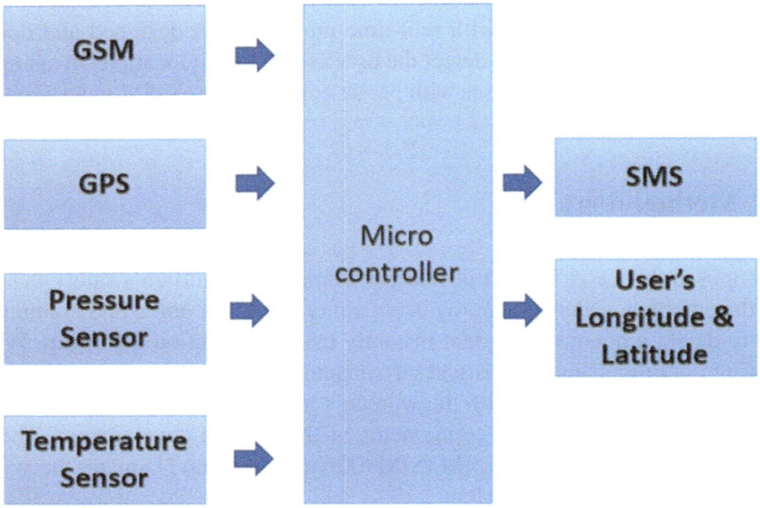

Fig. 27.1 The overall block diagram of the project

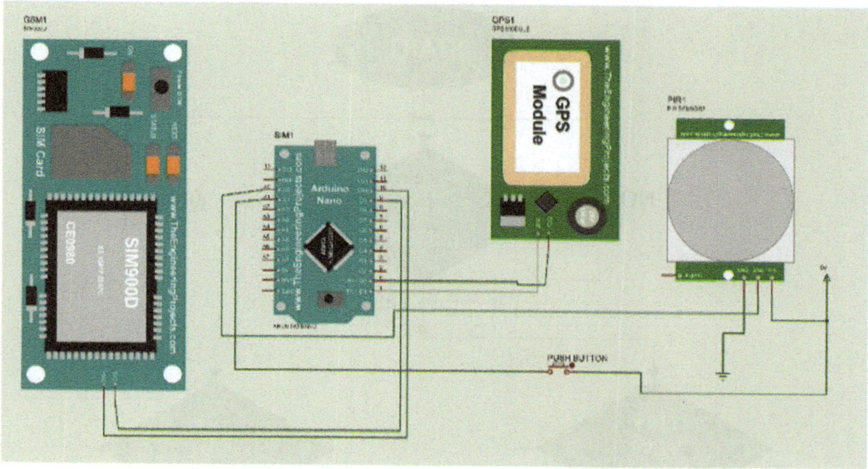

Fig. 27.2 The complete schematic diagram of the project

is created by using manual routing of the connections of each electronic component involved in developing the system.

27.4 Results and Discussion

Figure 27.4 depicts how all of the components used in this project are connected. The red light for the microcontroller board, GPS, and GSM blink when the power supply is turned on. As a result, the components appear to be in good working order. If the GPS red light is not blinking when the project is connected to the power supply, this GPS has not received any signal from the satellite. The GPS's ability to function properly can be affected by a variety of factors. One of the reasons is the user's location.

This project has been tested at three different places. Three other locations have been used to test this work. The first is in an open space area, where the GPS receives satellite signals and can track the user's location. GSM will then send SMS to pre-defined contacts where the message includes a link to the user's current longitude and latitude coordinates. This project was tested again at level 2 of the UniKL BMI building. According to the findings, the GPS received signals from satellites and tracked the user's location. The SMS is then sent to pre-defined contacts by GSM. A link to the user's current longitude and latitude coordinates is included in the message.

On the other hand, the GPS did not blink during the level 10 Residential Bayu Andaman test. The GSM sent SMS to pre-programmed contacts when the push button was pressed, but the latitude and longitude were set to "0, 0." As a result, the user's

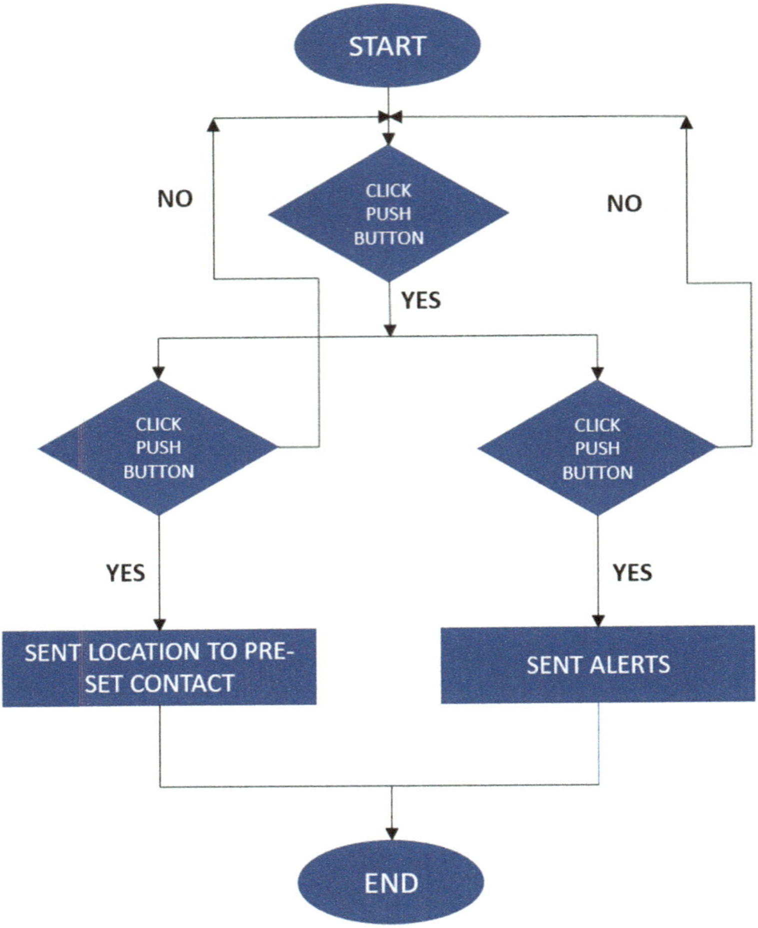

Fig. 27.3 Block diagram of the proposed system

location is hidden from the pre-set contacts. The above scenario demonstrates that the GPS did not work well in the higher elevations. The strong structure of a building's walls, as well as the weather, has an impact. Under condition two, when the project is powered up, the GPS red light blinks, and the GSM red light blinks fast every 0.5 s, indicating that the GPS receives satellite signals, but the GSM cannot read the GPS prepaid sim card. As a result, GSM did not send any SMS to the pre-defined contacts. Before using this project, it must be connected to a power source. Under condition three, the GPS red light blinks when the project is connected to the power supply. The GSM red light flashes a little slower than condition 2, every 3 s, indicating that the GPS has received satellite signals GSM can read the prepaid sim card. As a result,

Fig. 27.4 Prototype

after pressing the push button, pre-set contacts receive SMS from GSM. The time it takes for pre-set contacts to receive SMS is usually between 3 and 8 s.

Few factors, such as location issues, pre-set contact's mobile device issues, different networks used, and network traffic, can cause a delay after the project has been tested. Firstly, the sender's or recipient's location may cause a delay in text message delivery. There is a lag in transmission when all messages come at once on a mobile phone. It can be caused by moving at speeds above about 50 km/h on the border of two cell coverage areas. There are common reasons for this, including tall buildings and old houses with bad reception. Text message transmission can be delayed due to issues with mobile devices. The most obvious cause is a device turned off, but a weak or uncharged battery can also affect message delivery.

Moreover, due to network coordination or the carrier prioritizing their traffic, a text sender and receiver on separate networks may be more likely to experience texting delays than those on the same network. Finally, sending a text during peak network usage has an impact on text delivery speed. During peak traffic periods, the local network may become congested, delaying messages to mobile phones. Figure 27.5 shows the output at the receiver's phone screen. Suppose, the PIR detects any movement and GPS received a signal from the satellite. In that case, it will send a message to the receiver consisting of the user's locations in terms of longitude and latitude. In Fig. 27.6, the user's direction at the receiver's phone screen after clicking the link in the message is direct to google maps, as shown. Figure 27.7 shows the output at the receiver's phone screen. If GPS did not receive any signal from the satellite, but the GSM has functioned well, it will send a message to the receiver but consists of a "0, 0" coordinate for longitude and latitude.

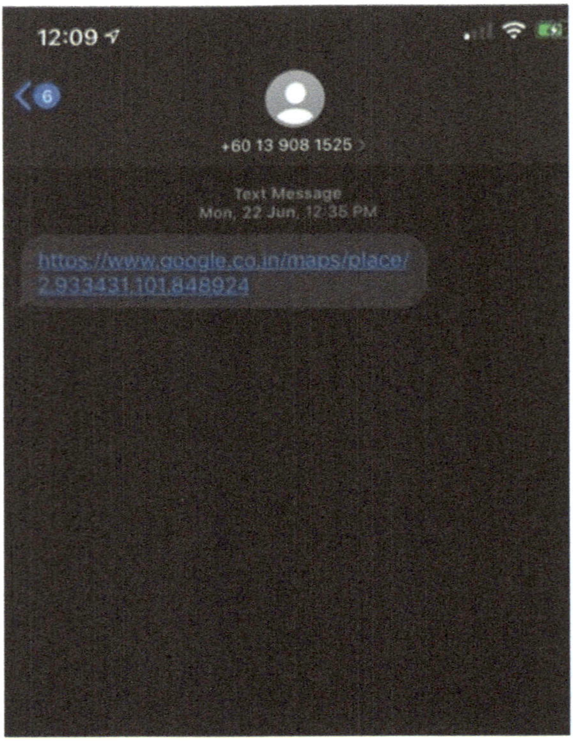

Fig. 27.5 Output 1 where the user sends a message

27.5 Conclusion

We proposed a fast-responding device that helps women during trouble, specifically in a public toilet and house parking. Using less power and based on an open-source single-chip ATmega328 microcontroller with 8-bit reduced instruction set computer RISC processor core based on Harvard architecture, this work aims to develop a wearable device for safety monitoring systems for women and children. This device identifies the situation and sends alerts to friends and families with the women's location coordinates without involving her ability to interact in critical situations. It immediately sends an alert notification to relatives and the nearest police station. The device is designed to be carried in any type of bag, including handbags and laptop bags. This work can be altered to small wearable devices and attach to an alarm system through customization development. An alarm or buzzer is good since it can alert nearby people if anything happens to the user. This device can play a big role in ensuring a safe and secure environment for women.

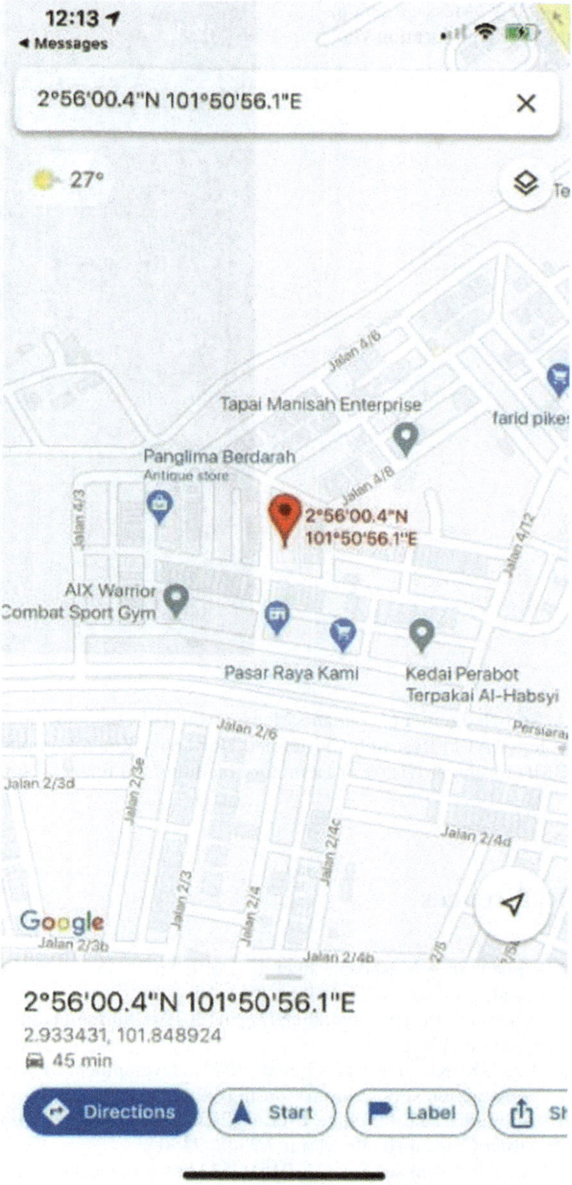

Fig. 27.6 Location of the user via GPS

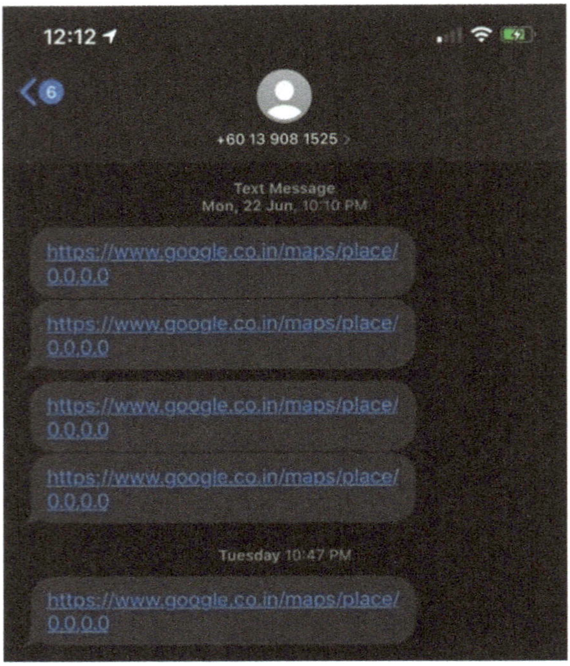

Fig. 27.7 Message sent to the user with location via GSM

Acknowledgements Zuhanis Mansor would like to thank the Advanced Telecommunication Technology (ATT) Research Cluster and University Kuala Lumpur British Malaysia Institute (UniKL BMI) for the provision of laboratory facilities and financial support.

References

1. Ranaware AA, Bhosale D et al (2019) Safety device for women with GPS, GSM and health monitoring system. JoMTRA 6:13–18
2. Senthamilarasi N, Bharathi ND et al (2019) Child safety monitoring system based on IoT. J Phys Conf Ser 1362:1–7
3. Lee UK, Kim JH, Cho H et al (2009) Development of a mobile safety monitoring system for construction sites. Autom Constr 18:258–264
4. Anuradha BD, Priyanka B et al (2020) Review paper woman security and position tracking. Struct Control Health Monit 7:2348–2350
5. Arijit P, Akanksha S et al (2015) GSM based home automation, safety and security system using android mobile phone. Int J Eng Res 4:490–494
6. Kiran M, Chaitra BV et al (2017) An intelligent safety system for individual's security. ICECDS. https://doi.org/10.1109/ICECDS.2017.8389614
7. Anand J, Madhvi K et al (2016) Design and development of an IOT based wearable device for the safety and security of women and girl children. RTEICT. https://doi.org/10.1109/RTEICT.2016.7808003
8. Sunil KP, Suvarna C et al (2018) Smart intelligent system for women and child security. IEMCON. https://doi.org/10.1109/IEMCON.2018.8614929

Chapter 28
Experiencing Layout Design Techniques from Highly Skilled IC Design Engineers

Mohd Amir Abas and Nur Intan Zazalinda

Abstract Our silicon industries are giving indicator that they are having shortage of expertise in IC design especially from local graduates. On top of that the employed graduates are also lacking of skills. IC design involves with wide topics due to wide application. Specialization is normally related to application such as RF design, memory design, analog design, and digital design. Expertise may end up with specializing in one or two applications only due to long effort to capture the whole skill and knowledge. It is known that the process to build up expertise in this field is hard and challenging. The degree structure needs to equip with high-end software, experience staff, and fund for fabrication. Most programs aim is to produce high level of graduate with good competitive skills that would help the silicon industries to employ them without retrain issue. Hence, in this article, a compilation of teaching process is shared which involves a highly skill mentor who is guiding the trainee on the design process. The sample results concluded that teaching advanced technology in IC design requires sophisticated facilities as well as experienced trainers.

Keywords Teaching · IC design · Specialization · Graduate and Industries

M. A. Abas (✉)
Electronic Section, Universiti Kuala Lunpur British Malaysian Institute, Gombak, Kuala Lumpur, Malaysia
e-mail: drmamir@unikl.edu.my

N. I. Zazalinda
T2-L05-2, Level 5, Tower 2, PJ33 Cybercentre, No.3 Jalan Prof. Khoo Kay Kim, Seksyen 13, 46200 Petaling Jaya, Selangor, Malaysia

A. Ismail et al. (eds.), *Advanced Materials and Engineering Technologies*,
Advanced Structured Materials 162,
https://doi.org/10.1007/978-3-030-92964-0_28

28.1 Introduction

Layout design is a process of translating circuits into silicon structures. The process of making the whole silicon chip involves huge efforts from the specialization team. The team must be led by a project engineer who has vast knowledge and experience. The team members should also have very good background in CAD simulation and silicon knowledge. Layout is a set of geometric patterns (typically in the form of polygons) which specifies the size and location of silicon layers, polysilicon, and metals to create all sorts of nano sized components. In fabrication process, the layout is used as mask the template for photolithographic process. The polygons in IC layout must satisfy the wafer fabrication's set of design rules. These rules must be applied to ensure the fab machine can fabricate accordingly against its minimum limitation.

Errors are not tolerable, therefore, all the geometric features must be precisely calculated. The impact of errors is failure. Failures are a huge topic in IC design. It can come from wrong geometric design, wrong techniques design, and wrong fabrication process. However, in this article, we are focusing on good design techniques which can safeguard on functional aspect of the silicon devices for increasing end yield.

An example of layout design has been identified and is used as a model in investigating all the proposed techniques from the expert. The layout model is the battery switchover IC (BSIC) [1–4] which consists of several functional blocks. In layout environment, there are two layout domains, analog, and digital which require different strategies in achieving a workable silicon structure. Some techniques and strategies are hardly written in any literature due to confidential issues. Despite the limitation, through guidance from an experience IC engineer, the work is still achievable. Therefore, BSIC is used as a platform to explore the implementation of good techniques under supervision of an IC engineer. After one year of hard work, BSIC is successfully designed and fabricated. The results in the last section show good justification for all the blocks within BSIC. It is evident that the knowledge and skill in IC design are well translated through guidance from experience engineer who has good knowledge in layout design and vast experience working in fabrication process. Without one of the two aspects, the works are potentially falling in failure mode.

28.2 Project Methodology

The aim of this project is to build up skill of IC making from layout until silicon prototype. The development of this project is carried out by a master student under guidance of an experienced IC engineer. It took one year to accomplish the whole design process. The team managed to complete the layout and submitted the design for fabrication. The end part of this study is to evaluate the successful rate or yield of 20 fabricated silicon dies. The results justify all the guidance from the engineer are well translated. Having this experience, the trainee gains very valuable knowledge and skills which are good for our future silicon industries.

In the design process, this article emphasizes on tips and guidance from the IC engineer. Other issues pertaining to design process such as simulation of CAD, silicon knowledge, and circuit operation is not elaborated due to content limitation by the conference organizer. The following section unfolds some of the good tips that make the investigation of layout making it a very successful story.

28.3 The Full Custom Layout Design Flow

The full custom design approach is the best option for a designer to obtain the most robust layout in terms of speed and power. The advantage of this approach is that designers have the ability to customize their own standard cells specific to their design project. This includes the gates positioning, the size, and shape of the circuit component as well as the connection paths among them. A typical full custom IC layout design process flow is illustrated in Fig. 28.1.

This process usually begins once a schematic design that fulfills all its design specification is fully functional at its system level. The next step is to define the floorplan which will take into consideration the special design requirements such as power grids, signal for critical area, symmetrical layout cells, noise, and latch-up protection as well as approximate total chip size. The important step during floorplanning is to decide the design for appropriate cells. Cells can vary from a single transistor to a cluster of combined transistors. Understanding the floorplan beforehand can help a designer to create a more effective cell library such that it will be easier for the designer to work on the top level design of the layout.

In layout design stage, there are three verification tools. Design rules check (DRC) is a tool to check the minimum dimension of components on its sizing, wires, connectors, etc., must be met. This setting must comply with the machines in fab process. The second tool is electrical rules check (ERC). ERC is a tool to verify the size of wire to avoid wire stress or known as electromigration. Wires which are thin but carry heavy current will make the wire stressed. This causes disconnection when the layout is under full testing procedure. The worst is when the wires intermittently failure after few months on operation. The third step is the layout versus schematic (LVS). The verification is carried out by comparing the netlist extraction from the schematic design with the layout drawing. This step is to ensure that the layout drawing is connected correctly corresponding to the schematic design as its reference. If the layout design does not match with the schematic, then the layout drawing has to be modified and later checked again until it matches perfectly with the schematic. Once the layout has completely passed with no errors for these three checking, the chip is now ready for tape-out and ready for the fabrication. The layout file is converted to graphic data system II (GDSII) format for vendor submission.

Fig. 28.1 SIC design flow

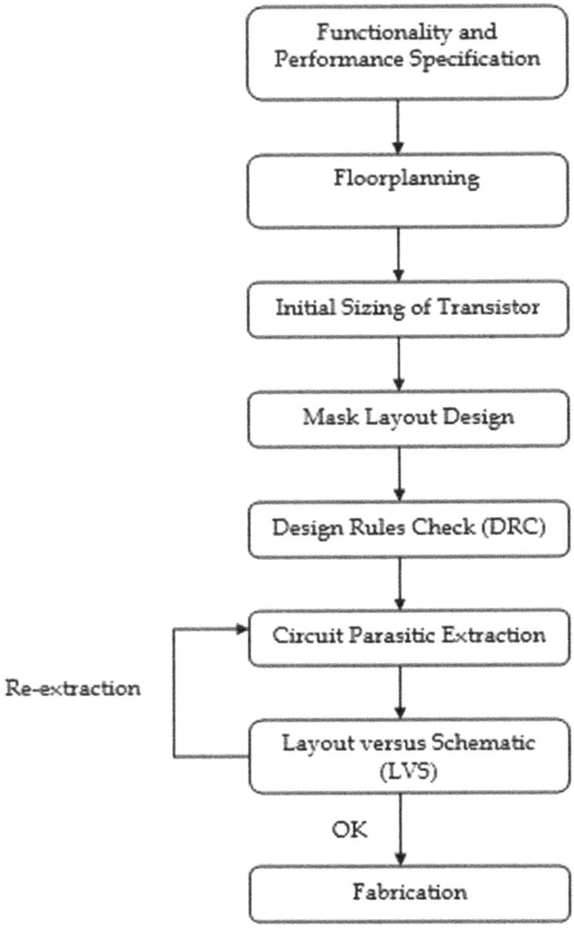

28.4 Layout Design Stages and Issues

IC design process is a game of war. It must be well planned and executed. A project leader who has very deep knowledge will lead the team. Team members are assigned to build up the block based on his capacity and skills. The work must accomplish within the time frame before tape-out dateline. The ultimate goal toward the end is the complete design must be functional, compact, and optimizing space intelligently.

28.4.1 Floorplanning

Floorplan is a draft of block placement associated with the total size of the silicon area. In certain circumstances, the total size of the floorplan could be in different shape depending on the silicon area allocated by the project manager. In this exercise, BSIC is the target of implementation. There are six blocks in BSIC and each block should occupy the space in different shape. Figure 28.2 shows the initial floorplan for BSIC. The six blocks are ring oscillator (ROSC), power on reset (POR), resistor divider, manual reset, power fail, and watchdog (WD).

The first decision to be considered in placing the block is the potential block that actively transits signals. The most active transition that occurs is the oscillator circuit. Oscillation causes active transition between two logics. This transition could transmit interference to other block due to the fact that all blocks are sharing the same substrate in silicon structure [5, 6]. The next decision is to identify block that responds to the interference. In confirming the theory and the advice of the expert to run away

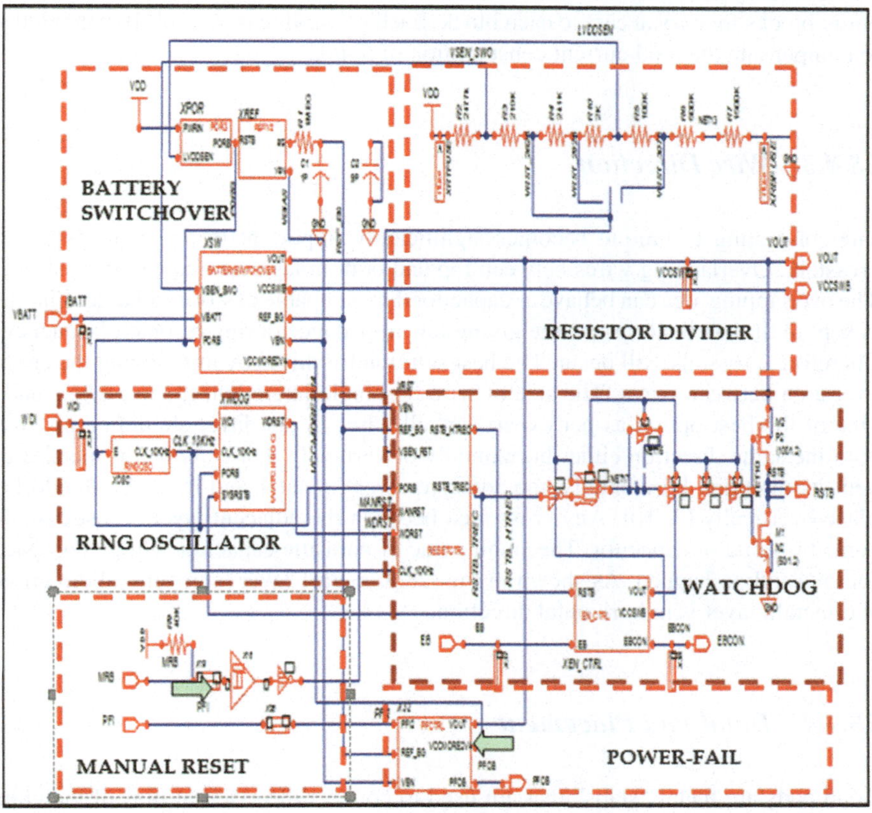

Fig. 28.2 Floorplan for BSIC

with this situation, the digital block must be separated with the analog block [7, 8]. Hence, the active block is the ring oscillator while the sensitive block is the battery switchover. Both must be separated with significant distance to avoid the transition of ring oscillator effecting the waiting signal mode of the battery switchover block.

28.4.2 Wire Size for Power Grid

Power issues in silicon need to be addressed efficiently with regards to simple concept the higher the number of transistor the bigger the wire. The most common issues in wiring is electromigration, whereby the wire is under stress when high current flows in the wire. According to the expert, the power grids should be the first issue that needs to be solved after floorplanning [5, 6]. Power supply lines and ground (VCC and GND) are large since both carry high amount of current, therefore they must be sized appropriately. Calculating the amount of current that flows in the power line will give easy estimation of size for the wire. The power lines are surrounding the entire blocks for easy access to each block. Each power line is set with 10 μm widths to compensate the total current consumption of 5 mA.

28.4.3 Wire Direction

Interconnecting technique becomes significantly important to avoid the issue of crosstalk. Overlapping wires between top and bottom layers could cause crosstalk. The overlapping area can behave as capacitor that is capable of storing charge. Charge is kept in that particular structure giving low impedance during resonant frequency. Therefore, crosstalk will occur. The best solution for this is by minimizing the cross over area between wires. The size of wire for single interconnect should be small. One of the best option as per expert say, all wires in one layer should emphasize with the same direction either horizontally or vertically. For example, if metal 2 is kept its direction horizontally, the adjacent layer (metal 1 and metal 2) should be placed vertically [9, 10]. Any cross area between the adjacent layers causes small parasitic value of capacitor. The small value of parasitic capacitor will give impact for crosstalk to happen. As shown in Fig. 28.3, metal1 layer is in vertical direction and metal2 layer is in horizontal direction.

28.4.4 Bond Pad Placement

Bond pads are metal "slabs" that are used to connect the external pins of the chip to the silicon layout. Generally, bond pads are located on the periphery of the chip as shown in Fig. 28.6. They are electrically connected to the devices within the

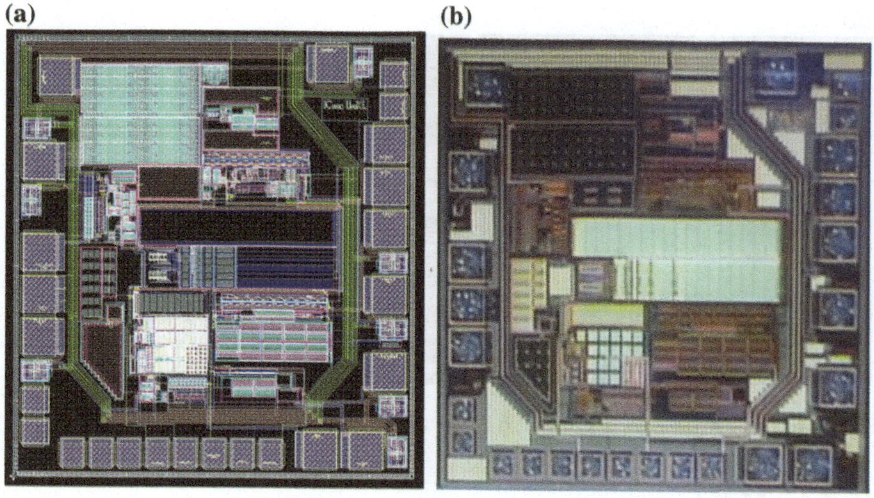

Fig. 28.3 **a** Layout from cadence. **b** Fabricated layout

ICs through buffers and electrically conducting interconnects. In the pad, a static discharge circuit is placed known as the electro-static discharge (ESD). The size for a bond pad for this chip is 80 μm × 80 μm.

Another pad known as probe pad is placed for testing purposes. A probe pad is placed when the node is not assessable and highly important to check the circuit's functional. In this case, ROSC output is one of the nodes to test its output for determining the frequency response. Probe pad size is much smaller; 50 μm × 50 μm.

28.4.5 Metal Density

Uniform metal layer is an important requirement for high yield in the fabrication process. Any empty space in metal layer should be filled up by dummy metal so that the total fill up area must be at the range of 30–70% [9, 10]. In BSIC structure, a set of dummy metal fills has been added to achieve minimum metal density as shown in Fig. 28.5.

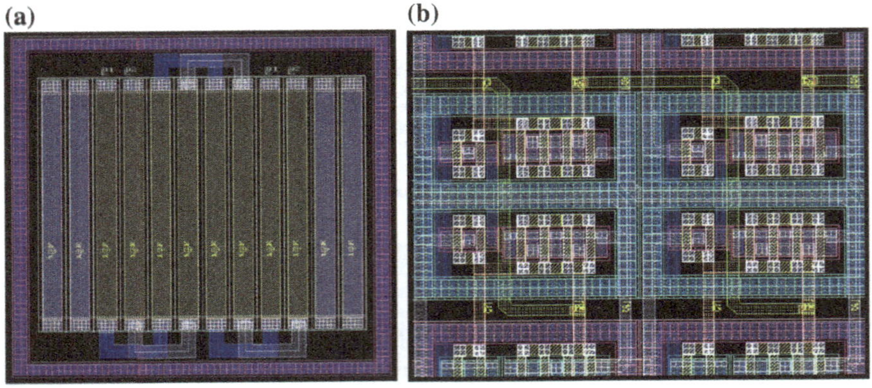

Fig. 28.4 **a** Dummy resistor 2 strip L/R **b** Polarization for p-substrate and n-well

28.4.6 Matching

In structuring devices on a chip, two identical resistors may not give the same performance. They are electrically different in term of carrier concentration. The neighboring components may also influence the performance due to parasitic element such as capacitor and resistor. In this case, a good solution is creating dummy devices. Dummy devices have identical structure with the devices under test. According to the expert, this can minimize the effect of process variations during fabrication. Figure 28.4 shows an example of resistor with twelve strips. Eight strips are used as resistor while two strips on left and right are functioned as dummy. The dummy strips are protecting the resistor from interference from neighboring components.

28.4.7 Transistor Arrangement

Polarization contact to substrate is a technique to eliminate latch-up effect in CMOS. The two transistors PMOS and NMOS are exposed to latch-up effect which causes damage when short circuit occurred. A solution is by shorting NWELL substrate for PMOS to VDD while P-substrate for NMOS is shorted to GND. This will secure the polarity of the substrate to permanently high and low. In addition to this, a maximum number of contacts depending on the space is recommended. Figure 28.4b shows the structure of transistor in the layout with four contacts for PMOS and NMOS, respectively.

28.4.8 Fingering

Fingering technique is used for creating large transistors. The finger structure is referring to the length of the polysilicon. Polysilicon is not a good conductor. It must be short to reduce the gate internal resistance. Therefore, for large transistor, the long polysilicon is chunk into number of fingers and each of them performs as a single transistor. All chunk transistors are connected in parallels for functioning as a single large transistor. There are two types of fingering technique; which are for wider W and larger L. PMOS with W = 60 μm/L = 0.6 μm and NMOS W = 30 μm/L = 0.6 μm as depicted in Fig. 28.6a, b shows the structure for width fingering. 12 fingers are applied to PMOS and NMOS.

While for larger L instead of polysilicon as fingering structure, diffusion is used as fingering element. Figure 28.6b shows the CMOS with the same size of NMOS and PMOS (W = 3 μm/L = 192 μm). The polysilicon size covering the diffusion fingers at the bottom. All the diffusion fingers are connected in series using METAL 1 and contact [9, 10].

(a) (b)

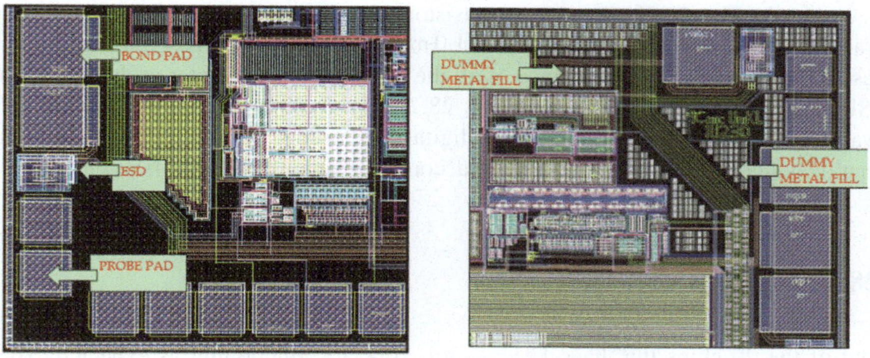

Fig. 28.5 **a** Bond and probe pad insertion. **b** Dummy fill up

(a) (b)

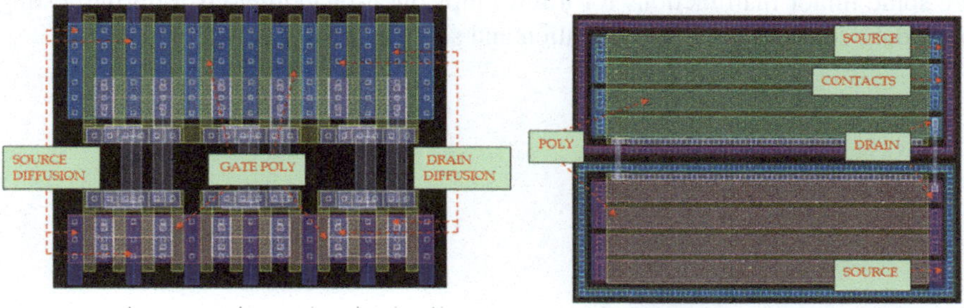

Fig. 28.6 **a** Fingering for width. **b** Fingering for length

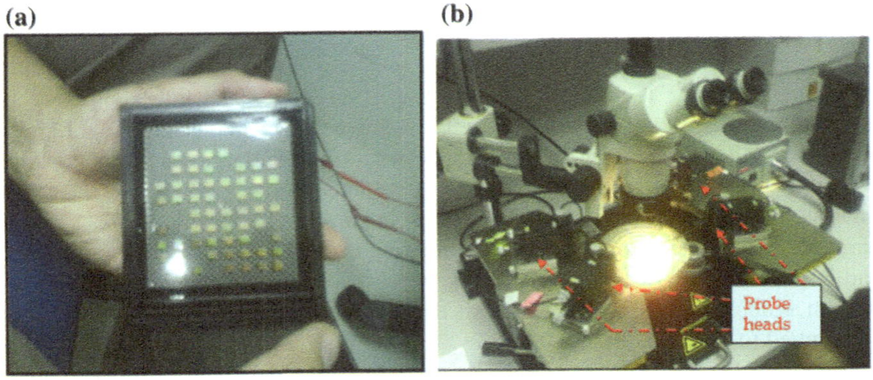

Fig. 28.7 **a** 20 Samples of silicon dies **b** Test bench and equipment

28.5 Layout of Battery Switchover IC (BSIC) Chip

The complete BSIC chip layout has been designed and fabricated. The silicon layout is shown in Fig. 28.2b. The total transistors that have been fabricated has 1091 transistors with the prototype die size of 1.0 mm × 1.0 mm. The prototype is manually tested to examine the performance of all the blocks using the wafer probing system (SUSS EP6 manual probe station, Fig. 28.7). Figure 28.8 shows three samples of waveform which are captured using the digital oscilloscope. All the waveform verifies that the silicon structure is performing accordingly against the specification.

28.6 Conclusion

Out of the 20 chips that have been tested, 70% are functioning as expected. The testing procedure using the manual probe station is targeting block for POR, ROSC, and WD. The testing results as shown in Fig. 28.8 verify the reset system is functioning accordingly and meets the specification at room temperature. However, there are some minor malfunctions for a few chips due to handling error during testing procedure such as chips contamination and scratched chips.

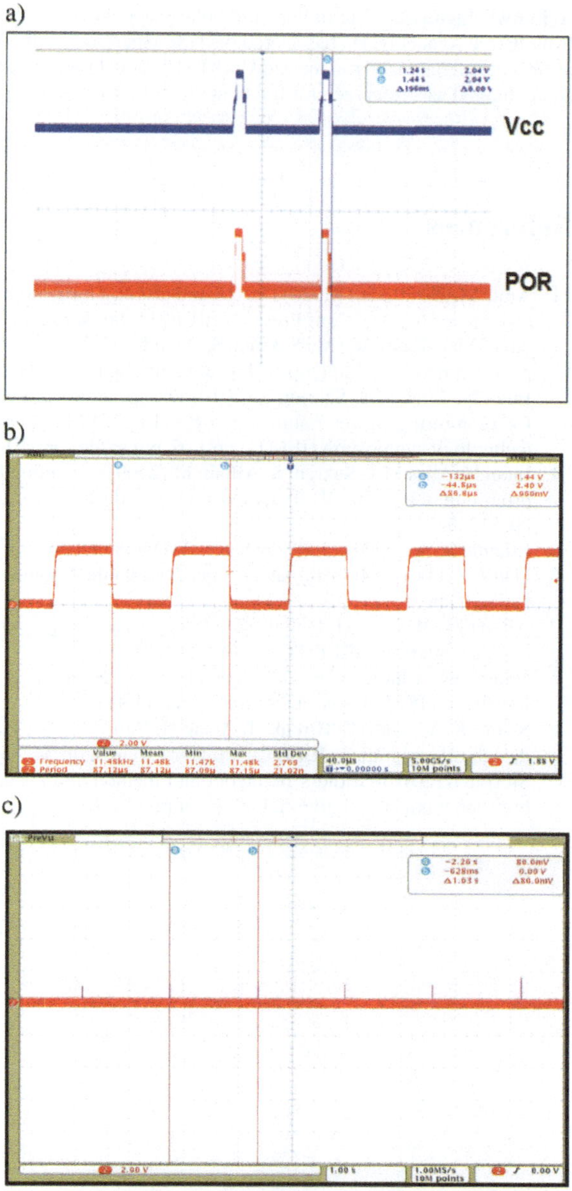

Fig. 28.8 **a** Battery switchover response **b** Ring oscillator output **c** Watchdog timer pulse trigger

Acknowledgements The author would like to applause to our highly experience mentor S. Sakrani from ICmic Academy. He has shared his vast experience in IC design which makes the flow of the project beneficial to the trainee and UniKL. His couching, skill, and knowledge have been captured and compiled as a reference for future study in fostering more graduate specializing in IC design.

References

1. Amir M, Zazalinda I (2020) Design and test on-chip power-on-reset (POR) function in battery switchover IC. AIP Conf Proc 2291:020011. https://doi.org/10.1063/5.0023517
2. Abas MA, Rusli MAM, Sakrani S, Aizudin NIZ, Syari AA (2012) Graduate enhancement programme for building up IC layout technologies. J Eng Technol 2:66–71
3. Intan N, Abas MA, Sakrani S (2010) Design an integrated microprocessor supervisory chip for monitoring power failure. In: 34th IEEE/CPMT international electronic manufacturing technology symposium (IEMT), pp 1–6. https://doi.org/10.1109/IEMT.2010.5746694
4. Intan N, Abas MA, Sakrani S, Afham M (2009) Grooming IC designer through apprenticeship programme. In: 2009 IEEE symposium on industrial electronics & applications, ISIEA 2009, pp 542–547
5. Etienne S, Sonia DB (2007) Advanced CMOS cell design. Tata McGraw-Hill, New Delhi
6. Dan C (2000) CMOS IC layout concepts, methodologies, and tools. Newnes, Butter-worth Heinemann
7. Allan GA, Walton AJ, Holwill RJ (1992) A yield improvement technique for IC layout using local design rules. IEEE TCAD 11(11):1355–1362
8. Ming DK, Chung YW, Tain SW (1997) Area-efficient layout design for CMOS output transistors. IEEE Trans Electron Devices 44(4):635–645
9. Satoru K, Yasunori S, Hiroaki T, Atsuo H, Isao O (1994) Transistor size optimization in layout design rule migration. IEEE Cust Integr Circuits, pp 542–544
10. Steven T (2010) Gridless IC layout and method and apparatus for generating such a layout. http://en.wikipedia.org/wiki. U.S. Patent 6 957 411

Chapter 29
Development of Blynk IoT Platform Weather Information Monitoring System

Muhammad Amiruddin Kamarudin, Noor Hidayah Mohd Yunus, Mohd Raziff Abd Razak, Mohd Shahrul Mohd Nadzir, and Kemal Maulana Alhasa

Abstract Internet of things (IoT) is a modern technology that connects between physical equipment and some other devices to the Internet network for the desired application. The demand for weather monitoring application leads to an initiative for smart IoT development. There is the issue of an urgent weather alert need for someone who wants to monitor environmental conditions before involving in outdoor activities. Thus, this paper proposes a device of weather information monitoring system to efficiently monitor weather conditions in real time and sending alert notifications via mobile Blynk application, SMS, and email. The device is composed of ESP8266 as microcontroller, temperature-humidity sensor (DHT11), rain drop, and carbon monoxide (CO) detection sensors with integrated real-time data transmit through GSM and Wi-Fi network. The proposed system is designed with a user-friendly GUI mobile application and Web based, accordingly the users can get the updates from the weather monitoring system at anytime and anywhere.

Keywords Weather monitoring · IoT · Sensor · Blynk application · NodeMCU · Web based

M. A. Kamarudin · N. H. M. Yunus (✉) · M. R. A. Razak
Communication Technology Section, Universiti Kuala Lumpur British Malaysian Institute, Batu 8, Jalan Sungai Pusu, 53100 Gombak, Selangor, Malaysia
e-mail: noorhidayahm@unikl.edu.my

M. A. Kamarudin
e-mail: amiruddin.kamarudin@s.unikl.edu.my

M. R. A. Razak
e-mail: mraziff@unikl.edu.my

M. S. M. Nadzir · K. M. Alhasa
Department of Earth Sciences and Environment, Faculty of Science and Technology, Universiti Kebangsaan Malaysia, 43600 Bangi, Selangor, Malaysia
e-mail: shahrulnadzir@ukm.edu.my

A. Ismail et al. (eds.), *Advanced Materials and Engineering Technologies*, Advanced Structured Materials 162,
https://doi.org/10.1007/978-3-030-92964-0_29

29.1 Introduction

The ancient peoples depended on their own experiences, animal actions, and photosynthetic fruiting to forecast the weather throughout the coming years. In the end of the nineteenth century, weather predictions are completely adaptable depending on observational principles, with no knowledge of weather and specific physical mechanisms [1, 2]. Every planned activity carried out in daily life is strongly influenced by the surrounding weather factors. Weather forecast is important to various sectors such as energy industries, agriculture, transportation department, and automation, and accordingly, weather forecasting is part of the economic progression. There are many situations which has been proven the importance of weather monitoring system in many aspects such as to ensure the stable healthy growth of plants, wind farms, movement in workplace, scheduling for travel trip, and so on. Certainty about weather conditions causes all activities could be run smoothly. There is weather forecast information reported in media, television, and radio; however, the information received is generally for the forecasting of a particular day in a district for the state. Therefore, real-time information updates on weather monitoring system is very affordable in many aspects.

In the world of modernization and technological growth, innovation is important to make ease in daily life [3–5]. Today, a growing demand for various desired applications over the Internet network keeps increasing. IoT enables a wide range of various application devices with identical protocols for coordinating the complete applications interaction [6]. Current technologies such as wireless sensor networks (WSNs), signal processing systems, smart home automation, controlling, and monitoring systems can be enhanced become smarter and more efficient with the association of IoT based [7].

For that reason, this paper aims to develop a moveable weather monitoring device linked with data storage cloud and displaying the information. An application is developed for the users to view in real-time weather information. The data from the connected sensors can be viewed by the users via smartphone with Blynk application, SMS, and email alerts. The sensors used to measure parameters such as humidity and temperature, rain drop, and carbon monoxide (CO). To demonstrate the practicability, the monitoring device is portable where it can be installed, mounted, and accessible from anywhere, to obtain such informational data.

29.2 Methodology

29.2.1 Project Development

Figure 29.1 shows the block diagram of the weather monitoring system. Figure 29.2 describes the flowchart functionality insights into the system. Firstly, the power supply at 5 V with 1200 mA is required to initialize the system. The connected

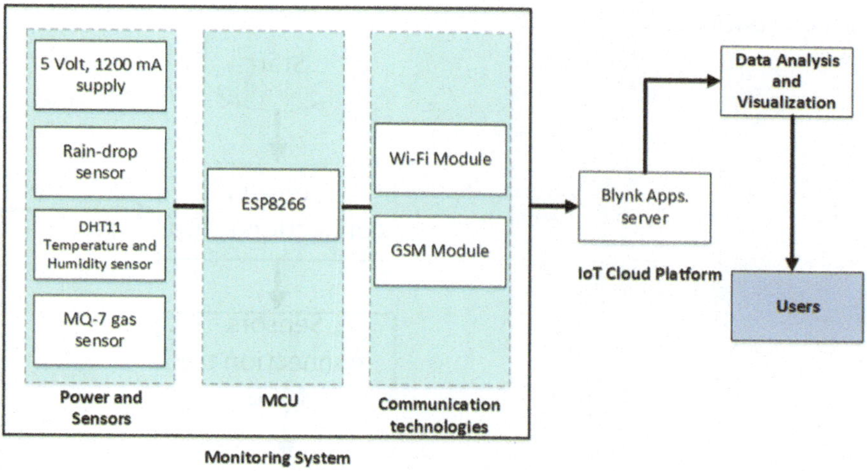

Fig. 29.1 Block diagram of the weather monitoring system

sensors are to ensure in active mode, so that the data from the sensors can be sent to the NodeMCU ESP8266 microcontroller. Arduino UNO SMD Rev3 is used to program the microcontroller. All the data from sensors being forwarded to Blynk server over a GSM and Wi-Fi network. Blynk is one of modern platforms that gives users to develop interfaces basically from iPhone operating system (iOS) and Android computer to track and monitor the desired applications [8]. The collected data are analyzed and employed on data visualization for the end users to receive information updates of the weather conditions.

29.2.2 Hardware Component

The developed monitoring system consists of NodeMCU ESP8266 microcontroller as a main controller unit (MCU), sensors such as rain drop sensor, CO sensor, temperature, and humidity sensor. In addition, there are connected devices to the MCU which are GSM and Wi-Fi that are widely used.

The Arduino UNO SMD Rev3 is used an ATmega328-based microcontroller module with 14 optical input/output (I/O) pins. There are also six analog input pins available and a crystal oscillator which has a bandwidth 16 MHz. Some functionality also includes several I/O which consist of an ICSP header, a reset button, a USB link, and a power port [9]. The Arduino UNO SMD Rev3 provides all the microcontroller needs, by connecting to a charging computer with a USB cable to get started with an AC-to-DC adapter or battery.

The rain drop sensor is the most important part as a rain sensing tool. There are two components in the sensor, which are a weather board sensing a control module and the

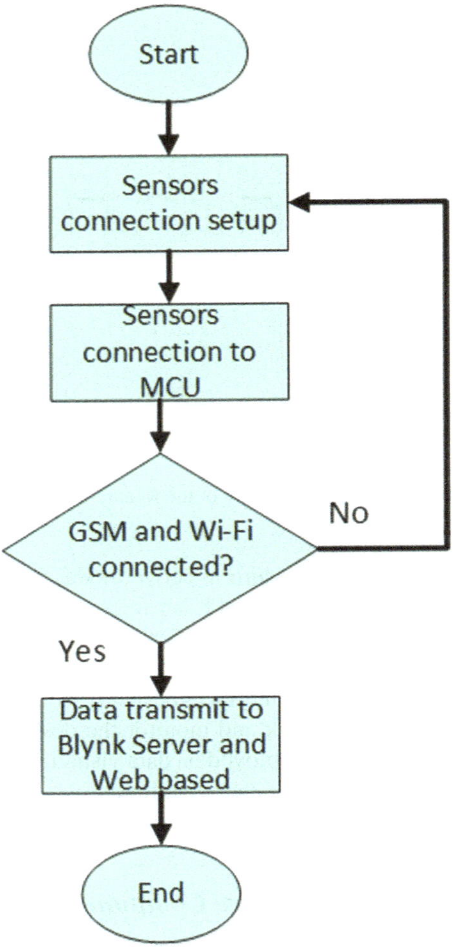

Fig. 29.2 Flowchart of system

rain comparing and translating the analog value to a digital value. In the automobile industry, rain drop sensors can also be used to remotely control windscreen wipers [10]. In the agricultural sector and in home automation systems, rain drop sensors can be used to automatically track the rain water distribution.

The CO sensor used can be applied to detect various CO containing gases and is suitable for various applications. Besides the sensor detects low and high temperatures by the cycle process and detects CO at low temperatures which is heated by 1.5 V. The conductivity of MQ-7 sensor is greater as the concentration of gas increases.

The DHT11 is a sensor for temperature and humidity that is widely used in sensing weather parameters. The sensor is also able to be mounted to other microprocessors easily. The sensor provides the features of an 8-bit microcontroller and temperature measurement for the output of serial relative humidity and temperature value data.

29.2.3 Software Development

The functionality of the system includes integrating the peripheral devices with software. Figure 29.3 shows the circuit diagram of the weather monitoring system. The monitoring system includes several sensors, communication connectivity, and MCU programmable by Arduino UNO SMD Rev3 as the main part of the weather monitoring system. All of the virtual connections and hardware components are to be synchronized with the Blynk application. Besides, Arduino IDE application was used as an editor to develop the coding called sketch that allows the system to work, whenever a signal for a hazardous weather condition is detected. Once coding is successfully verified and compiled, the codes will be uploaded to the NodeMCU via the serial interface of the MCU. The sketch will allow all the sensors to send the sensed data to the NodeMCU. NodeMCU will process all the data collected and upload it to Blynk application via the Wi-Fi connection. The data are delivered to the OLED display module, and Blynk application. Furthermore, the data are analyzed

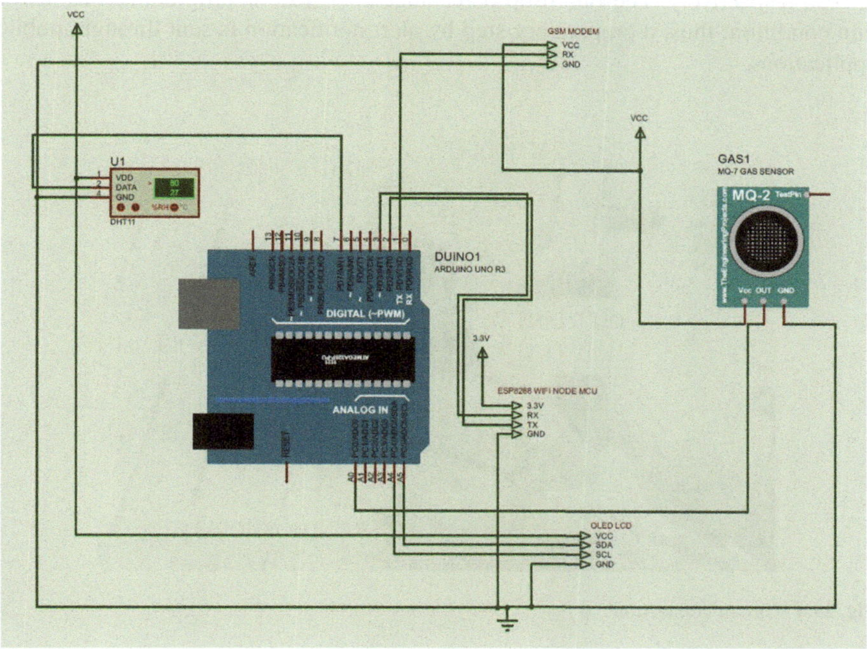

Fig. 29.3 Circuit diagram of system

and employed on data visualization for the end users to receive information updates of the weather condition based on notification alert.

The reliability data measurements for temperature, humidity, percentage of rain, and CO concentration from the sensors during the monitoring application were observed. The information data such as time, temperature in degree Celsius (°C), percentage humidity, percentage rain drop, and air quality index (AQI) by CO percentage were extracted from Blynk application. Then, the data are sent to OLED, SMS, and email notifications.

29.3 Results and Discussion

Figures 29.4 and 29.5 show the circuit connection for the device prototype of the monitoring system. For data collection, the device is placed at outdoor environment. All of the data were recorded by the Blynk application. Figure 29.6 shows OLED display module in which displaying the data collected from all sensors. In case the air quality index (AQI) value is $\geq$ 300, the surrounding is considered hazardous, and an emergency alert is likely to be issued. In normal room temperature conditions as the rain drop sensor is dry and not covered by any rain drop, the average values of temperature, humidity, and rain drop percentage are at 31 °C, 71%, and 1.07%, respectively. The rain drop percentage at $\geq$ 40% is programmed as heavy rain condition; thus, a preparatory step by alert notification is sent through mobile applications.

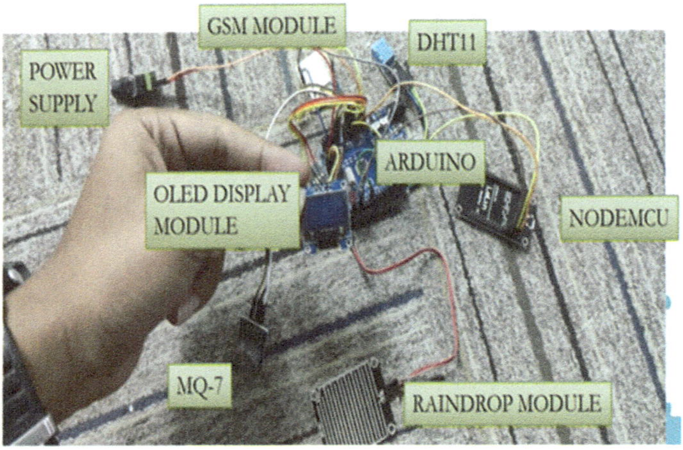

Fig. 29.4 Circuit connection

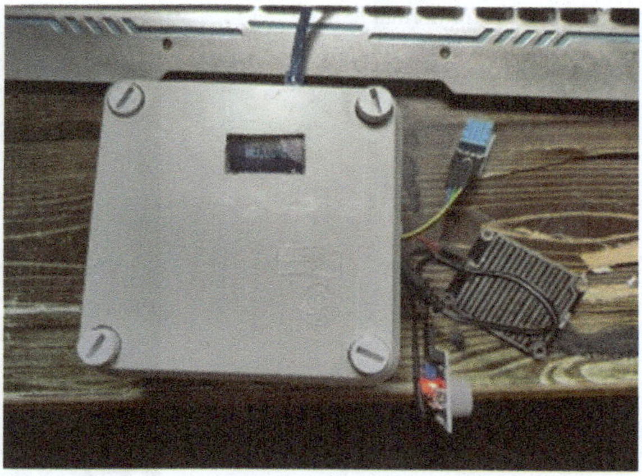

Fig. 29.5 Project prototype

Fig. 29.6 Information data display on OLED

Figures 29.7 and 29.8 show the graphical user interface (GUI) of the real-time data collected by all the sensors based on normal and high AQI value situations, respectively. The Blynk application interface displayed the updated data instantaneously.

Figure 29.9 shows the alert notification message from the weather monitoring system from GSM module. The notification of abnormal conditions such as 'It's raining heavily!!' and 'Abnormal air quality index!!' will pop-up whenever a certain parameter is not in agreement with the normal conditions. Figure 29.10 shows the

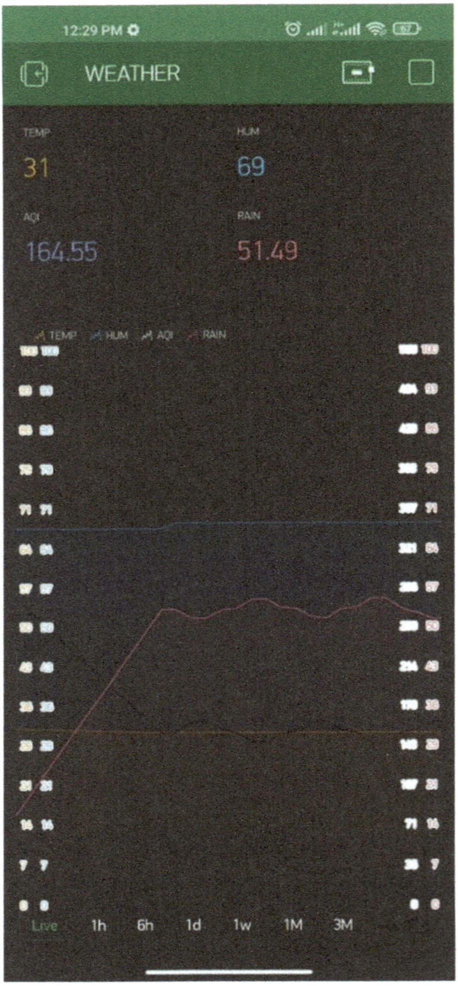

Fig. 29.7 GUI for normal data

notification by email from the Blynk application. The message displayed is similar to the alert notification from SMS via GSM.

29.4 Conclusion

In this paper, the development of an IoT-based weather monitoring device system using Arduino UNO SMD Rev3 is presented. The device does not require to have continuing maintenance for the system functionality, thus save labor cost. Several achievements of the development were accomplished, including (1) the information

Fig. 29.8 GUI for high AQI data

data such as temperature, humidity, rain, and AQI of CO was successfully transmitted from all the sensors to the OLED and mobile application. By this, the user could manage any activities properly, (2) the system used Arduino as a microcontroller associated with Blynk application was able to effectively display all the data information in a real-time basis, and (3) the GUI development through mobile application allowing easy alert notification platform by the user to monitor the weather conditions at anywhere.

According to the final model of the device system, significant advancements in hardware design can be implemented in order to produce better results and commercialized on a worldwide scale. Future work should be conducted to research on the

Fig. 29.9 SMS alert notification

Fig. 29.10 Email alert notification

sensitivity enhancement of the sensors and adapting long range (LoRa) communication technology connectivity as low power and long range technology capabilities preference.

Acknowledgements The authors wish to thank UniKL and Earth Sciences and Environment research team of UKM for the support given in facilitating and contributing to the success of this project.

References

1. Wiston M, Mphale KM (2018) Weather forecasting: from the early weather wizards to modern-day weather predictions. J Climatol Weather Forecast 6(2):1–9
2. Zhu AW, Pi H (2014) A method for improving the accuracy of weather forecasts based on a comprehensive statistical analysis of historical data for the contiguous United States. J Climatol Weather Forecast 2(1):1–4
3. Sampe J, Yunus NHM, Yunas J et al (2019) Architecture of an efficient dual band 1.8/2.5 GHz rectenna for RF energy harvesting. Telkomnika 17(6):3137–3144
4. Sampe J, Yunus NHM, Yunas J et al (2017) Ultra-low power RF energy harvesting of 1.9 GHz and 2.45 GHz narrow-band rectenna for battery-less remote control. IJIEE 7(3):118–122
5. Yunas J, Yunus NHM, Sampe J et al (2020) Design and fabrication of glass based mems patch antenna for energy harvester. In: 2020 IEEE international conference on power and Energy (PECon), pp 362–365
6. Al-Sarawi S, Anbar M, Alieyan K et al (2017) Internet of things (IoT) communication protocols. In: 2017 8th International conference on information technology (ICIT), pp 685–690
7. Kumar NM, Mallick PK (2018) The Internet of things: insights into the building blocks, component interactions, and architecture layers. Procedia Compu. Sci 132:109–117
8. Serikul P, Nakpong N, Nakjuatong N (2018) Smart farm monitoring via the Blynk IoT platform: case study: humidity monitoring and data recording. In: 2018 16th International conference on ICT and knowledge engineering (ICT&KE), pp 1–6
9. Seneviratne P (2017) Getting ready for the development environment. In: Building Arduino PLCs, Apress, Berkeley, CA, pp 1–22
10. Rabiei E, Haberlandt U, Sester M et al (2013) Rainfall estimation using moving cars as rain gauges–laboratory experiments. Hydrol Earth Syst Sci 17(11):4701–4712

Chapter 30
IoT Based Home Security for Housing Areas

Suraya Mohammad, Muhammad Fadhil Mohamad Bakri, Mohd Raziff Abdul Razak, Dzulhadi Sapari, and Mohamed Shahril Abdul Rashid

Abstract Internet of things (IoT) refers to the infrastructure of connected physical devices which is growing at a rapid rate as a huge number of devices and objects are getting associated with the Internet. Home security is one of the useful applications of IoT. This paper describes a project for a home security system. The system will inform the owner about any unauthorized entry or whenever the door is opened or if there is any gas leakage by sending the notification to the user through a smartphone. The system consists of an Arduino Uno as the main microcontroller, a PIR sensor to detect human movement, a buzzer to sound an alarm, an MQ-6 gas sensor to detect gas leakage, and an ESP8266 Wi-Fi module or NodeMCU to connect to the Internet. The main advantages of the system are ease of use, lower cost, and low maintenance.

Keywords Internet of things · Home security · ESP8266 Wifi module · PIR sensors · And MQ-6 sensors

---

S. Mohammad (✉) · M. F. M. Bakri · M. R. A. Razak
Advanced Telecommunication Technology, Communication Technology Section,, Universiti Kuala Lumpur British Malaysian Institute, Gombak, Malaysia
e-mail: surayamohamad@unikl.edu.my

M. R. A. Razak
e-mail: mraziff@unikl.edu.my

D. Sapari
Unit Perunding ICT, Cawangan Kejuruteraan Elektrik, Ibu Pejabat JKR Malaysia, Kuala Lumpur, Malaysia
e-mail: dzulhadi@jkr.gov.my

M. S. A. Rashid
Jabatan Kerja Raya Malaysia, Cawangan Kejuruteraan Elektrik, Ibu Pejabat JKR Malaysia, Kuala Lumpur, Malaysia
e-mail: mshahrilar@jkr.gov.my

A. Ismail et al. (eds.), *Advanced Materials and Engineering Technologies*, Advanced Structured Materials 162,
https://doi.org/10.1007/978-3-030-92964-0_30

30.1 Introduction

Everyone wants to be as much safe as possible that is why security is a prime concern in our day-to-day life. Home security systems are sensible in cost and can be added to practically any house or building. There are a few significant motivations to consider getting a home security system. To start with, when criminals break into a house, they are not liable to be worried about the individual well-being of anybody in the house. The internet of thing (IoT) based home security system can be embraced at home or office as it has different sorts of sensors. This project was utilized to identify two things (intrusion and fire). The reason why fire detection is used is that based on research, most of the fire cases that occur in Malaysia in a year started with gas leakage. Usually, when the police/firefighter arrives, it is too late, and the damage has been done.

This project is easy to use as it can be operated wirelessly to turn it ON or OFF via a web app. Any intrusion or gas leakage will promptly be notified to the user via web app and all the info will be stored inside the database for future reference and analysis. If there is either an intrusion or gas leakage, the alarm will go off indicating there is danger, that way the neighbor can somehow be notified about it which maybe can save the user home with quick action. By detecting the intrusion and sounding an alarm, a security system also will often scare an intruder away.

Several works have been proposed for home security systems. Work by [1] proposed two approaches for home security. The first approach depends on GSM innovation, and the second one uses a web camera to detect an intruder. For the first approach, the smart home prototype was designed using various sensors such as infrared (IR), temperature, gas, and light-dependent resistor sensor. LED lights and fans were also included in the design to make it more resemble actual home appliances. The proposed system is controlled by an Atmega644p microcontroller. It collects information from the sensors, makes decisions, and sends an SMS notification to the corresponding number or homeowner by using a GSM modem. Some examples of a message being sent to the homeowner are "Fire at home" and "Gas leakage." These two messages will be sent if the temperature is higher than the set threshold value or if the gas sensor is activated. As for intruder detection, the security system uses the Yawcam (Yet Another WebCAM Software) software to interface with the web camera. Yawcam is a Multilingual Java software for viewing and transferring images or videos from a webcam. The system detects the motion of the intruder, sounds an alarm, and sends an email to the homeowner.

The second work is described in [2]. The prototype system is also designed to detect an intruder. When an intruder is detected, the system will send a voice call notification to the home owner's smartphone. Unlike the previous approach, the alarm activation is optional and is based on homeowner discretion. In addition, the provision for sending alert messages to concerned security personnel in case of critical situations is also built into the prototype system. In the case of the owner identifies that the person entering his house is not an intruder but an unexpected guest, the user/owner can make arrangements such as opening the door or switching on various appliances inside the house by using his/her smartphone. This system is designed using TI CC3200 LaunchPad as the main controller, PIR motion detector sensor, alarm, relays for connecting home appliances, electromagnetically controlled doors or windows, smartphone phone to receive Voice Call and Energia software.

Another work is described in [3]. This paper describes the development of a home security system comprising of two different sensors, a passive infrared sensor, and an infrared sensor to monitor the home condition, in this case, motion, ESPresso Lite V2.0 to connect all the sensors and act as the main controller, Blynk application as the main interface with the smartphone and FAVORIOT platform to receive the data sent from the ESPresso Lite V2.0. Sarkar et al. [4] proposed an android based home security system using IoT and firebase. The system used two types of sensors, PIR sensors and flame sensors. Whenever the sensors get triggered, a notification will be sent to an Android app installed on a smartphone.

30.2 Methodology

30.2.1 Hardware Setup

The main block diagram is as shown in Fig. 30.1. Based on Fig. 30.1, the main components are NodeMCU ESP8266, web application, MQ 6 gas sensor, infrared passive sensor (PIR) sensor, a buzzer, and a Firebase database. NodeMCU is a micro-controller board that comes together with a Wi-Fi module. In this project, NodeMCU is the main controller which acts as a communication hub for sending and receiving

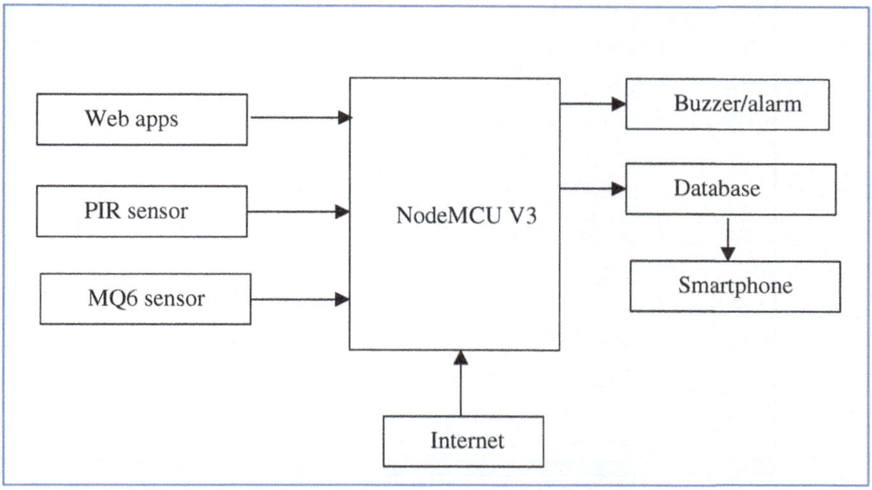

Fig. 30.1 System block diagram

data over the Internet and relaying it to the rest of the module accordingly. Arduino IDE software is used to provide the necessary coding to ensure the NodeMCU can interact with all connected sensors, the firebase database, and the web application [5]. A sample code is shown in Fig. 30.2.

The PIR sensor stands for infrared passive sensor. It is a low-cost sensor, capable of detecting human or animal activity by detecting infrared radiation emitted from objects with a temperature above zero. As the name implies, the MQ-6 gas sensor is a gas sensor. It was chosen for this project because the sensor can detect or measure common household gases such as LPG and butane. The type of buzzer used is an active passive buzzer. It is a small yet efficient component to add sound features to our project/system. In this case, to provide an alarm for the home security system.

This project will be using Firebase [6] as the database to store and send the information needed to the components or web application. Firebase also provides a

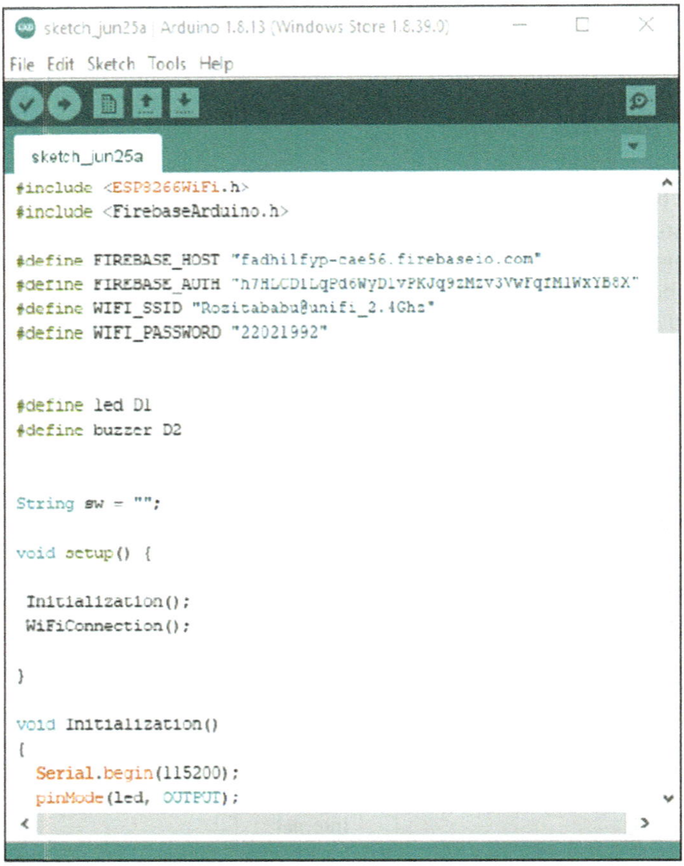

Fig. 30.2 Sample of an Arduino sketch

real-time database and back-end services which makes it very good to be used in an IoT project like this. Firebase is free to use, and it is also one of the easier databases to operate which makes it the database of choice in this project.

Like any other IoT-based home security system, the system also comes with a web application. In this project, 000webhost is used to create a working website which is the web application to monitor the state of the system. 000webhost is a free website hosting solution that provides a variety of features, including a website builder and WordPress support, all of them without any ads. Users can upgrade to a paid plan to get even more features and support, but based on the reviews, 000webhost is the best free web hosting solution for those who are truly on a budget. The HTML and JavaScript codes in the development of the web application are written using the Notepad++ editor.

To convert the web application to an android application for smartphones, a software called MIT App Inventor is used [7]. The software lets the user develop their application for Androids smartphone by using a web browser or either a connected phone or emulator. The software server will store the user's work and helps to keep track of the project. This software will convert the URL of the web application received from 000webhost into the apk file which can be downloaded and installed on any Android smartphone or device.

30.2.2 Circuit Diagram

The circuit connection for the system is as shown in Fig. 30.3. Firstly, for the PIR sensor, the output pin (orange) is connected to the D0 pin of the NodeMCU while the VCC (red) and GND (black) pin is connected to the ground and 3.3 V pin of the NodeMCU, respectively. Next, the gas sensor analog pin (yellow) is connected to the A0 pin on the NodeMCU, and similar to the PIR sensor connection; the GND and VCC pin on the gas sensor is connected to the ground and 3.3 V of the NodeMCU. The LED anode (blue) is connected to the D1 pin on the NodeMCU and the cathode is connected to the ground. Lastly for the buzzer, its positive pin is connected to the D2 pin of NodeMCU, and its negative pin is connected to the ground.

30.2.3 Flow Chart

Next, Fig. 30.4 shows the flow chart for the system. Based on the figure, the system starts with the home security system web application in the smartphone is open. Initially, the system will be in an idle state in which it will wait until either the infrared passive sensor (PIR) sensor or the MQ-6 sensor detect something or is activated. If any of the sensors is triggered, the buzzer will go off and information will be sent to the database. At the same time, a notification will be sent to the user

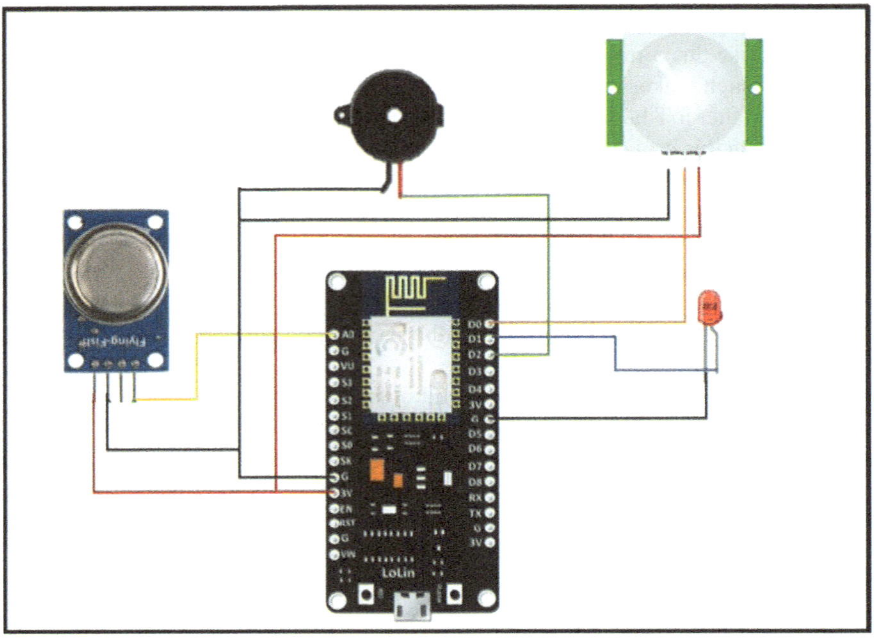

Fig. 30.3 Circuit diagram

or homeowner through the web application indicating either an intruder or a gas leakage is detected, depending on which sensor is triggered.

30.3 Results and Discussion

Figure 30.5 and 30.6 show the completed hardware and web interface, respectively. The Android app installed on an Android smartphone also looks the same as the web interface and it works the same as the web version. This project is designed in two modes of operation. The first is the user or homeowner view and the second mode is the administrative or admin view. Admin in this context can be the guard in the guardhouse or even the administrative people at a condominium. It refers to the people who have to watch over the user's house in a housing area.

30.3.1 User View Mode

To start using the system, a user need to sign in by entering their username and password. This way the system is made secure against unauthorized access. Upon

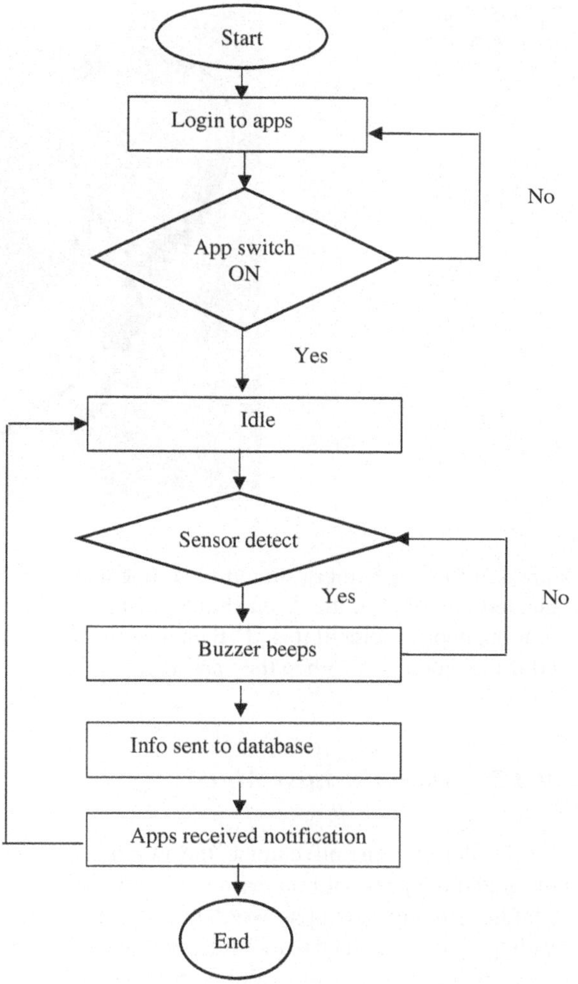

Fig. 30.4 Flow chart

entering the correct username and password, access will be granted and a user monitoring page will appear. To activate the system, the button SWITCH must be turned "ON." The page will display the status of the PIR sensor and the MQ-6 gas sensor in real-time. Supposedly when there is an intrusion or gas leakage, the sensor will be triggered and a notification will be sent to the web application and the android application. At the same time, an alarm will also be activated. Sample of notification received by the applications is as shown in Fig. 30.7.

In addition, the system is also able to generate a status report for both the PIR and the gas sensors. This report will be generated when a user pushes the "Click to see PIR sensor report" button for the PIR sensor and the "Click to see Gas sensor report" button for the gas sensor. A sample of the PIR report is as shown in Fig. 30.8. As

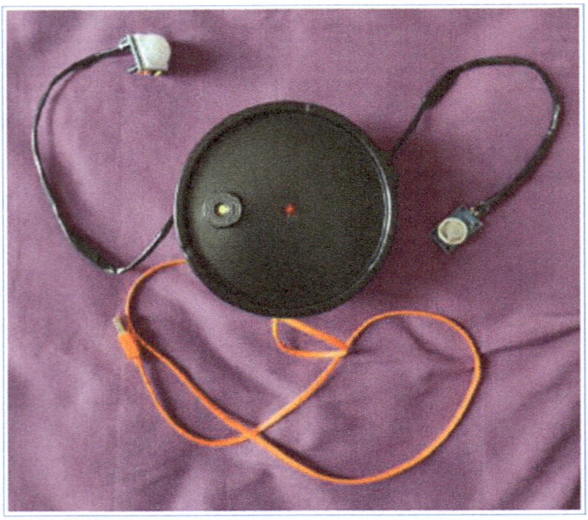

Fig. 30.5 Completed system hardware

shown in Fig. 30.8, there should be a time and date for the movement in the house detected recorded in the "Detachment history." The reading value will indicate the component parameter, if it is "1" then the sensors/components are currently activated and if the value is "0," then they are idle.

30.3.2 Administrative Mode

To enter the administrative mode, the admin needs to click on the "Log in as Admin" button and the page will be redirected to the admin login page as seen in Fig. 30.9. After inserting the right password, the admin can then proceed to the admin monitor by clicking the "Proceed to monitor" button. As seen in Fig. 30.10, the admin monitor can view the state of many houses. Similar to user view, the admin will get a notification if the house is breached or there is a gas leakage (refer to Fig. 30.11). The admin view however does not has the option to see the report of either gas leakage or intruder detection.

30.4 Conclusion

This paper described a development for a home security system for a housing area. The system incorporates PIR sensors to detect an intruder and an MQ-6 sensor to detect gas leakage. If any sensors are triggered, a notification will be sent to the homeowner and security personal. A log report of the event can also be generated

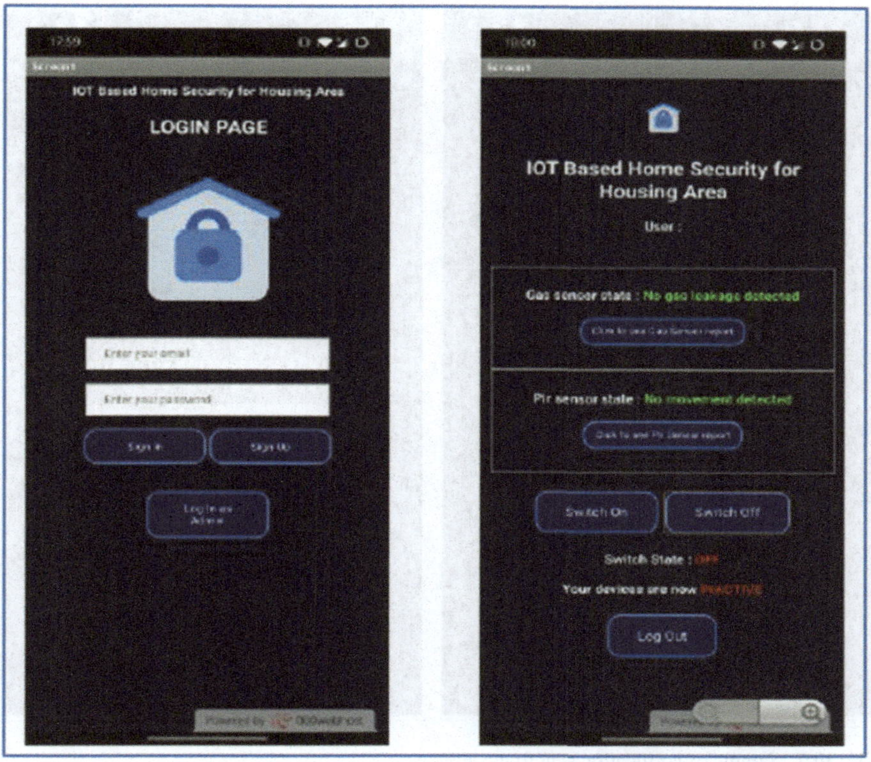

(a) (b)

Fig. 30.6 Android app interface. **a** User login page, **b** User monitoring page

upon request. For ease of use, a web interface application was also designed. Although the system is shown to be working as expected, further improvement can still be made to the system.

Other monitoring devices such as a camera can be added to the system. This way users can identify the intruder as the real intruder or simply an unexpected guest. Another improvement concerned the fact that the system is highly dependent on the Wi-Fi connection. Network latency will affect the system performance. The time taken to transfer data to NodeMCU and then to firebase will be increased as the network latency increased. The system must also have a way to notify the user that the Wi-Fi at their home is down as the system just won't work if there is no internet connection.

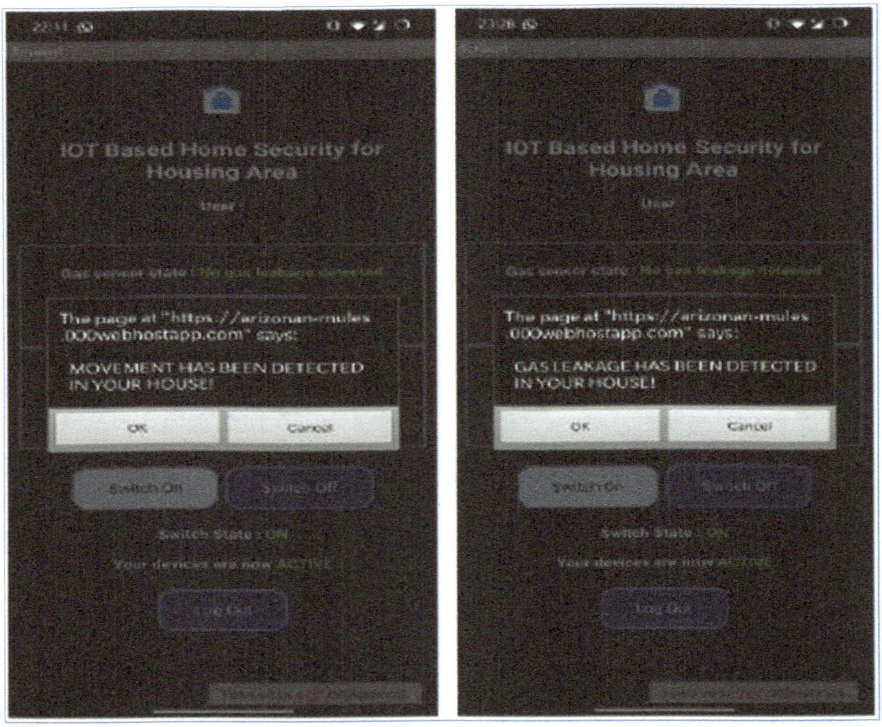

Fig. 30.7 Sample of notification sent to the app

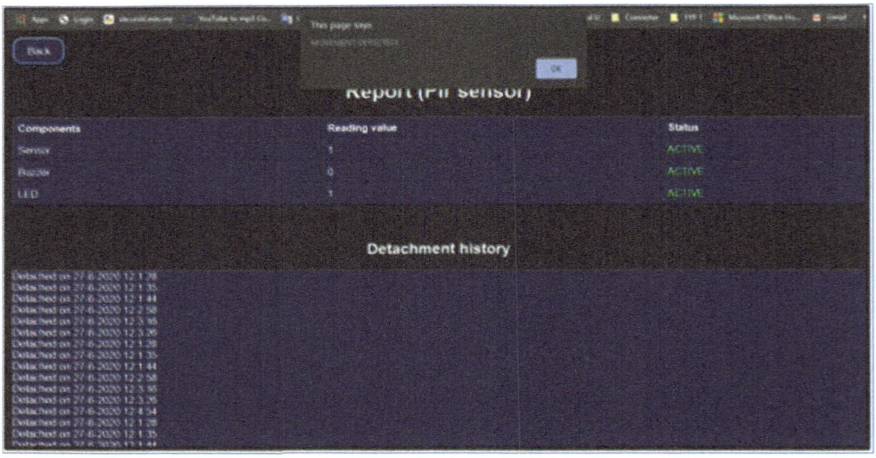

Fig. 30.8 Sample of PIR sensor report

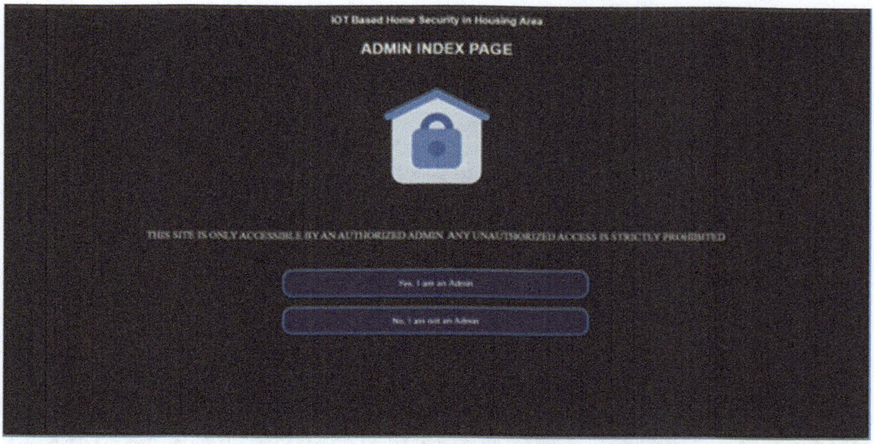

Fig. 30.9 Admin main page

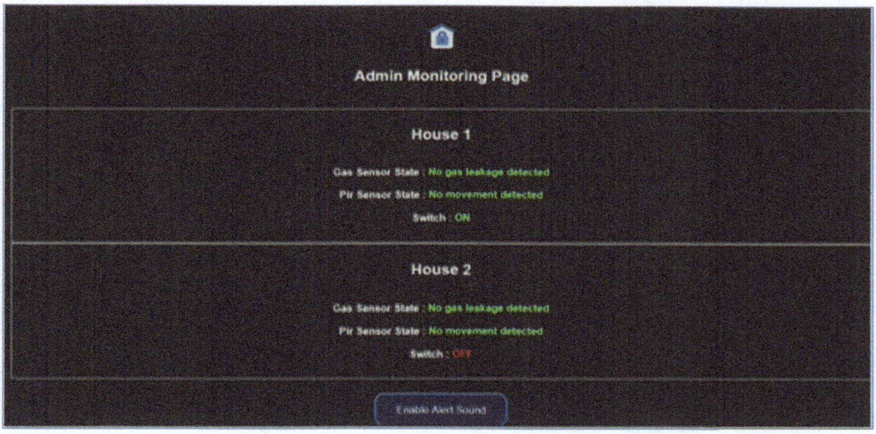

Fig. 30.10 Admin monitoring Page

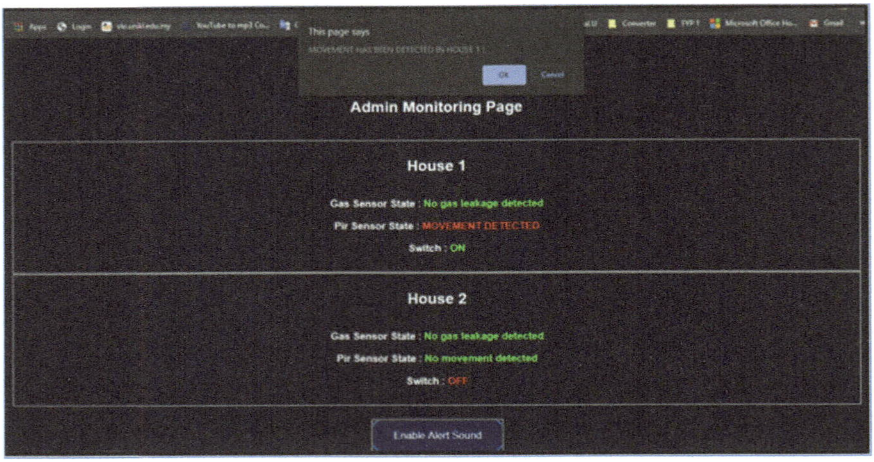

Fig. 30.11 Sample of admin alert notification

References

1. Bangali J, Shaligram A (2013) Design and implementation of security systems for smart home based on GSM technology. Int J Smart Home 7(6):201–208
2. Kodali RK, Jain V, Bose et al (2016) IoT based smart security and home automation system. In: 2016 international conference on computing, communication and automation (ICCCA), IEEE, pp 1286–1289
3. Abu MA, Nordin SF, Suboh et al (2018) Design and development of home security systems based on internet of things via favoriot platform. Int J Appl Eng Res 13(2):1253–1260
4. Sarkar S, Gayen S et al (2018) Android based home security systems using internet of things (IoT) and firebase. In: 2018 International conference on inventive research in computing applications (ICIRCA) IEEE, pp 102–105
5. Connecting NodeMCU and Arduino. https://github.com/FirebaseExtended/firebase-arduino. Accessed 20 July 2021
6. Firebase. https://firebase.google.com/docs. Accessed 20 July 2021
7. MIT App Inventor. https://appinventor.mit.edu/. Accessed 20 July 2021

Chapter 31
Malay Cued Speech Recognition Using Image Analysis: A Review

Muhammad Ghazali Twahir, Zulkhairi Mohd Yusof, and Izanoordina Ahmad

Abstract Automatic real-time translation of gestured languages for hearing-impaired would be a major advancement on disabled integration path. Cued speech (CS) is a specific visual hand gesture that complements oral languages lip-reading. Cued speech in Bahasa Malaysia (CSBM) is an adaptation of cued speech for use in the Malay language. The cued speech recognition system is capable of detecting all necessary parameters of CS (handshape, hand position, and hand movement) and translate it to text equivalent. The aim is to help the deaf learn and practice the basic of cued speech consonant and vowel using hand gesture. This paper looks into the existing researches involved in this area and also the sensors and methods they used. Due to the limited number of researches for cued speech, related researches such as sign language (SL) translator systems and hand gesture recognition are also reviewed. This paper gives a general overview on the implementation of cued speech recognition system that automatically recognize a succession of cued speech hand gestures in real time. A Malay cued speech recognition system using image analysis is proposed.

Keywords Cued speech · Machine learning · Transliterate system · Image analysis · Hand gesture

---

M. G. Twahir (✉)
R4R Research Cluster, Communication Technology Section, Universiti Kuala Lumpur British Malaysian Institute, Batu 8, Jalan Sungai Pusu, 53100 Gombak, Selangor, Malaysia
e-mail: ghazali@unikl.edu.my

Z. M. Yusof · I. Ahmad
R4R Research Cluster, Electronics Technology Section, Universiti Kuala Lumpur British Malaysian Institute, Batu 8, Jalan Sungai Pusu, 53100 Gombak, Selangor, Malaysia
e-mail: zulkhairi@unikl.edu.my

I. Ahmad
e-mail: izanoordina@unikl.edu.my

A. Ismail et al. (eds.), *Advanced Materials and Engineering Technologies*, Advanced Structured Materials 162,
https://doi.org/10.1007/978-3-030-92964-0_31

31.1 Introduction

As someone speaks, a hearing-impaired can try to guess the oral message by lip-reading. This is a difficult task for different phonemes which correspond to identical mouth shapes. Therefore, Dr. Cornett developed the cued speech in order to improve the lip-reading efficiency [1]. The manual gestures to lip shapes are proposed to ensure that each sound has an original visual aspect. Thus, the "hand and lip-reading" becomes as meaningful as the oral message.

A significant difference between these two communication systems is that SL is a complete language, while CS is not a language at all. For example, cued speech in Bahasa Malaysia (CSBM) is a visual representation of the spoken Malay language itself. Due to that, CS may have an advantage over SL in an environment where a translation is being made [2]. The deaf will also learn to actually speak with their mouth while learning cued speech with their hand. However, it will take a lot of time and practice for them to be able to speak proficiently and fluently. In order to improve the communication of deaf cure, there is a need of an automatic system that recognizes the CS and translates it to text equivalent.

A new approach called CSBM is based on a syllabic decomposition: The message is formatted into a list of "consonant–vowel syllable" (a CV list). Each CV is coded with a specific gesture, which is combined to its lip shape, so that the whole looks unique and understandable. A gesture contains two pieces of information which is the area handshape (for the consonant coding as shown in Fig. 31.1) and a location around the face (for the vowel as shown in Fig. 31.2). Hand coding brings the same quantity of information than the lips movement. It is as difficult to lip-read without gestures as to understand the hand coding without lips movements. This symmetry explains why a single gesture codes of numerous phonemes, which correlate to different lip shapes. Thus, there are only eight handshapes and four positions for a combination of 32 CV-gestures.

The contribution of this research is to review methods of automatically recognize a succession of cued speech hand gestures in real time. In the future, a complete hearing-impaired translator could be feasible by coupling such a device with an automatic lip-reading module and others various automates.

31.2 Literature Review on Cued Speech

A cued speech recognition system is a system that automatically recognizes the hand shape, movement, and position and then display the text equivalent in the computer screen. In order for the system to do the recognition, a camera that can capture necessary parameters of CS such as handshape, hand position, and hand movement is needed. All these parameters will be processed by certain method in order for the gesture to be accurately determined. Some of the methods have been developed for

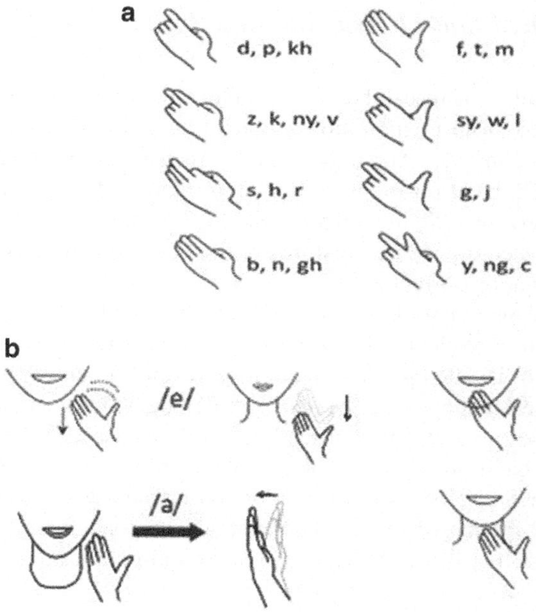

Fig. 31.1 **a** Eight handshapes for the 24 Malay consonants. **b** Four position with respect to the face for the six Malay vowels

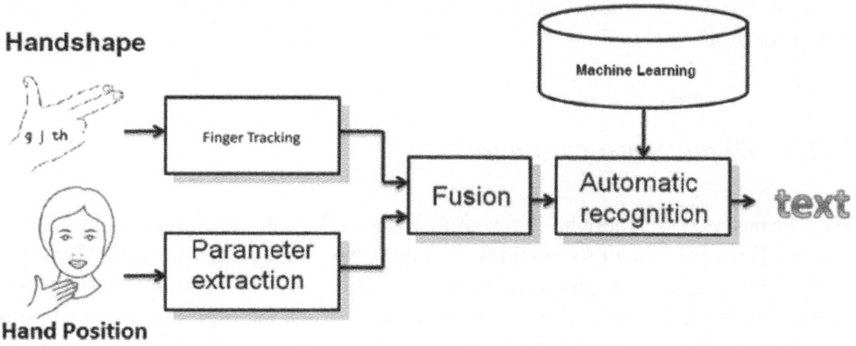

Fig. 31.2 System diagram of Malay cued speech recognition

recognizing the hand gesture such as using computer vision, depth cameras, glove-based system, hidden Markov models, adaptive boosting (AdaBoost), random forest regression (RFR) Progress, finger tracking algorithm, and glove-based system. These methods will be discussed in the following sub-section.

31.2.1 Existing Cued Speech Transliterate System

There are a limited number of researches for an automated cued speech recognition system. Some of the earliest explorations into CS recognition were conducted using computer vision techniques. Aboutabit et al. [3] focused on the identification of vowels by merging CS hand positions and lips information. Hand position was conducted using the Gaussian classifier which took the 2D hand positions as input. The vowel recognition used the merged features of the lips and hand position, and obtained 77.6% identification correctness.

Given the success that hidden Markov models (HMM) have had in the field of automatic speech recognition, Heracleous et al. [4] used the context-independent HMM-GMMs to decode a set of isolated phonemes extracted from CS sentences, i.e., the temporal boundaries of each phoneme to be recognized in the video was given at the test stage. In fact, the audio-based temporal segmentation was used for the temporal alignments of the lips, hand position, and shape. The corpus was derived from a video recording of the CS speaker with blue colors on lips and hand pronouncing and coding a set of 262 French sentences. The experiments concerning the vowel, consonant, and phoneme recognitions were presented and obtain an accuracy of 78.9%.

In the previous work on CS recognition, the video images were recorded with artifices applied to the CS speaker before the recording (blue sticks on the lips, blue marks on the hand and forehead) in order to mark the pertinent information and make their further extraction easier. However, recent researches are struggled to provide a robust and real-time solution that can adequately track handshapes against varying backgrounds and occlusions.

31.2.2 Depth Sensor Camera

Over the past decade, there has been significant exploration into using depth cameras such as Microsoft Kinect for tracking the hands gesture. The device itself features an RGB camera, a depth sensor, and a multi-array microphone and is capable of tracking the users' body movement. Since Kinect is able to track the user's full body, it seems natural to build a framework for sign language recognition. Much of the work has focused on generalized hand tracking with a priority on real-time processing and arbitrary camera angles. Ahmed et al. [5] develop Deaftalk, a sign language interpreter using Microsoft's Kinect depth camera that provides 84% accuracy detection.

The gestures recognition technologies in Kinect visual gesture builder implemented are the AdaBoost and RFRProgress. AdaBoost is a trigger which gives us a true Boolean value, while the person is performing a particular gesture; it uses the adaptive boosting machine learning algorithm. RFRProgress on the other hand produces continuous results giving us an analog data of progress. The user is performing the gesture, thus, enabling the system to detect how much of the gesture is completed and how much is the hit rate at the particular frame of the gesture. This approach uses the random forest regression machine learning algorithm. Although it facilitates body and hands tracking and creates the depth image directly, it does not support hand shape recognition. Since sign language and CS generally features different hand shapes, similar signs cannot be distinguished.

31.2.3 Finger Tracking

A number of approaches have been explored to get around the problem of hand shapes detection, but no clear consensus as to which holds the most promise has formed. Some of the approaches, such as finger counting using convex hull algorithm were shown to work in a particular case. Gurav et al. [6] developed a method using background subtraction and HSV segmentation together to create a mask. After the hand is segmented, the number of fingers raised could be detected. The largest contour in the image which is assumed to be the hand is then found. Then, the convex hull and convexity defects which are most probably the space between fingers is classified. All this algorithm such as convex hull, background subtraction, and HSV segmentation are provided in the open computer vision (OpenCV) library. OpenCV is used in HCI, robotics, biometrics, image processing, and other areas where visualization is important and includes an implementation of Haar classifier detection and training [7]. Its finger counting algorithm achieved 92% accuracy with convex hull technique. This is a manual way of finding the number of fingers and a necessary step to identify each different finger.

31.2.4 Glove-Based

Glove-based systems have achieved the most impressive SL results in terms of vocabulary size, with over 90% accuracy being obtained in continuous sign detection across more than 5000 Chinese signs [8]. However, such systems are both expensive and require the user to wear unnatural devices.

31.3 Requirement of Cued Speech Recognition

31.3.1 System Overview

The ultimate goal of this research is to develop a Malay cued speech hand gesture recognition system. The system diagram in Fig. 31.2 shows the basic flow of the proposed system. There are a number of subsystems within the proposed system that will be explained in more detail in the section B.

In this proposed system, the user will perform desired hand gesture; thus, the Kinect camera will capture this hand shape and hand position. The handshape image is processed by the finger-tracking algorithm to determine how many fingers are engaged. Then, each finger could be classified into its respective consonant. Similarly, hand position and movement video frame are captured in parameter extraction module where skin, depth, contour points, and movement are detected. The images or video streams are next sent to the machine learning functions where they are matched and compared with stored images templates to identify the correct consonant. Lastly, the consonant is fused with the vowel to produce text of Malay syllable in the computer screen.

31.3.2 Proposed Method

Hand shape recognition by using finger gesture tracking is proposed for this research. The hand shape will be captured by using the Kinect camera which has sensors of both RGB and depth data. The hand data will be processed by convex hull methods. This method will count the number of engaged fingers, and then, each finger name is classified using fingertip tracking methods. This will be made possible by implementing the OpenCV library in the system. In result, the algorithm will be able to translate eight different hand shapes into specific consonant.

Microsoft Kinect provides Visual Gesture Builder software to detect hand position and movement using machine learning methods. VGB has built in detection technology such as discrete gesture and continuous gesture. Hand position and movement recognition can be achieved by using the AdaboostTrigger and RFRProgress algorithm. In result, the system will be able to translate four different hand positions and movements into specific vowel.

The cued speech recognition software is build using C# in MS Visual Studio. This software will be used to integrate the finger tracking algorithm and machine learning method that could translate the consonant and vowel. This method will produce a syllable which resulting a Malay words.

31.3.3 Dataset, Test, and Evaluation

In order to measure the accuracy of the system, the average hit rates of particular gesture from test runs by different users will be compared. In addition, the data validation will be counted based on false positive and false negative against the correct result. A confusion matrix will summarize the result of the testing algorithm for further inspection.

31.4 Conclusion

After the survey on the approaches used in various CS and SL recognition systems, the methodologies and algorithms involved in this research could be justified. Most of the times, a combination of different methods and algorithms has to be used to achieve a moderate to acceptable rate of recognition. This will allow the system to offer complete phonetic representation of cued speech hand gesture recognition. In the future, a complete hearing-impaired translator could be feasible by coupling such a device with an automatic lip-reading module and others various automates. From a technical point of view, there is a vast scope for the future research and implementation in this very field. The ultimate gain of the proposed study is enormous.

References

1. Cornett RO (1994) Adapting cued speech to additional languages: procedures followed in assigning phonemes to cue groups in the development of cued speech in its original form. Cued Speech J 19–29
2. Nicholls GH, Ling D (1982) Cued speech and the reception of spoken language. J Speech Hear Res 25:262–269
3. Heracleous P, Aboutabit N, Beautemps D (2009) Lip shape and hand position fusion for automatic vowel recognition in cued speech for French. IEEE Signal Process Lett 16:339–342
4. Heracleous P, Hagita N, Beautemps D (2010) Gestures and lip shape integration for cued speech recognition. In: 2010 20th International conference on pattern recognition, pp 2238–2241
5. Ahmed M, Idrees M, Abideen ZUl, Mumtaz R, Khalique S (2016) Deaf talk using 3D animated sign language: a sign language interpreter using Microsoft's kinect v2. In: 2016 SAI Computing Conference (SAI), pp 330–335
6. Gurav RM, Kadbe PK (2015) Real time finger tracking and contour detection for gesture recognition using OpenCV. In: 2015 International Conference on Industrial Instrumentation and Control (ICIC), pp 974–977
7. Perimal M, Basah SN, Safar MJA, Yazid H (2018) Hand-gesture recognition-algorithm based on finger counting. J Telecommun Electron Comput Eng 10:19–24
8. Farooq U, Asmat A, Rahim MSBM, Khan NS, Abid A (2019) A comparison of hardware based approaches for sign language gesture recognition systems. In: 2019 3rd International Conference on Innovative Computing (ICIC). https://doi.org/10.1109/ICIC48496.2019.8966714

Chapter 32
The Analysis on Haze Attenuation in 5G Millimeter Wave: A Case Study

Siti Fatimah Nordin, Zuhanis Mansor, and Aizat Faiz Ramli

Abstract Haze is a natural atmospheric phenomenon that happens in Malaysia yearly. Since haze mainly consists of dust and soot, there are high chances that haze particles can be caused by attenuation in millimetre waves. This paper is mainly focused on the haze attenuation rate in Batu Muda, Kuala Lumpur area and analyzed the results if haze particles can affect attenuation. The double Debye formula is used to analyze the haze attenuation rate since there is no concise mathematical modeling for haze. Few variables need to be considered for haze attenuation, such as the API values, size of the particles and particle concentrations. The lower the API values, the concentration of the particles is lower. The size of particles can be divided into two, which are 10 and 2.5 microns. Results show that the size of haze particles also plays an essential role in causing attenuation in millimetre waves. Bigger particles such as PM10 will cause more significant problems in implementing millimetre waves in a real-life scenario.

Keywords 5G · Haze attenuation · Millimeter wave · Radio propagation

S. F. Nordin · Z. Mansor (✉)
Advanced Telecommunication Technology Research Cluster, Communication Technology Section, Universiti Kuala Lumpur British Malaysian Institute, Batu 8 Jalan Sungai Pusu, 53100 Gombak, Selangor, Malaysia
e-mail: zuhanis@unikl.edu.my

S. F. Nordin
e-mail: sfatimah.nordin@s.unikl.edu.my

A. F. Ramli
Electronics Technology Section, Universiti Kuala Lumpur British Malaysian Institute, Batu 8 Jalan Sungai Pusu, 53100 Gombak, Selangor, Malaysia
e-mail: aizatfaiz@unikl.edu.my

A. Ismail et al. (eds.), *Advanced Materials and Engineering Technologies*,
Advanced Structured Materials 162,
https://doi.org/10.1007/978-3-030-92964-0_32

32.1 Introduction

Mobile data usage has been increased up to 50% and is overgrowing as the technological advances [1]–[3]. But, wireless service providers have faced a significant challenge with mobile traffic growth due to the global bandwidth shortage [4]–[6]. Therefore, researching on fifth-generation (5G) mobile communication systems to cope with the demand. Few problems can be caused by attenuation for millimetre waves (mmwave), such as rain and fog. However, haze also is one of the problems causing a disturbance in mmwave signals.

Haze can be considered a common atmospheric phenomenon in South-East Asian countries, including Malaysia. This paper examines and analyses the case study on the attenuation rate for haze in the Batu Muda station location in Kuala Lumpur. This research started as there is less research performed until this day on haze attenuation. Besides that, the primary research focus currently is on rain attenuation. Many kinds of research are made on rain attenuation because it causes more disturbance in the mmwave signal. The rest of the paper is organized as follows. Section 32.2 explains the previous works on haze attenuation. Section 32.3 consists of the methodology of the research. Section 32.4 discussed the results on haze attenuation based on the Batu Muda station's data. And finally, Sect. 32.5 concluded this paper and suggested future work.

32.2 Analysis of Previous Works

Previous work in [7] stated that haze is a very complex particle due to its diversity. The researchers in [8] and [9] have no specific recommendations released by the ITU-R to tackle haze attenuation. The researcher in [8] suggested the Mie scattering theory to estimate the haze attenuation rate. This is because haze particles can be either soluble or insoluble. Haze particles can be categorized into six; soluble particles, sea salt particles and sulfate particles, insoluble particles, dust particles and soot. The majority of haze particles compromise 60% of dust and soot.

Scattering can be caused by dust and soot due to it has a high effect on electromagnetic wave propagation. Combining the scattering cross-section equation, $Q_s(r)$, and the number of particles equation in a unit volume, $n_m(r)$, as shown in Eq. (32.1) can be obtained.

$$\alpha_m = 169.2 \times \frac{\mu D^3}{\rho \lambda^4} \left| \frac{\varepsilon - 1}{\varepsilon + 2} \right|^2 \qquad (32.1)$$

However, further research needs to be done to produce more concise mathematical modeling for haze to help and prevent any disturbance on the mmwave signal.

32.3 Methodology

This work focuses on the attenuation rate caused by haze particles due to the scattering and absorption of the particles. The research concentrated on the haze attenuation rate in a station located at Batu Muda, Kuala Lumpur. The data is obtained from the Department of Environment (DOE) Malaysia under the Ministry of Environment and Water. The air pollution index, also known as API, is a standard calculation that uses the 24 h data retrieved from the air quality monitoring network. The API data is updated every 1 h.

As shown in Fig. 32.1, the system modeling for haze is illustrated. First, the requested data from DOE is extracted and sorted using Excel. It is sorted based on the API values, date and time and also particles concentration $\left(\mu g/\text{m}^3\right)$ In particulate matter 10 (pm10) and particulate matter 2.5 (PM2.5). Next, the variables such as latitude and longitude, signal path length and frequencies are determined. Since the data extracted for temperature is in Celsius, it then is converted into Kelvin. In addition, temperature, speed of light, θ and wavelength will also need to be determined. The dielectric permittivity of water, $\varepsilon(f)$, is obtained by using the double Debye equation. A few things need to be considered: the diameters of the particles and particles concentrations. Finally, the attenuation rate of haze was obtained in terms of a decibel per kilometer (dB/km).

32.4 Results and Discussion

The differences between a year with and without a severe haze crisis are pretty apparent. Figure 32.2 until Fig. 32.4 illustrate the results for the haze attenuation in the station located at Batu Muda, Kuala Lumpur, from the year 2015 until 2018. However, the year 2017 is not presented due to the unavailability of the haze data for that year. Based on the results, it can be said that there is a clear relationship between the API values and attenuation. Higher API values consisting of high particles concentration may cause mmwave attenuation. Secondly, the size of the particles plays a vital role in causing attenuation. This is because larger size particles can quickly scatter the wavelength.

The year 2015 produced the highest recorded API values, which was between 201 until 300, as shown in Fig. 32.2. This is due to the severe haze crisis that happened during that year. The results show that as the API values increase, the attenuation rate will also be increased. For API values between 201 and 300, the mmwave signal starts to have a slight disturbance as the frequency reaches 74 GHz. Meanwhile, for frequency above 84 GHz, it can affect the mmwave signal as the API value is between 101 until 200. The attenuation rate for these three API values is around 0.1 dB/km. As the frequency range is more than 100 GHz, the attenuation for API less than 100 will become more prominent. The attenuation rate for API values between 51 until 100 and less than 50 were 0.09–0.04 dB/km, respectively.

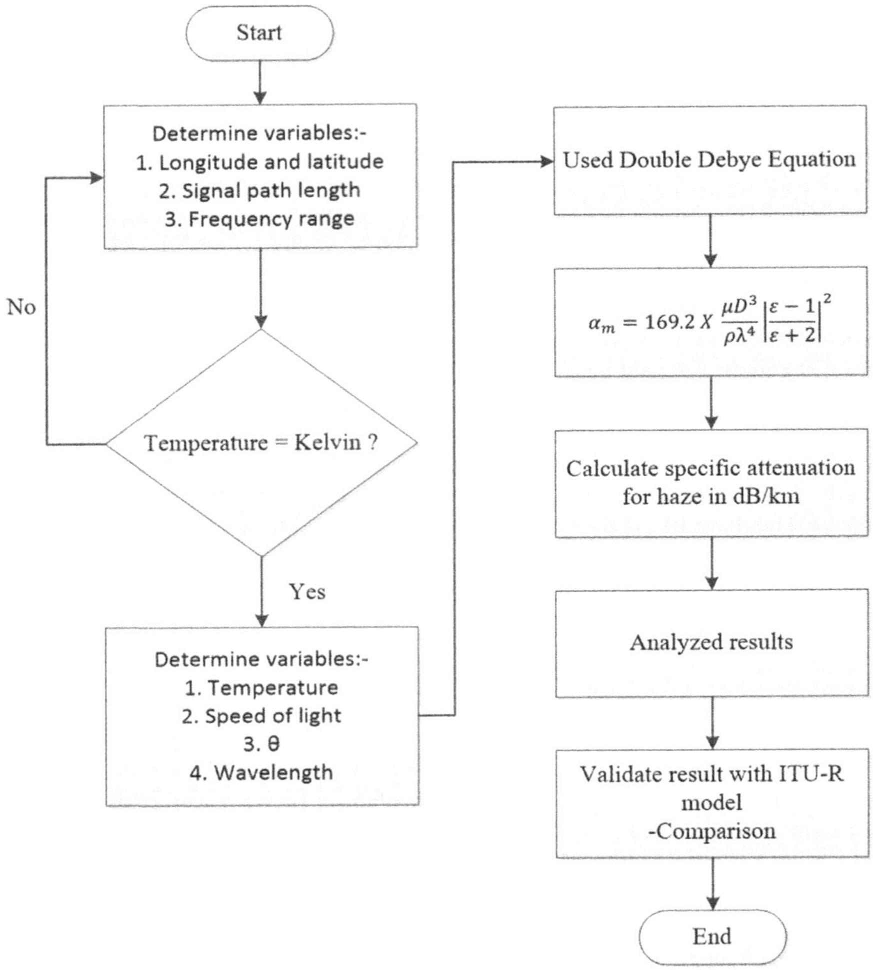

Fig. 32.1 The overall system modeling for haze

Figure 32.3 demonstrates the haze attenuation rate in 2016 for the Batu Muda station. During this year, there is no severe haze crisis that affected the area. The highest API values recorded for this year is between 101 until 200. The disturbance of mmwave signal is more pronounced when the frequency reached 100 GHz with an attenuation rate of 0.09 dB/km. Meanwhile, the mmwave signal is affected by interference as the frequency is more than 100 GHz for API values between 51 until 100 and less than 50 with the attenuation rate was around 0.08–0.04 dB/km, respectively.

There are two results for each API value for haze attenuation rate in 2018, as shown in Fig. 32.4. During this year, there is also no severe haze crisis happened in

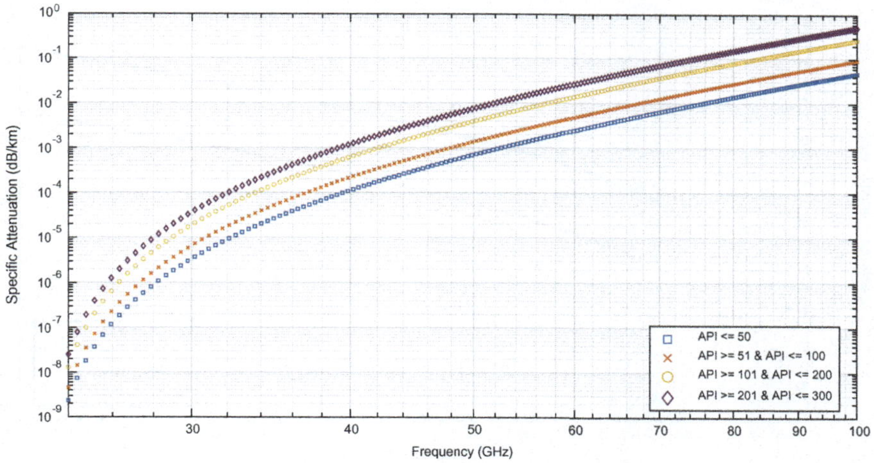

Fig. 32.2 The haze attenuation rate at Batu Muda station in 2015

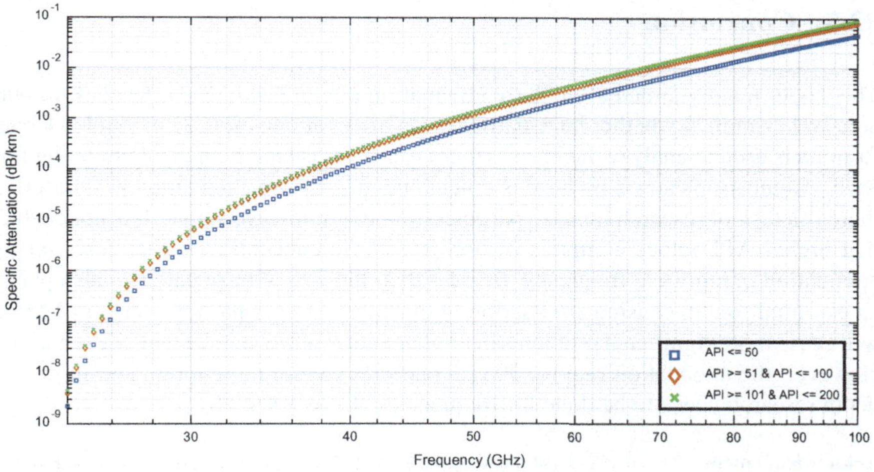

Fig. 32.3 The haze attenuation rate at Batu Muda station in 2016

Batu Muda. Based on the result, there are two different sizes of haze particles which are PM10 and PM2.5. The haze attenuation rate caused by PM10 particles is around 0.04 dB/km and 0.02 dB/km for API values between 51 until 100 and less than 50, respectively. However, the attenuation rate for PM2.5 for frequency below 100 GHz is too low, with an attenuation rate between 0.4×10^{-3} and 0.2×10^{-3} dB/km. However, the attenuation caused by haze cannot be negligible as it may still affect mmwave signal for frequencies above 100 GHz.

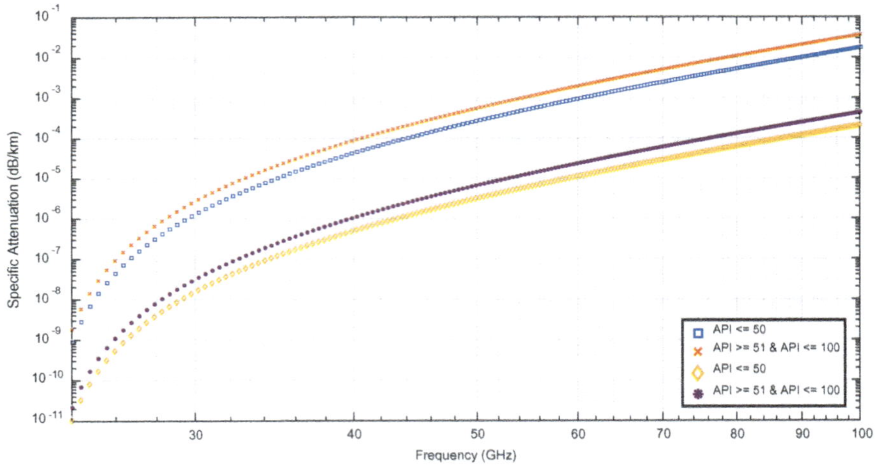

Fig. 32.4 The haze attenuation rate at Batu Muda station in 2018

32.5 Conclusion

This paper mainly aimed at the haze attenuation rate in Batu Muda, Kuala Lumpur area and examined whether haze particles can affect attenuation. The double Debye formula is used to analyze the haze as there is no concise mathematical modeling for haze. Attenuation caused by haze is crucial and need more attenuation as currently, there is a lack of research on haze particles affecting mmwave signal. It is noticeable that higher API values comprise of higher particle concentration that can lead to attenuation in mmwave frequency. Besides that, the size of haze particles also plays an essential role in causing attenuation in mmwave. Bigger particles such as PM10 will cause more significant problems in implementing mmwave in a real-life scenario. However, the attenuation rate for PM2.5 particles cannot be ignored, and more in-depth research needs to be done for frequencies above 100 GHz.

Acknowledgements This research is funded by the Ministry of Higher Education under the Fundamental Research Grant Scheme (FRGS) with reference code FRGS/1/2017/TK04/UNIKL/02/12. The authors would like to thank the Universiti Kuala Lumpur British Malaysian Institute for the provision of the laboratory.

References

1. Theodore SR, Yunchou X et al (2017) Overview of millimeter wave communications for fifth-generation (5G) wireless networks-with a focus on propagation models. IEEE Trans Antennas Propag 65(12):6213–6230

2. Jayavardhana G, Rajkumar B et al (2013) Internet of things (IoT): a vision, architectural elements and future directions. Future Gener Comput Syst 29(7):1645–1660
3. Theodore TR (2016) Spectrum frontiers: the new world of millimeter-wave mobile communication. New York University, Tandon School of Engineering
4. Theodore SRSS et al (2013) Millimeter wave mobile communications for 5G cellular: it will work! IEEE Access. 1:335–349
5. Theodore SR, James NM (2011) State of the art in 60-GHz integrated circuits and systems for wireless communications. Proc IEEE 99(8):1390–1436
6. Zhouyue P, Farooq K (2011) An introduction to millimeter wave mobile broadband systems. IEEE Commun Mag 49(6):101–107
7. Chen CC (1975) Attenuation of electromagnetic radiation by haze, fog, clouds, and rain. DTIC: 42
8. Lin L, Li Z (2016) Study of transmission effects of millimeter wave through fog and haze. In: Proc. 2016 IEEE Adv. Inf. Manag. Commun. Electron. Autom. Control Conf. IMCEC, pp. 1591–1596
9. Siti FN, Zuhanis M et al (2019) Propagation challenges in 5G millimeter wave implementation. Indones. J Electr Eng Comput Sci 15(1):274–282

Chapter 33
Performance of Propagation for Millimetre-Wave Frequency: Case Study

Nurul Husna Mohd Rais and Zuhanis Mansor

Abstract Most works in antenna design are focussed on the results for the fundamental antenna parameter that can be achieved in the antenna simulator. For further investigation, the antenna must be fabricated, measured and installed to observe the propagation results. This work involves antenna characterisation and a 3D propagation model. Characterisation of the antenna for the transmitter and receiver is for 28 GHz frequency. The input for the computation of the propagation model is using the radiation pattern from the designed antenna. The antenna performance is analysed in different mobile user antenna orientations in vertical and horizontal array configurations. Results show that the propagation performance for the horizontal orientation for the user equipment antenna is outperformed compared better as the downlink SNIR receive power is 45 dBm, 10 dBm higher than the vertical orientation.

Keywords Millimetre wave · Antenna rotation · 3D propagation model · Channel model

33.1 Introduction

Recently, the ray-tracing approach to investigate propagation characteristics is widely used. Ray-tracing techniques that are commonly used are 3D ray tracing as it can employ propagation environment in a 3D geometrical representation [1, 2].

Network planning for the large area can be done by using the existing geometrical map from the internet. Zhang in [3] used 3D ray tracing to investigate the coverage and characteristic of the millimetre-wave band. The utilisation of ray tracing also provides more capability to analyse the propagation performance, as the input transmitter and

N. H. M. Rais (✉) · Z. Mansor
Advanced Telecommunication Technology Research Cluster, Communication Technology Section, Universiti Kuala Lumpur British Malaysian Institute, Batu 8 Jalan Sungai Pusu, 53100 Gombak, Selangor, Malaysia
e-mail: nurulhusnamr@unikl.edu.my

Z. Mansor
e-mail: zuhanis@unikl.edu.my

A. Ismail et al. (eds.), *Advanced Materials and Engineering Technologies*, Advanced Structured Materials 162,
https://doi.org/10.1007/978-3-030-92964-0_33

receiver can be set to a different configuration. This work performs the analysis of the millimetre-wave frequency performance of mobile user equipment (UE) within the urban scenario. The study aims to analyse the performance on the user side with different base station (BS) placement. The BS is placed on the rooftop with a diverse distance range of blockage. The propagation model of 28 GHz was created using the WinProp tool. The type of propagation model chosen for the analysis is the 3D intelligent ray tracing (IRT). The 3D IRT technique computes the propagation path in three dimensions, including reflections at building walls and diffractions around the building wedges. Input for the propagation model is the radiation pattern of the BS and UE. The radiation pattern exported from the simulated antenna design using the computer simulator technology (CST) antenna simulator software. This work focuses on the downlink (DL) transmission of the model from the BS to UE. Computation of the simulation for this research is as follows:

(a) Two development files were created with a similar pre-processed map and antenna pattern as the input.
(b) One design file is for the horizontal orientation of the mobile user antenna.
(c) Another design file is for the vertical orientation of the mobile user antenna.

Data from both simulation development files are compiled and plot in the same graph for the analysis.

33.2 Methodology

33.2.1 Simulation Workflow

The first stage of this research is the antenna design for the mobile phone and the BS. The propagation model needs an antenna pattern as the source input for the simulation at the transmitter and receiver. As for this research project, the antenna pattern is required for the BS as the transmitter and the user mobile phone. The WinProp software tool is used to perform propagation modelling. There are three software components with different purposes. Figure 33.1 shows the workflow for the urban propagation model analysis. A geometry database simulation in 3D model is implemented using Wallman. The geometry database for the area of investigation is produced using Wallman. The urban database is pre-processed using Wallman from the imported map file from OpenStreetMap. In addition, CST is used to design and simulate the antenna for BS and UE. The antenna manager (AMAN) is used to export the antenna pattern for the propagation computation. The radiation pattern from the antenna design is used as the input for the computation for the 3D IRT propagation model. The radiation pattern from another antenna simulator can be used by converting the file format using AMAN. Finally, based on the Wallman database, the propagation manager (ProMan) is used, including the radiation pattern of the BS and UE to create the urban propagation. The BS's placement at the desired

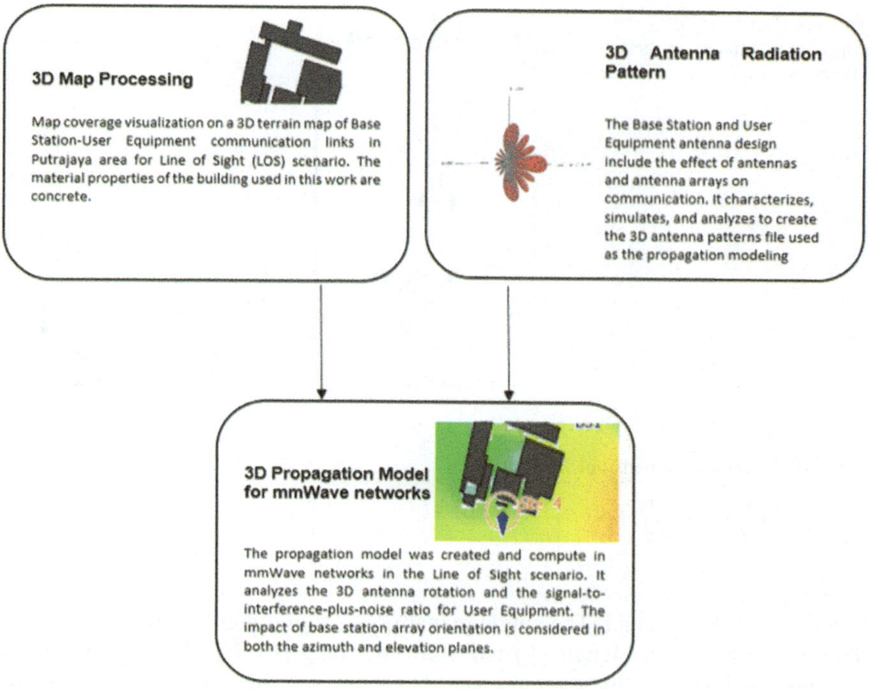

Fig. 33.1 Graphical methodology of the 3D propagation modelling

site location and inclusion of UE were done for the computation of propagation and network analysis. Each BS and the network results are analysed by extracting the graph for each type of the implementation's result.

33.2.2 Antenna Characterisation for the Mobile User Antenna

Characterisation of planar quasi Yagi antenna design is proposed for the mobile user antenna. The radiation pattern of the planar quasi Yagi antenna will be included as the receiver (mobile user) in the ray tracing computation. The substrate used for the antenna design is Rogers 5880 substrate with 2.2 permittivities and 0.0009 conductor loss with a 0.254 mm thickness. The planar quasi Yagi antenna consists of a dipole driver and three director elements. The antenna fed by micro-strip feed line with 4.8 mm length (L_f) and 0.72 mm width (W_f) match with 50 Ω input impedance. Partial ground planes with 3 mm height with parasitic elements act as reflectors to reflect the electromagnetic waves to produce a directional beam pattern [4]. The geometry of the proposed antenna is shown in Fig. 33.2. Table 33.1 describes the

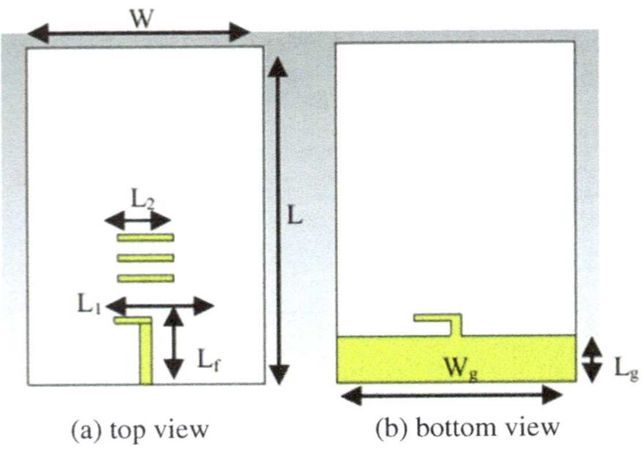

Fig. 33.2 Geometry of proposed antenna **a** top view **b** bottom view

Table 33.1 Antenna dimensions in mm

L	W	L_f	W_f	L_g	W_g	L_1	L_2
20	15	4	0.72	3	20	4.8	3.6

dimension of the antenna. The single planar quasi Yagi dimension for the length is 20 mm × 15 mm for the length (L) and width (W), respectively, where L_g and W_g are the length and width of the ground plane. The lengths of the driver (L_1) and director (L_2) for the proposed quasi Yagi are used to calculate the antenna. The equation for the length of the driver is $0.5\lambda_{g,}$ and the length of the director is $0.455\lambda_g$ [5]. The effective wavelength (λ_g) of the antenna can be calculated by using the equation as follows:

$$\lambda_g = \frac{c}{f_r\sqrt{\varepsilon_{r+1}}} \tag{33.1}$$

where c is the velocity of light and ε_r is the permittivity of the substrate, and f_r is the resonant frequency of the antenna. The calculated length is optimised through the simulation. The optimised length of L_1 and L_2 is 4.8 mm and 3.6 mm, respectively.

33.2.3 *Antenna Characterisation for the Base Station*

The antenna substrate is similar to the one used in Sect. 33.2.2 for the rectangular U-slot antenna design. Figure 33.3 shows the proposed antenna design for the BS, which is a rectangular U-slot antenna. The dimension of the radiating patch is 6.5 mm × 5 mm placed on a 10 mm × 8 mm substrate. The rectangular patch's length and width were chosen from several optimisations, with the initial value derived from the calculation described in [6]. To take advantage of the directional antenna pattern, the

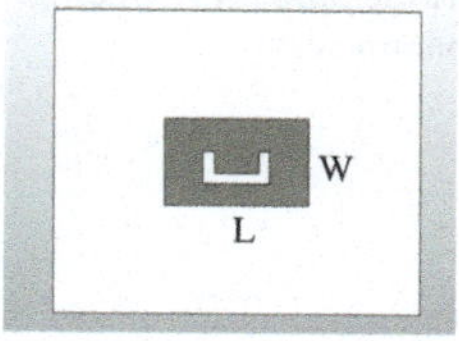

Fig. 33.3 Rectangular U-slot antenna design

antenna design used probe feed antenna input. Creating a slot on the antenna patch [7] is one impedance matching method suitable for the probe feed patch antenna. The U-shape slot is promising to enhance the antenna matching as it has the dominant effect on the return loss of the antenna.

33.2.4 Prediction Parameter for the Propagation Simulation

Figure 33.4 depicts the study area, which includes the location and distance of each BS from other geometry blockages (1.3×1.5 km^2) urban area with a building surrounding is selected for the investigation. All buildings in the map are modelled as solid blocks with electrical properties of $\varepsilon_r = 4$, $\mu_r = 1$ and $\sigma = 0.01$ S/m. These electrical prosperities correspond to concrete commonly found in buildings. In the simulations, the signals are unable to penetrate the buildings. There are four BSs located at different locations. Each location has a different distance to the blockage.

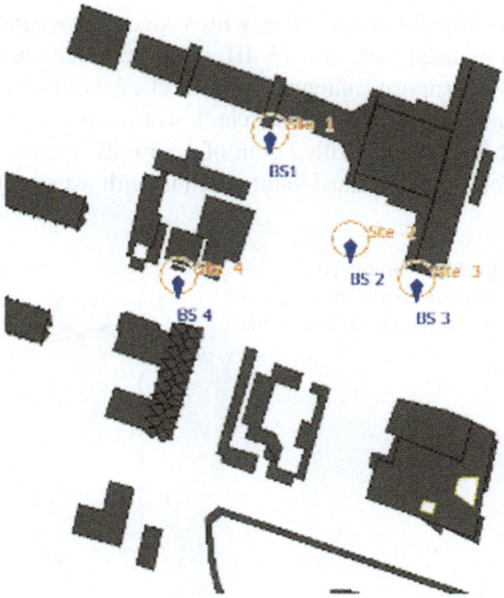

Fig. 33.4 Location and distance of each BS to other geometry blockage 1.3×1.5 km^2

Table 33.2 Ray-tracing parameter

Parameters	Values
Operating frequency	28 GHz
Number of BS	4
BS transmitter antenna height (m)	5 m
Bs transmitter power	40 dBm
BS antenna down tilt	10^0
UE receiver antenna height	1.5 m
UE transmitter power	0.2 dBm

The radiation pattern of the BS and the UE antenna is used as the input for the propagation model. The BS sets to be 5 m in height, and the UE height is set to 1.5 m. Table 33.2 summarise the ray tracer parameter for the BS and the UE antenna.

33.3 Results and Discussion

33.3.1 Return Loss and Antenna Pattern for the Antenna Design

The simulated results for the planar quasi Yagi for the UE and rectangular U-slot antenna for the BS are presented in this section. The simulated reflection coefficient ($S11$) is shown in Fig. 33.5. As can be seen, the proposed antenna has a wide frequency bandwidth of 4.37 GHz, which covers the frequency range of 24.76–29.13 GHz. With a realised gain of 8.15 dB, the single element of the antenna performs remarkably. The proposed antenna has a directional radiation pattern along the xy plane, as shown in Fig. 33.6. For this research work, a planar Yagi antenna was used to create a 1 × 2 linear array with a gain of 10.56 dB. Figure 33.7 shows the simulated return loss for a rectangular U-slot antenna with a frequency bandwidth ranging from 27.2 to

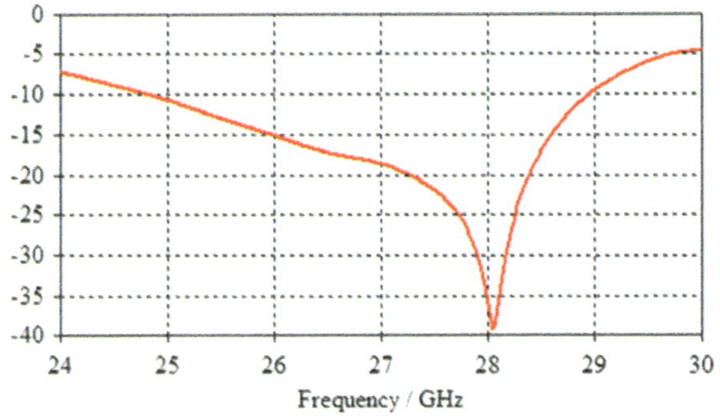

Fig. 33.5 Simulated reflection coefficient (S_{11}) of the single quasi Yagi antenna

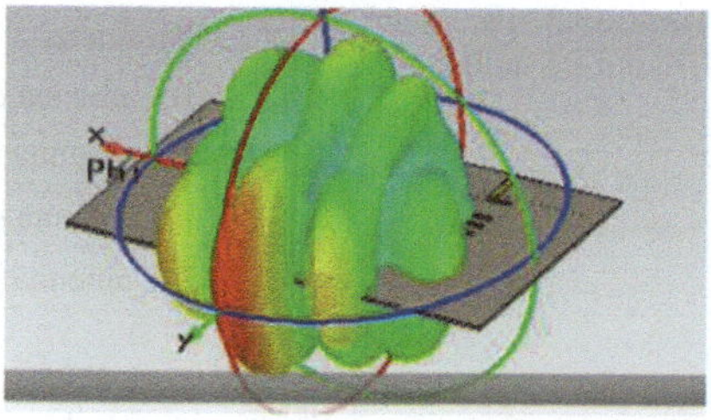

Fig. 33.6 3D radiation pattern for 1 × 2 linear array

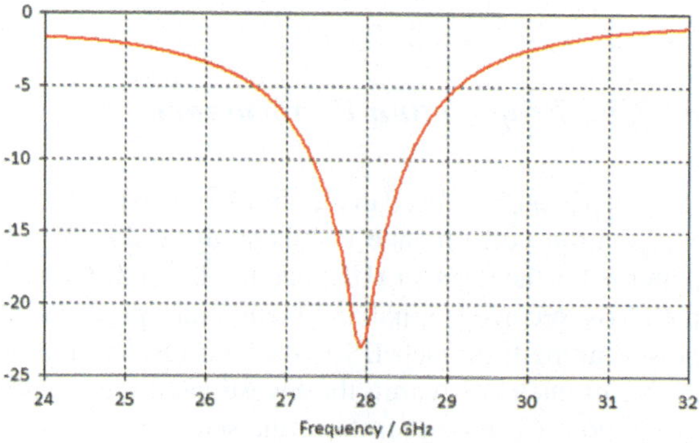

Fig. 33.7 Return loss for rectangular U-slot antenna

28.6 GHz. The BS antenna is designed to an 8–3 antenna array. Figure 33.8 shows the 3D radiation pattern for the 8 × 3 antenna array to meet the high gain requirement for BSs. The radiation pattern for the BS and mobile user antenna is then exported to a suitable format for use in WInProp Proman for propagation prediction. The antenna

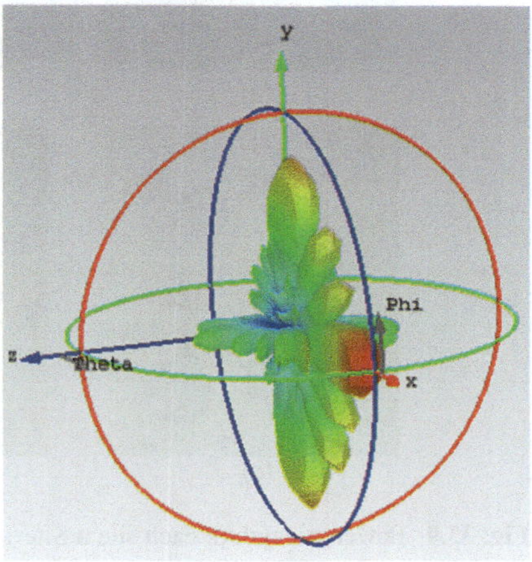

Fig. 33.8 3D antenna pattern for the rectangular U-slot antenna

Table 33.3 Parameter for the UE and BS antenna

Parameter	Value
BS antenna type	8 × 3 linear array
UE antenna type	1 × 2 linear array
BS antenna gain	20.15 dB
UE antenna gain	10.56 dB

patterns parameters are summarised in Table 33.3. In this paper, the radiation pattern is used in ray tracing.

33.3.2 *Propagation Performance*

WinProp Proman is used in the simulation to predict the millimetre-wave's frequency propagation performance in this study. Using both inputs, which are the antenna patterns for the receiver (UE) and the BS at different locations, the power of the BS, path loss, received signal and many other propagation results can be obtained. The power strength of each BS is the first observation of the ProMan simulation. The power of each BS within the investigated area is depicted in Fig. 33.9. Indication of the power can be seen from the scale on the right side of each Figure measured

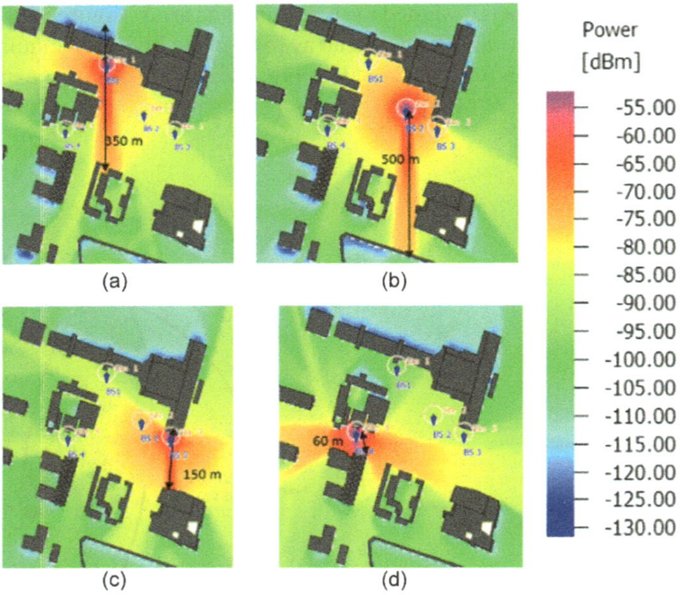

Fig. 33.9 Power strength of each site **a** Site 1, **b** Site 2, **c** Site 3 and **d** Site 4

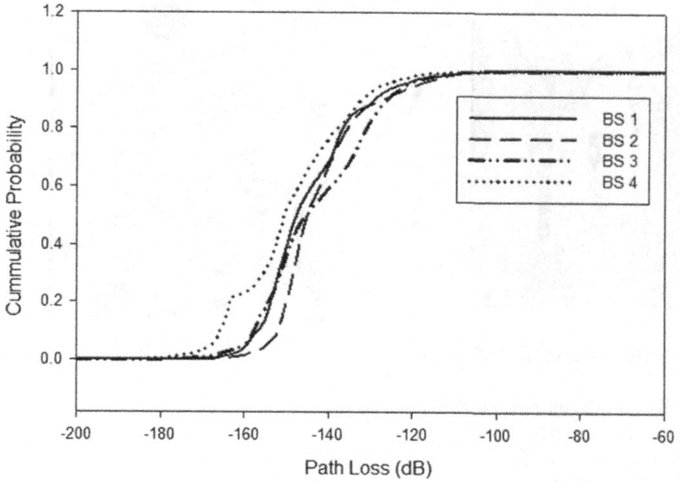

Fig. 33.10 Path loss of each BS

in dBm. The pink area shows the highest connectivity, whilst the yellow and green regions show areas with good signal strength, and the blue area shows the weakest signal strength. Based on the simulation prediction, the strongest power reached up to −55 dBm. Data on path loss are compared for each BS with different distance range blockage. Figure 33.10 shows the cumulative distribution of each BS in a comparative manner. The plots of BS1 and BS2 are almost identical because the two BSs cover a large area. CDF curve for BS4 shows the best results as it covers the small area as it covers a shorter range of blockage. The signal-to-noise-and-interference-ratio (SNIR) results at the user are determined, taking into account the maximum possible power of the transmitting station and the system's maximum potential noise interference power. Figure 33.11 shows the DL SNIR for both simulations with horizontal and vertical mobile user antenna orientation. From the maximum scale value of both results, it can be seen that the value for vertical orientation has 10 dBm difference compared to horizontal orientation. Even though the SNR value for vertical decreases, it is still considered suitable for performance since it is greater than 20 dB. The 3D ray-tracing propagation model also makes use of the UE's maximum receive power. The UE received power resulted in absolute power received directly from the BSs. These values are considered by taking into account the ultimate power of the transmitting BS. A comparison of the UE DL received power in vertical and horizontal is shown in Fig. 33.12. The results show that the maximum received power for the horizontal array configuration is higher, at—40 dBm. The most significant gain on the antenna pattern facing directly to the BS causes a 5 dBm loss in received power. At point 0.28, the cumulative distribution function plot also shows that the horizontal antenna orientation outperforms the vertical antenna orientation.

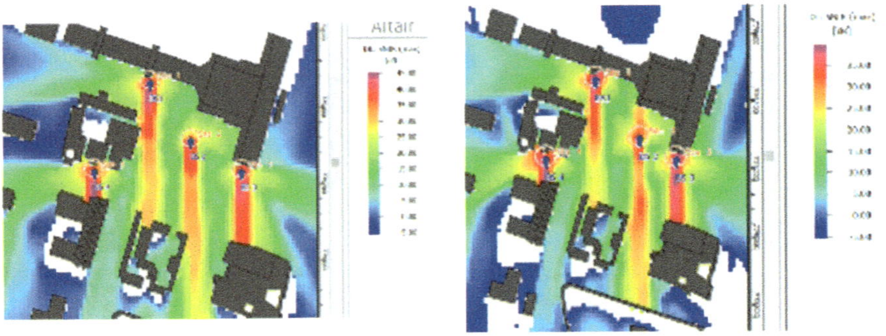

(a) DL SNIR horizontal 45 dBm (b) DL SNIR vertical 35 dBm

Fig. 33.11 2D visualisation of DL SNIR

33.4 Conclusion

This paper examines the millimetre-wave frequency performance of mobile UE in an urban environment. The WinProp tool was used to create the 28 GHz propagation model. A similar BS antenna location and various UE positions are used to simulate the ray-tracing propagation model. The performance is compared by plotting both the horizontal and vertical results for the UE in the same cumulative distribution function (CDF) graph. It can be seen that if the receiving antenna is in a different orientation, the propagation performance from the analysis will differ. In this case, the mobile user antenna is analysed with two antenna array configuration orientations, horizontal and vertical, respectively. In addition, most CDF plots show no noticeable differences between the two orientations. The results show that the planar quasi Yagi antenna designed for mobile users has good performance even when used in two different orientations. Based on these results, the horizontal array orientation has a receive power, at—40 dBm, 5 dBm higher than the vertical antenna orientation, which has a receive power of—45 dBm. The 5 dBm degrade of the received power is caused by the most substantial gain on the antenna pattern facing away from the BS. Thus, the results show that the planar quasi Yagi antenna designed for mobile users has good performance even when used in two different orientations.

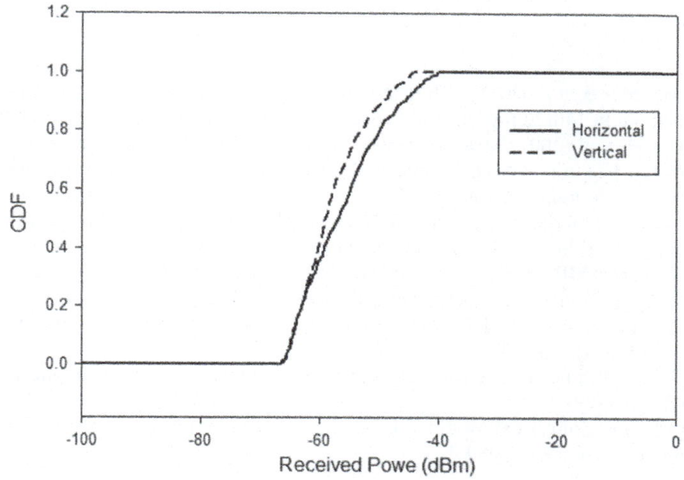

(a)

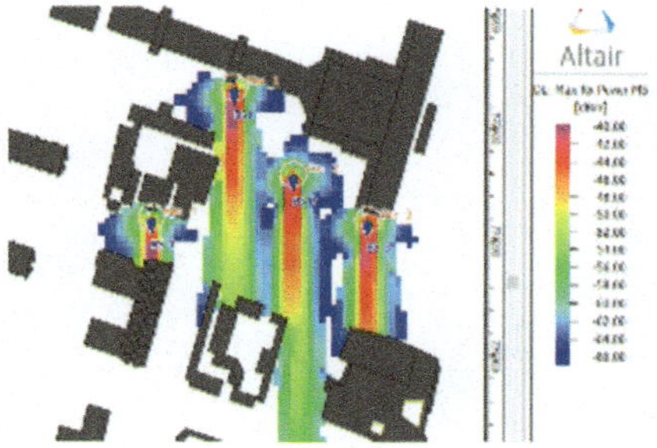

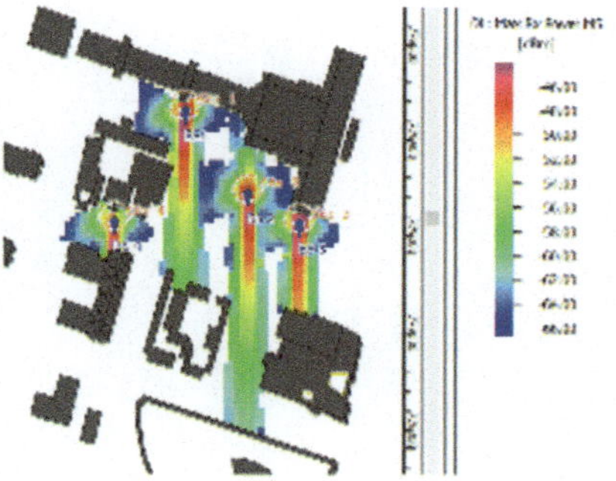

(b) horizontal (c) vertical

Fig. 33.12 DL received power of mobile users in different antenna orientation **a** CDF graph **b** 2D visualisation for horizontal **c** 2D visualisation for vertical

Acknowledgements This research is funded by the Ministry of Higher Education under the Fundamental Research Grant Scheme (FRGS) with reference code FRGS/1/2017/TK04/UNIKL/02/12. Nurul Husna would like to thank the Majlis Amanah Rakyat (MARA) and Universiti Kuala Lumpur British Malaysian Institute for her postgraduate scholarship and the supervision of laboratory facilities.

References

1. Higo TP, Rafael MD et al (2020) Cell-free at millimeter wave frequency simulation using the ray tracing method. In 14th European conference on antennas and propagation (EuCAP).
2. Terhi R, Reiner H et al (2007) Measurements and 3D ray tracing propagation predictions of channel characteristics in indoor environments. In: IEEE international symposium persistence indoor mobile radio communication, pp 1–5
3. Zhenliang Z, Jung R et al (2015) Coverage and channel characteristics of millimeter wave band using ray tracing. IEEE Int. Conf. Commun. 9:1380–1385
4. Ramadan AA, Gabriel MR (2009) High-Gain Yagi-Uda antennas for millimeter-wave switched-beam systems. IEEE Trans Antennas Propag 57(11):1356–1359
5. Sreelakshmi K, Pronami B et al (2017) Linear array Yagi-Uda 5G antenna for vehicular application. Int J Eng Technol 7(1):513
6. Adamu MJ (2017) Millimeter wave patch antenna design antenna for future 5g applications. Int J Eng Res 6(02):289–291
7. Amit AD, Aarti GA et al (2018) Modified U-slot cut rectangular patch antenna for wideband response. Appl. Electromagn Conf 1:1–2

Chapter 34
Uninterruptible Power Supply System Configuration Reliability Studies

Mohd Khairil Rahmat, Mohd Akmal Hadi Mazlan, Abd Halim Jaafar, Wan Abdul Azir Wan Musa, and Mohd Nizam Mat Ros

Abstract The purpose of this paper is to predict the reliability parameters of the DC uninterruptible power supply (UPS) by using the reliability block diagram (RBD) method. Application of RBD in predicting reliability of the DC UPS is capable to produce the important quantitative reliability indices such as the system's failure rates, mean time between failures (MTBF), availability, and unavailability, which will be useful to the UPS designer, manufacturer and finally the user to decide for the best DC UPS configurations. In this paper, two configurations of the DC UPS (with and without generator) are considered, and comparisons on their resultant reliability parameters by using the proposed RBD method are discussed in detail. Sensitivity analysis on the major components of the DC UPS is performed to investigate the effect on the overall reliability of the power systems. Field data from industrial best practice are used to validate the results of the analytical model.

Keywords Uninterruptible power supply · Reliability block diagram · Mean time between failures

---

M. K. Rahmat (✉) · M. A. H. Mazlan
Renewable Energy Research Laboratory, Universiti Kuala Lumpur British Malaysian Institute, Gombak, Selangor, Malaysia
e-mail: mkhairil@unikl.edu.my

M. A. H. Mazlan
e-mail: akmal.mazlan@s.unikl.edu.my

A. H. Jaafar
Cawangan Kejuruteraan Elektrik, Ibu Pejabat JKR Malaysia, Jabatan Kerja Raya, Kuala Lumpur, Malaysia
e-mail: abdhalimj@jkr.gov.my

W. A. A. W. Musa
Cawangan Kejuruteraan Elektrik, Jabatan Kerja Raya Negeri Kedah, Kedah, Malaysia
e-mail: wazir@jkr.gov.my

M. N. M. Ros
Unit Kejuruteraan Elektrik, Kementerian Perumahan & Kerajaan Tempatan, Putrajaya, Malaysia
e-mail: nizammr@jkr.gov.my

A. Ismail et al. (eds.), *Advanced Materials and Engineering Technologies*,
Advanced Structured Materials 162,
https://doi.org/10.1007/978-3-030-92964-0_34

34.1 Introduction

Most electrical installations require uninterruptible power, virtually free of frequency excursions and voltage dips, surges and transients. Reliability demands become much more important, especially under emergency conditions such as natural catastrophes and accidents in which a public power failure is likely to happen [1].

Conventionally, reliability parameters of the DC uninterruptible power systems are presented by using the state-space method [2]. In this approach, firstly, all the possible system states have to be identified. Then, the state-transition diagram has to be constructed in order to show the interdependencies between the states. The system states probabilities and the inter-state transition rates (from state i to j) have to be determined. The simulation result suggested that the reliability indices improve considerably with increase in battery discharge time. The results agreed that a standby generator should be considered if high reliability of critical DC load supply is required. The difficulty in constructing the state-transition diagram for the redundant and bigger system has made the method become less popular as a method in determining the standby power system's reliability. Furthermore, the inter-state transition rates are quite complicated to identify.

A study by [3] proposed the minimal cut set method. This method is very similar to the state-space approach where the state-transition diagram has to be constructed, and the inter-state transition rates and the steady-state probabilities have to determine first. The failure rates of some major components used for this simulation also being applied in this paper. The author suggested that the reliability of DC UPS could be improved by a generator and by introducing accelerated maintenance. Again, the difficulty in constructing the state-transition diagram of the overall system has made this method less preferable in determining the reliability indices of power systems.

Papers [4, 5] describe the methodology to analyse the availability and reliability performance of the DC uninterruptible power supply (UPS) based on Markov chains. A typical telecom DC UPS system is taken into consideration in this paper. The results obtained were useful as the reliability parameters of each major component can be achieved individually, and the outcomes of the overall system can be compared easily. However, to construct the system's Markov graphs, a great understanding of the system operations is essential. Like the other two methods before, this method is less likely to be used in the UPS reliability analysis.

Due to the difficulty and complicatedness in constructing the model for the uninterruptible power systems, UPS manufacturers rely on the existing field data [6, 7]. The method consists of tracking a sample population of a product and gathering the failure data. With this data, the failure rate and MTBF can be calculated.

This paper proposed the reliability block diagram (RBD) method for predicting the failure rates (λ), mean time between failure (MTBF), availability (A), and unavailability (W) for the two types of DC UPS configurations, with and without generator. Reliability indices of the major components that made up the DC UPS are considered. Finally, comparisons between the DC UPS will be made by considering the overall system's reliability.

The RBD method is widely used in software engineering. It is found to be very popular in determining the reliability of the computer networking systems and in the telecommunication network industry. However, this method is never being applied in predicting reliability parameters in power engineering studies. The RBD for the electrical power system is usually difficult to prepare, and in some cases, a unique diagram may not exist.

The advantages of using the RBD method to estimate the reliability parameters of the UPS system manifested in several ways:

1. Simple to construct the reliability model as it is quite close to the system single-line diagram/layout.
2. RBD modelling require less component's input data compared to all other method to obtain reliability indices.
3. RBD model can clearly show the interdependencies of the components in the system as they will be arranged in series or parallel between the system input and output nodes.
4. As the model is quite similar to the single-line diagram of the system, a thorough understanding of the system's working operation is not essential as the user can always refer to the single-line diagram.
5. Every components input data are treated independently. Thus, it is possible to perform a sensitivity analysis of every component in the system to investigate 'its effect on the overall system.

34.2 Reliability Block Diagram (RBD) Method

In the RBD method, the UPS reliability model must be build first, and a failure rate, λ (failures/h) must be assigned to each block. From the system's single-line diagrams, the resultant reliability model will be constructed [8]. For simplicity, in this UPS reliability modelling, the major components were assumed to be connected either in series or parallel, as shown in Table 34.1. For the overall system's unavailability calculations, the unavailability (W) and repair rate (μ) of each component are

Table 34.1 Component failure rates and unavailability equation

Series configuration	Parallel configuration
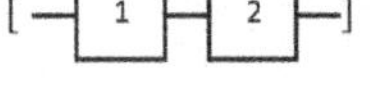	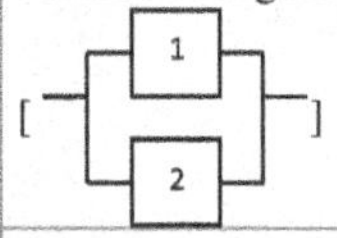
Failure rates, $\lambda_S = \lambda_1 + \lambda_2$ (34.2)	Failure rates, $\lambda_P = \underline{\lambda_1, \lambda_2}$ (34.4)
Unavailability, $W_S = W_1 \exp(-\mu_1.T) + W_2 \exp(-\mu_2.T)$ (34.3)	$\lambda_1 + \lambda_2$ Unavailability, $W_P = W_1 \exp(-\mu_1.T), W_2 \exp(-\mu_2.T)$ (34.5)

required. The battery reserve time, T is also needed in the calculation. The equations for failure rates and unavailability in series and parallel configurations are presented by Eqs. 34.1–34.4.

34.2.1 DC UPS system—With Generator

In this DC UPS system, the rectifier will receive its input AC power either from utility or the generator unit. Figure 34.1 shows the single-line diagram of the DC UPS system with generator [9]. Under normal operation, the utility power will supply AC power to the rectifier through the mechanical static switch. Rectifier will convert the AC supply to DC supply in order to feed to the critical DC loads. At the same time, the DC output from rectifier will flow to the battery charger to charge the battery system. In the event of utility power failure, there is no AC power supplied to the rectifier. As the battery system is connected to the critical load bus, the battery will be discharged to supply DC power to the DC loads. At the meantime, the generator set is ready to start and supplying the AC power to the rectifier. The generator will need some time (few minutes) to initialize before supplying AC power. For this reason, a battery system is needed to cater during the generator initialising time. The reliability model that has been built by the system configuration diagram is presented in Fig. 34.2, which comprises two parallel-connected blocks, block 1 and block 2.

After the reliability model has been constructed, the failure rate and unavailability of the UPS system can be calculated using (Eqs. 34.1–34.4) in Table 34.1. In this paper, for simplicity, the battery reserve time, T of 1 h is considered for both UPS configurations. The models for the failure rate and unavailability measurements of the UPS system are presented in Table 34.2.

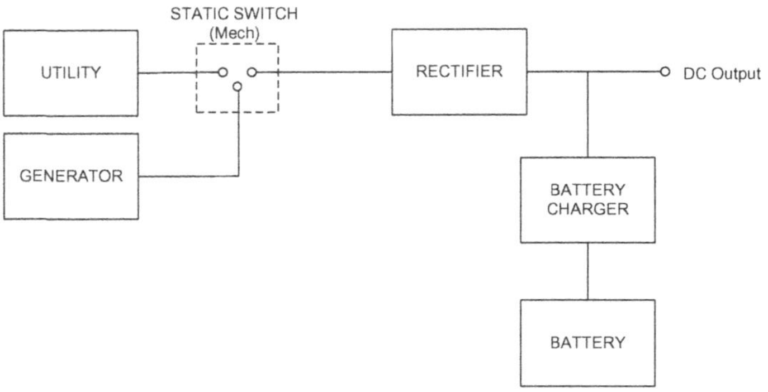

Fig. 34.1 DC UPS (with generator)—system configuration

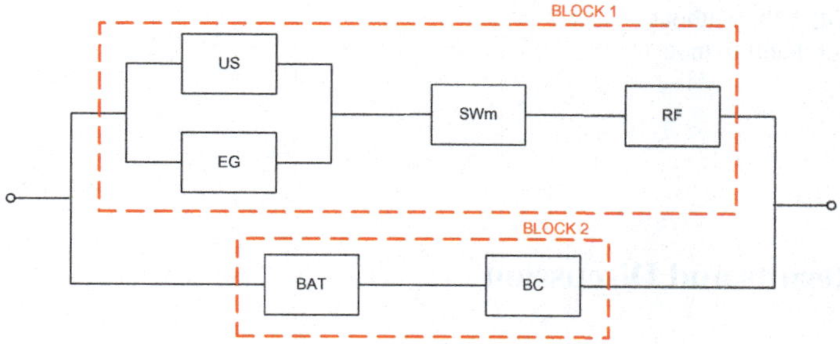

Fig. 34.2 DC UPS (with generator)—reliability model

Table 34.2 Failure rates and unavailability calculations for DC UPS with generator

Failure rate calculation	Unavailability calculation
$\frac{1}{\lambda_1} = \frac{1}{\lambda_{US}} + \frac{1}{\lambda_{EG}}$ $\lambda_1 = \frac{\lambda_{US} \cdot \lambda_{EG}}{\lambda_{US} + \lambda_{EG}}$	$W = W_1 + W_2 = \begin{bmatrix} W_{US/EG} \cdot \exp(-\mu_{US/EG} \cdot \mathrm{T}) \\ + W_{SWm} \cdot \exp(-\mu_{SWm} \cdot T) + \\ W_{RF} \cdot \exp(-\mu_{RF} \cdot T) \end{bmatrix}$ $+ \quad [(W_{US/EG} + W_{SWm} + W_{RF}] \cdot [W_{BAT} + W_{BC}]$
$\lambda_2 = \lambda_1 + \lambda_{SWm} + \lambda_{RF}$	
$\lambda_3 = \lambda_{BAT} + \lambda_{BC}$	
$\lambda_T = \frac{\lambda_2 \cdot \lambda_3}{\lambda_2 + \lambda_3}$	

34.2.2 DC UPS System—Without Generator

This is the most basic configuration of the DC UPS system [9]. From the system configuration in Fig. 34.3, it is shown that under the utility failure condition, the back-up time depends solely on the battery system. Once the utility power resumed, the rectifier will feed DC supply to the critical DC loads and charge the battery. The reliability model in Fig. 34.4 confirms that the critical load will only be supplied by either the utility or the battery supply.

Fig. 34.3 DC UPS (without generator)—system configuration

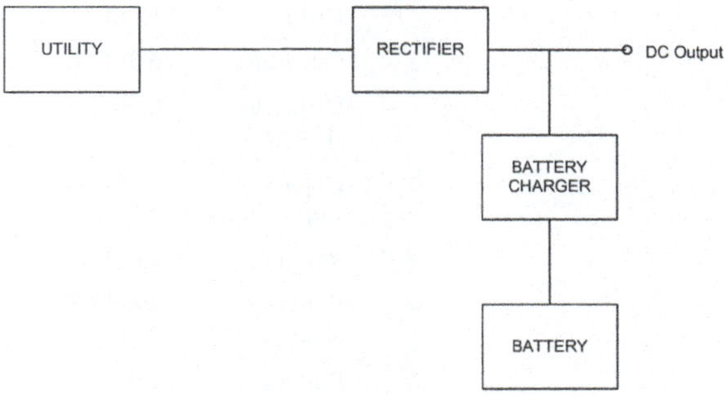

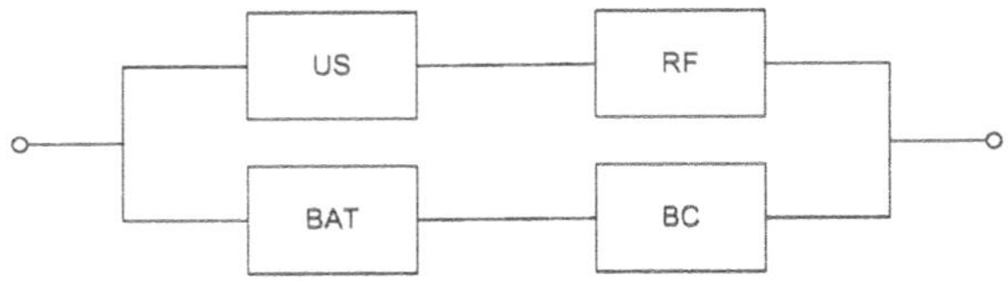

Fig. 34.4 DC UPS (without generator)—reliability model

34.3 Results and Discussion

Table 34.4 shows the reliability values of the major components for the UPS being studied indicating the values for failure rate (λ), unavailability (W), and repair rate (μ) [1, 3, 10]. Table 34.3 shows how the failure rate and unavailability of the UPS system were modelled by using the RBD modelling method.

Three major indicators for UPS system reliability were considered here. Unavailability (W) is defined as the inability that an entity is not in a state to perform a required function, under a given conditions, at a given instant of time [1]. Failure rate (λ) is the probability that an entity loses its ability to accomplish a function during the interval $[t, t + dt]$, knowing that it is not failed between $[0, t]$. The mean

Table 34.3 Failure rates—unavailability calculations for DC UPS without generator

Failure rate calculation	Unavailability calculation
$\lambda_1 = \lambda_{US} + \lambda_{RF}$	$W_{DC2} = \left[W_{US} \,.\, \exp(-\mu_{US} \cdot T) + W_{RF} \,.\, \exp(-\mu_{RF} \cdot T) \right] + [(W_{US} + W_{RF}] \cdot [W_{BAT} + W_{BC}]$
$\lambda_2 = \lambda_{BAT} + \lambda_{BC}$	
$\frac{1}{\lambda_T} = \frac{1}{\lambda_1} + \frac{1}{\lambda_2}$ $\lambda_T = \frac{\lambda_1 \cdot \lambda_2}{\lambda_1 + \lambda_2}$	

Table 34.4 Reliability data used for the reliability block diagram modelling

Component	Unavailability W	Failure rate, λ (failures/h)	Repair rate, μ (h)
Utility	1.08E–5	1.0000E–3	0.232
Generator	6.30E–6	3.6597E–6	0.054
Utility/gen (US/EG)	6.80E–8	1.5000E–7	2.2
Bypass switch, SWm	1.50E–6	1.0000E–5	0.5
Rectifier	4.60E–8	4.3478E–6	0.5
Battery charger	4.60E–8	4.3478E–6	0.5
Batteries	3.90E–6	1.6393E–7	0.042

Table 34.5 Results from reliability block diagram modelling

	Unavailability, W	Failure rate, λ	MTBF (h)	MTBF(y)
DC UPS with generator	9.452E–7	3.432E–6	2.913E–5	31.65
DC UPS without generator	8.592E–6	4.492E–6	2.226E–5	25.42

time between failures (MTBF) is the expected operating time between two failures [1]. MTBF is the reciprocal of failure rate (i.e. MTBF = $1/\lambda$)

34.3.1 DC UPS MTBF Comparison

The result in Table 34.5 shows that DC UPS with a generator will give higher MTBF value. The generator will provide another path of the back-up power supply in the event of utility power failure. Furthermore, the battery system will not be "deep-discharged" during power failure as the critical loads only on-battery for a short time (i.e. initialising time of generator). The failure rate of DC UPS with generator is lower compared to the system without generator. The failure rate of the generator is found to be low enough to affect the overall system failure rates. The unavailability of the system without generator is lower, and this can suggest that inclusion of generator unit in the DC UPS configuration can increase the availability of the overall system.

34.3.2 Battery Back-Up Time (T)

By increasing the battery reserve or back-up time, T the unavailability of the DC UPS will be reduced, as shown in Fig. 34.5. The slope of the two straight lines obtained for both DC UPS configurations, clearly shows that DC UPS with generator unavailability will reduced in faster rate compared with the DC UPS system without generator. For the DC UPS without generator, during the mains failure, the input power to the critical loads depends solely on to the battery supply and the battery supply depends on the battery's back-up or reserve time. For the DC UPS with generator, during mains failure, the critical loads will mainly receive input supply from generator. The load will only take power from battery system during the initialising of the generator unit.

34.3.3 Sensitivity Analysis of DC UPS MTBF

Sensitivity analysis was performed to investigate the effect of varying the MTBF of major components in the DC UPS on the overall reliability of the power systems.

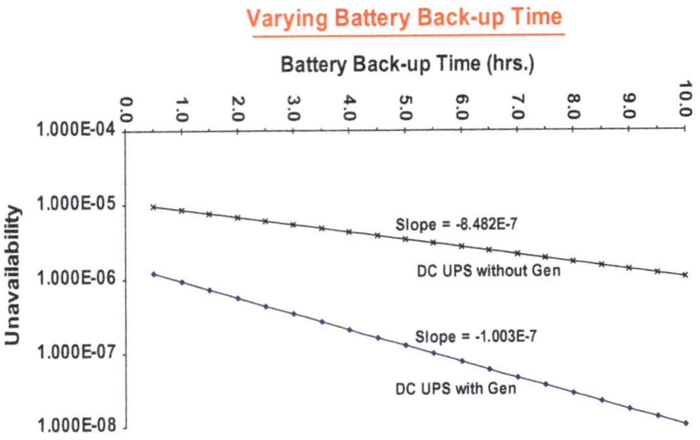

Fig. 34.5 Varying battery back-up time on DC UPS unavailability

The sensitivity analysis was done by increasing the components' MTBF value, and the effect of this value to the overall system MTBF was indicated by the slope of the graph obtained.

By varying the utility MTBF, from Fig. 34.6, the result suggested that for the DC UPS without generator, there is a high dependability on the utility's MTBF. Utility's MTBF has a very small effect on the MTBF of DC UPS with generator. Figure 34.7 shows that for DC UPS with generator the rectifier MTBF has a significant effect on the overall system MTBF. Higher rectifier MTBF will result in an improved system MTBF. For the DC UPS without generator, an increase in the rectifier MTBF has no effect on the overall system MTBF.

By varying the battery MTBF, from Fig. 34.8, both DC UPS produced almost the same curve. The DC UPS with generator has a higher system's MTBF compared

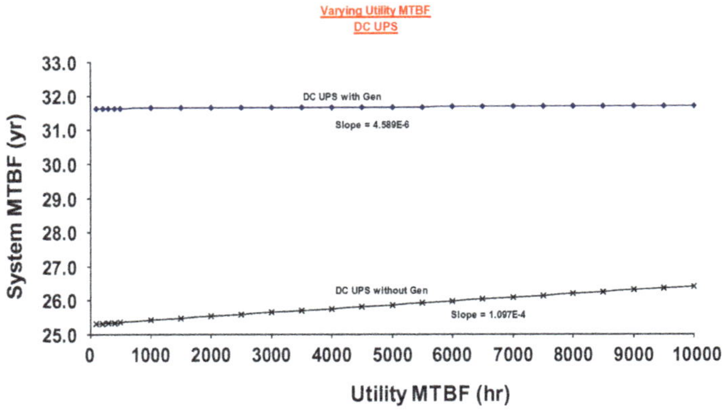

Fig. 34.6 Varying utility MTBF on DC UPS system MTBF

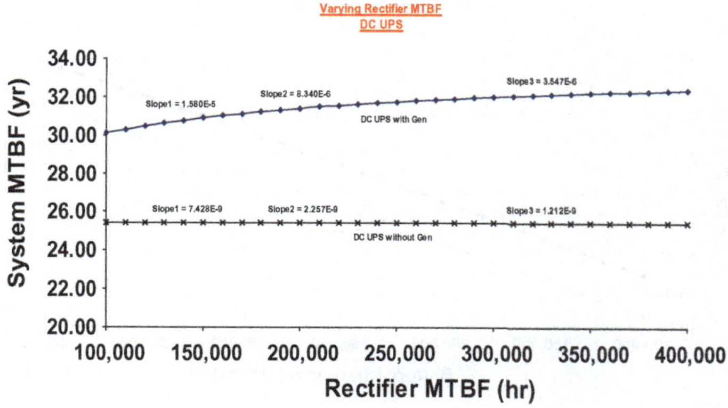

Fig. 34.7 Varying rectifier MTBF on DC UPS system MTBF

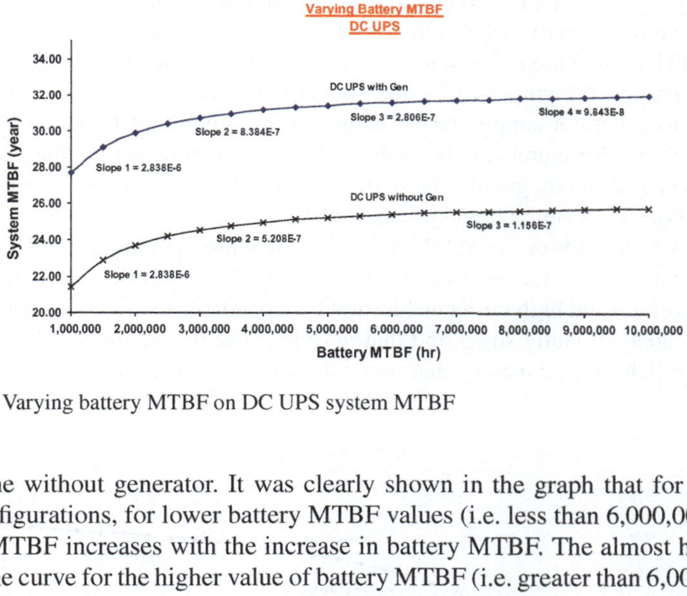

Fig. 34.8 Varying battery MTBF on DC UPS system MTBF

to the one without generator. It was clearly shown in the graph that for both DC UPS configurations, for lower battery MTBF values (i.e. less than 6,000,000 h), the system MTBF increases with the increase in battery MTBF. The almost horizontal line on the curve for the higher value of battery MTBF (i.e. greater than 6,000,000 h), suggested that the increment in battery MTBF values have no effect on the overall system MTBF.

Figure 34.9 shows the two straight lines with the same slope for both DC UPS systems. Similarly, the DC UPS with generator will give a better system MTBF value compared to the DC UPS without generator. The graph also suggested that the system MTBF is highly dependent on the battery charger MTBF. The result also suggested that the battery charger module is the most important component in the system to ensure its reliability. During the normal operation of the UPS, battery charger will float charge the battery, and during power failure, battery charger will be the link between the battery supply and the critical load.

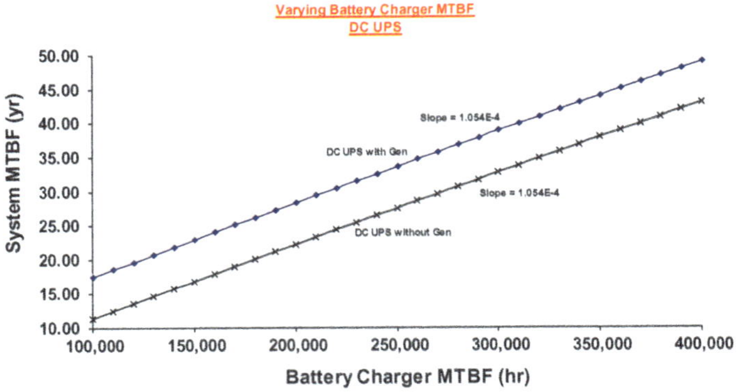

Fig. 34.9 Varying battery charger MTBF on DC UPS system MTBF

By varying the generator MTBF up till 200,000 h, it was found that the system MTBF increases greatly. However, for the higher value of generator MTBF, the system MTBF increment became less significant, as shown in Fig. 34.10. As discussed earlier, the purpose of static switch is to transfer the input supply from the utility to generator supply during mains failure. Figure 34.11 shows that as the static switch MTBF increases, the system MTBF also increase. In other words, the system MTBF depends greatly on static switch MTBF and higher value of static switch MTBF will result a high system MTBF.

Table 34.6 shows how the MTBF values of the major components in UPS system being rated with respect to its effect on the overall system's MTBF values. Low rating means less effect and high rating means highly dependable to the system MTBF. The sensitivity analysis study suggested that in order to achieve a higher DC UPS with generator reliability, the battery charger reliability has to be improved. As for the DC

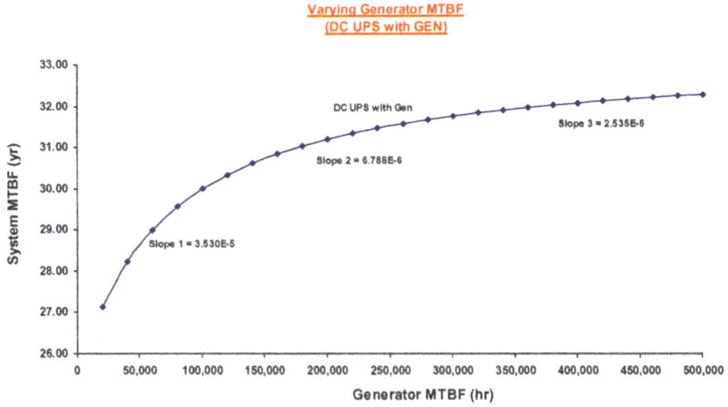

Fig. 34.10 Varying generator MTBF on DC UPS system MTBF

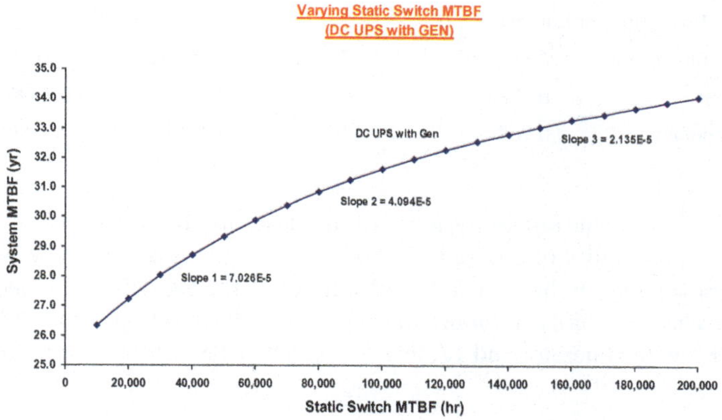

Fig. 34.11 Varying static switch MTBF on DC UPS system MTBF

Table 34.6 Rating on determining factors on overall UPS system's MTBF value

UPS configuration	Determining factors (on system MTBF)	Rating
DC UPS (with generator)	Utility	Low
	Generator	Low
	Static switch	Medium
	Rectifier	Low
	DC battery	Low
	Battery charger	High
DC UPS (without generator)	Utility	High
	Rectifier	Low
	DC battery	Low
	Battery charger	High

UPS without generator, the overall system's reliability will increase when the utility and the battery charger reliabilities increase.

34.3.4 Probability of Failures

The failure rates (λ) of each UPS configuration can be used to calculate the reliability and the probability of failure of a system with respect to time (t), using the equation:

$$\text{Reliability, } R = \exp^{-\lambda . t} \quad (34.1)$$

Table 34.7 Probability of failures

DC UPS configuration	*R* (1 year)	Fail (1 year)	*R* (5 years)	Fail (5 years)
With generator	0.9704	0.0296	0.8604	0.1396
Without generator	0.9614	0.0386	0.8214	0.1786

Table 34.7 shows the reliability, *R* or in a simple terms, the probability of surviving $R(t)$ and the probability of failure Fail(t) of the DC UPS systems. There was a very slight improvement on the probability of failure for the DC UPS with and without generator. The probability of failure to happen with the 5-year period is 13.96% for the DC UPS with generator and 17.86% for the DC UPS without generator.

34.3.5 Unavailability Comparison

Unavailability measures with respect to time per year for both UPS systems are presented in Table 34.8. The DC UPS with generator will be unavailable only for 29.81 s/year, whereas DC UPS without generator will be unavailable for 4.516 min/year.

In order to verify the result obtained by using the RBD method, the actual field data from various UPS companies were used. Table 34.9 shows the availability and MTBF figures for the two types of DC UPS considered. As expected, the results from the RBD method were close to the field data.

Table 34.10 shows that there was a very slight difference between the result obtained from RBD and field data for both availability and MTBF. For availability of the DC UPS with generator, the field data are $3.01 \times 10^{-5}\%$ more than the RBD method. The field data are $1.38 \times 10^{-5}\%$ more than RBD method for availability of DC UPS without generator configuration. In the MTBF results, field data give

Table 34.8 UPS unavailability

	Unavailability			
	%	h/year	min/year	s/year
DC UPS with generator	9.452E–05	0.0083	0.497	29.81
DC UPS without generator	8.592E–04	0.0753	4.516	270.95

Table 34.9 Results verification (RBD vs. field data)

	Calculated (RBD)		Field data	
	availability, *A*	MTBF (h)	availability, *A*	MTBF (h)
DC UPS (with generator)	99.9999054%	2.772E5	99.9999355%	2.600E5
DC UPS (without generator)	99.9991408%	2.226E5	99.9992788%	2.142E5

Table 34.10 Percentage differences between RBD and filed data results

	Availability % difference	MTBF % difference
DC UPS (with generator)	3.010E–5	– 6.61538
DC UPS (without generator)	1.380E–4	– 3.92157

6.62% less value compared to the RBD for the DC UPS with generator and 3.92% less value to RBD for the configuration without generator.

It was observed that for the DC UPS, the inclusion of generator will result:

- Lower failure rates
- Higher mean time between failures
- Increased system's availability
- Reduced unavailability

Performing sensitivity analysis on the major components of both DC UPS configuration clearly shows that the system without generator reliability depends highly on utility power reliability. For the configuration without generator, the battery system will be the only line of defence against the critical load shutdown during utility failure. Thus, the DC supply to the critical loads depends solely on the back-up time of the battery system. However, for the configuration with generator, the battery system will be the secondary back-up power supply (i.e. during the initialising of the generator) as primarily the back-up supply comes from the generator unit.

The sensitivity analysis also highlights the importance of the battery charger reliability on both the DC UPS system's reliability. Improvements in the battery charger's MTBF will result a higher value of overall system's MTBF.

Although the availability and failure rate is certainly attractive for such a simple scheme, the DC system without generator cannot be considered as a viable option for customers with critical loads (essential and vital customers) that require long-back-up time, as this configuration depends solely on-battery system reliability during power outage. Unlike DC UPS with generator, the battery system is used to feed the loads only before the generator has started up.

34.4 Conclusion

This paper proposed the RBD for the reliability analysis of uninterruptible power supplies (UPS). RBD is found to be a simple and effective method to predict the important reliability parameters of UPS systems.

The main advantage of this method is its simplicity in constructing the RBD reliability model compared to other methods. The ability of this method to produce the reliability indices of the overall UPS system allows user to perform a sensitivity analysis to investigate the effect of varying the MTBF of major components in the DC UPS on the overall reliability of the power systems. Eventually, the result obtained can be used to establish the area where improvements have to be made to achieve the highest system reliability.

The major finding from this study is for the critical loads, with a long-back-up time, DC UPS system with generator will be the best option. In terms of the system reliability, it will give a lower failure rates (λ), higher MTBF, and finally better availability (A).

References

1. Heising C (2007) Recommended practice for the design of reliable industrial and commercial power systems. IEEE Inc, New York
2. Singh CA, Patton AD (2015) Reliability evaluation of emergency and standby power systems. IEEE Trans Ind Appl 21(2):222–234
3. Singh C, Gubbala NV (1993) Reliability analysis of electric supply including standby generators and uninterruptible power supply system. IEEE industry applications conference twenty-eighth IAS, pp 1454–1460
4. Suntio T (2018) Novel approaches to ensure redundant rectifier system's reliability performance. INTELEC Int. Telecommun. Energy Conf, pp 414–420
5. Suntio T, Askola S (2017) DC UPS System's reliability performance: facts and fiction. TELESCON, pp 212–220
6. Singh C, Patton AD (1985) Reliability evaluation of emergency and standby power systems. IEEE Trans Ind Appl 2:476–480
7. Singh CJ, Mitra (2007) Reliability analysis of emergency and standby power systems. IEEE Ind Appl Mag pp 25–35
8. Bentley J (2006) Introduction to reliability and quality engineering, 2nd Edn. Addison-Wesley
9. King A, Knight W (2013) Uninterruptible power supplies and standby power systems. McGraw-Hill Publishing
10. Military-Handbook (1988) Electronic reliability design handbook. In: MIL-HDBK-338. Department of Defense United States of America

Chapter 35
Mock-Up Green Building Thermal Comfort Studies

Mohd Khairil Rahmat, Muhamad Fauzi Abdull Aziz, Abdul Muhaimin Mahmud, Iskandar Haszuan Ismail, and Rosnizah Ghazali

Abstract Buildings account for about 40% of the global energy consumption and contribute over 30% of carbon emission, globally. A large proportion of this energy is used to achieve a thermal comfort within the building. To overcome this problem, green building elements in the construction reduce excessive energy consumption of the building. This paper reviews the effect of thermal comfort to electrical energy consumption through wall and roof material selection of each four mock-up green building in UniKL Green Building test-site. All the data are collected and descriptive analysis has been done to investigate the potential of green building as a method to reduce electrical energy consumption. Test results have been obtained and analysis has been performed to identify the characteristics of the four different materials of the wall (building envelope). The result from this study showed that mock-up green building that is built from green material could provide a better thermal environment than mock-up green building that is built from non-green material. The result also showed the amount of energy consumption could be reduced as less active cooling than the mock-up green building that is built from non-green material.

M. K. Rahmat (✉) · M. F. A. Aziz
Renewable Energy Research Laboratory, Universiti Kuala Lumpur British Malaysian Institute, Gombak, Selangor, Malaysia
e-mail: mkhairil@unikl.edu.my

M. F. A. Aziz
e-mail: fauzi.abdullaziz@s.unikl.edu.my

A. M. Mahmud · I. H. Ismail
Cawangan Kejuruteraan Elektrik, Ibu Pejabat JKR Malaysia, Jabatan Kerja Raya, Kuala Lumpur, Malaysia
e-mail: muhaimin@jkr.gov.my

I. H. Ismail
e-mail: haszuan@jkr.gov.my

R. Ghazali
Cawangan Kerja Kesihatan, Jabatan Kerja Raya, Kuala Lumpur, Malaysia
e-mail: rosnizahg@jkr.gov.my

A. Ismail et al. (eds.), *Advanced Materials and Engineering Technologies*, Advanced Structured Materials 162,
https://doi.org/10.1007/978-3-030-92964-0_35

Keywords Green building · Thermal comfort · Energy efficiency · Electrical energy consumption studies

35.1 Introduction

In this globalization era, the demand of electrical energy is increasing every year especially within the building sector. This raises concern on fossil fuels, which has been used widely in every country across the world as a source of energy and the implication for the environment. Global warming and air pollution are some of the effects that come from fossil fuels. Buildings account for about 40% of the global energy consumption and contribute over 30% of CO_2 emission with a large proportion of this energy is used for thermal comfort building [1]. It shows that there is a lot of excessive energy consumption is meant for achieving the thermal comfort within the building. To overcome this problem, green building is one of many methods used to reduce excessive energy consumption through thermal comfort. All specific requirements to achieve thermal comfort level should be considered. It is included the material of the building envelope/wall, type of roof, air ventilation and circulation, building positioning toward sun path, window glazing, and many others.

Green buildings will contribute to a healthy lifestyle, long lasting of the building, and cost effective. It also means to maintain natural resources and improve the built environment for the people, communities, and ecosystems to grow and live together in good condition [2]. There are a lot of benefits of green building such as they can save the environment and help in economic growth. Malaysia is one of the countries that are very committed to utilize the potential of green building. As an example, there are many green buildings that have been built by government, such as the Diamond Building of Suruhanjaya Tenaga at Putrajaya and Green Energy Office (GEO) building in Pusat Tenaga Malaysia, Bangi. Energy consumption in Suruhanjaya Tenaga building has reduced by 45%, and one of most contribution is from thermal comfort that requires space heating, cooling, and ventilation. It shows that thermal comfort in a building is very important because it consumes electrical energy.

Thermal comfort is defined as that condition of the human mind that shows satisfaction with the thermal environment surrounding and is determined by subjective evaluation, by ASHRAE Standard 55 (2013) [3]. It was found that building thermal comfort could be influenced by a combination of two factors; the environmental factors and personal factor. Environmental factor is the condition inside the building such as air temperature, air velocity, radiant temperature, and relative humidity. Personal factor is the condition of human body such as clothing, metabolic rate and also the behavior of human. There are many studies that have been done before to determine actual thermal comfort temperature in Malaysia climate. Department of Standards Malaysia, 2007 had established MS 1525: 2007 (Code of Practice on Energy Efficiency and Use of Renewable for Non-Residential Building) for non-residential building as shown in Table 35.1.

Table 35.1 The recommended indoor condition of MS: 1525: 2007

Recommended dry bulb temperature	23 °C–26 °C
Recommended relative humidity (RH)	55–70%
Recommended air movement	0.15–0.5 m/s
Minimum dry bulb temperature	22 °C
Maximum air movement	0.7 m/s

35.1.1 Impact of Building Material on Building Thermal Comfort

This study was made in Gombak, Malaysia on the hot and humid climate; a study on the thermal comfort in all residential buildings or dwellings. Specifically for hot and humid climates, the design of the building must be focused on lowering the indoor climate of the building to achieve thermal comfort standard. In order to achieve this, major design considerations must focus on building orientations, which can affect solar and thermal conditions and potential ventilation of buildings.

Besides building orientation, material of building used to design is also important as it can change the indoor climate inside the building. Selection of material of the building is necessary, especially for walls, roofs, and glazing along with the shape of the building. Paper [4] has highlighted that the thermal properties of building materials can govern the relationship between average indoor climate of the building and the outdoor air temperature pattern where thermal conductivity, specific heat capacity, and density of the building material play an important role. For non-residential place, such as office buildings and schools, the use of insulation material is most effective to change the indoor the climate inside buildings in order to achieve thermal comfort temperature over various external conditions [5]. Although, it might not achieve comfort condition, suitable material for insulated wall can minimize thermal discomfort and also reduce energy consumption where air conditioning used for space heating and cooling [6].

35.2 Methodology

Experimental investigation of thermal comfort and energy consumption was conducted on four (4) mock-up green buildings that have been built with various different types of building material for wall/building envelopes. These four mock-up green buildings are named as $M1$, $M2$, $M3$, and $M4$. All of the mock-up green buildings have the same size, which is $3 \times 3 \times 3$ cubic meter and have the same type of windows which is casement window have is insulated with straw board. There are exhaust fans and air conditioner installed in $M3$ and $M4$, so energy consumption only measured for $M3$ and $M4$ only. For each mock-up green building, there is one door and three windows for each side, as seen in Fig. 35.1. Figure 35.2 shows the plan location of the Mock-up Buildings (*M*1, *M*2, *M*3, and *M*4), respectively.

All the four MUBs have an identical floor area of msq. (3 m × 3 m) and constructively different in terms of its wall material and compound, roof design, door, and window features—to investigate all these distinctive variables toward the internal thermal comfort of each MUBs. Table 35.2 depicts characteristics for each MUB.

(a) Specification of Building Material (wall) of Mock-Up Green Building

All the mock-up green buildings have been built from different types of materials for the wall. All types of construction materials used is conventional according to Malaysian building standard. There are three mock-up green buildings that have been built from green material that is *M*1, *M*2, and *M*3 and one mock-up green building that has been built from non-green material that is *M*4. The wall is made of 20 cm thick material block with 2–3 cm thickness of cement plaster and paint for both

Fig. 35.1 *M*1 Mock-up buildings on-site

Fig. 35.2 *M*2 Mock-up buildings on-site

Table 35.2 MUBs—building characteristics

	Wall	Roof	Roof design
Mock-up building 1 (MUB 1) 	Wool wood cement board	Composite cladding roofing element	5° inclination roofing design
Mock-up building 2 (MUB 2) 	Concrete-polystyrene lightweight cement mix	Metal decking without insulation	40° inclination with overhang roofing design
Mock-up building 3 (MUB 3) 	Clay cavity brick wall	Metal decking without insulation	20° inclination with overhang roofing design
Mock-up building 4 (MUB 4) 	White sand brick wall	Metal decking without insulation	20° inclination with overhang roofing design

interior and exterior side of the building. The floor is made from cement with 42 cm of thickness. The specification of each main construction material is shown in Table 35.3.

(b) Thermal Comfort Temperature and Relative Humidity Parameter

In this research, thermal comfort temperature and relative humidity standard are necessary in order to analyze data. Thermal comfort temperature is very important as a set point temperature. If the temperature taken is higher than the set point temperature, it means that the environmental condition inside the building will cause discomfort. From the literature review, the most suitable thermal comfort temperature and relative humidity standard were taken as a parameter during the research. Table 35.4 shows the parameters that will be used while doing this research.

Table 35.3 Specification of each main construction material

Type of material	Density (10^3 kg/m^3)	Specific heat capacity (J/kg °C)	Thermal conductivity (W/(m °C)
Polystyrene concrete block (*M*1)	0.8–1.0	1300–1500	0.1–0.5
Straw slab insulation, compressed (*M*2)	1.1–1.3	900–1000	0.09
Fired clay bricks (*M*3)	1.8–2.6	1400	0.60–1.30
Concrete block (*M*4)	1.3–1.7	880	1.7

Table 35.4 Thermal comfort temperature and relative humidity parameters

Thermal comfort temperature (°C)	23 °C–28.6 °C
Relative humidity	40–60%

35.3 Results and Discussion

(a) Selection of Data Sample

Data collection for this research is divided by two, which is data collected without using mechanical ventilation and air conditioner and data collected using the mechanical ventilation and air conditioner. The data that have been collected is temperature surrounding in all 4 mock-up green buildings and the power consumption.

For data collected without using the mechanical ventilation and air conditioner, it started on March 12, 2019 until March 14, 2019. Then, it continued on March 16, 2019 until March 19, 2019 which total up to 7 days of collecting data. During 7 days of collecting data, the weather condition was mostly sunny and partly sunny. Partly sunny means that the area received 40–70% of sunlight during daytime while mostly sunny means that the area received 70–90% of sunlight during daytime. This type of weather condition makes the outside temperature of the mock-up green building were very hot with a maximum temperatures of 37°C. The details about weather description, maximum and minimum temperature of all 7 observed days are shown in Table 35.5.

For data collected while using mechanical ventilation and air conditioner, 3 days have been chosen which started on March 15, 2019 until March 17, 2019. Energy consumption and temperature were measured for both mock-up green buildings to find out the relationship between energy consumption and thermal comfort temperature. During 3 days of collecting data, the weather condition was partly sunny and had a light rain. This type of weather condition makes the outside temperature of the mock-up green building was not very hot with maximum temperature of 34°C. The details about weather description, maximum and minimum of all three observed days show as in Table 35.5.

Table 35.5 Weather description of 1 observed week

Date	Weather condition	Temperature Hi/Low (°C)
12 March 2019	06.00–12.00 partly sunny	35/24
	12.00–18.00 mostly sunny	
13 March 2019	06.00–12.00 partly sunny	34/26
	12.00–18.00 mostly sunny	
14 March 2019	06.00–12.00 partly sunny	36/26
	12.00–18.00 partly sunny	
16 March 2019	06.00–12.00 partly sunny	37/25
	12.00–18.00 partly sunny	
17 March 2019	06.00–12.00 partly sunny	36/25
	12.00–18.00 mostly sunny	
18 March 2019	06.00–12.00 partly sunny	36/26
	12.00–18.00 partly sunny	
19 March 2019	06.00–12.00 partly sunny	36/26
	12.00–18.00 partly sunny	

(b) Findings of Data Collected Without Using Mechanical Ventilation and Air Conditioner

The finding is confirmed in the boxplot in Fig. 35.3 below which shows the difference in median and standard deviation between mock-up green building $M1$, $M2$, $M3$, and $M4$. It can be seen that, the $M3$ has the lowest median temperature that almost reaches the thermal comfort temperature standard (28.6°C) that is 28.9 °C while $M4$ has the highest median that is 32.5 °C. For standard deviation, the thickness of the boxplot indicates the difference where $M3$ has the lowest standard deviation, and $M4$ has the highest standard deviation. In this boxplot, it can be seen that all the mock-up green building that have been built from green material ($M1$, $M2$, and $M3$) have a lower median temperature and standard deviation compared to mock-up green building $M4$ that is built from non-green material ($M1$).

By determining the detailed sigma level (z benchmark for overall capability), the percentage of occupant satisfaction can be determined as an indicator for the relative overheating from the perspective of the occupants. Below is the process capability report for all the mock-up green building by using the highest level of 28.6 for residential area without A/C or mechanical ventilation. This will show the percentage of occupant satisfaction and the percentage of active cooling needed. From the result of process capability of $M1$ in Fig. 35.4 below, Z bench value is −0.31 means that only in 38.21% the occupants can be satisfied which makes almost 61.79% more active cooling is necessary.

From the result of process capability of $M2$ below in Fig. 35.5, Z bench value is −0.62 means that only in 26.76% the occupants can be satisfied which makes almost 73.24% more active cooling is necessary. From the result of process capability $M3$

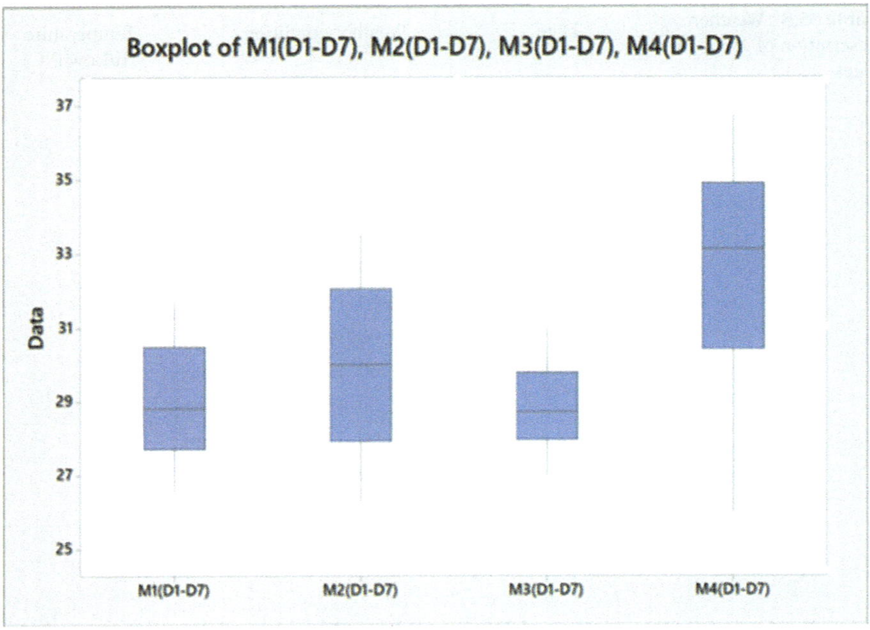

Fig. 35.3 Boxplot of $M1$, $M2$, $M3$ and $M4$

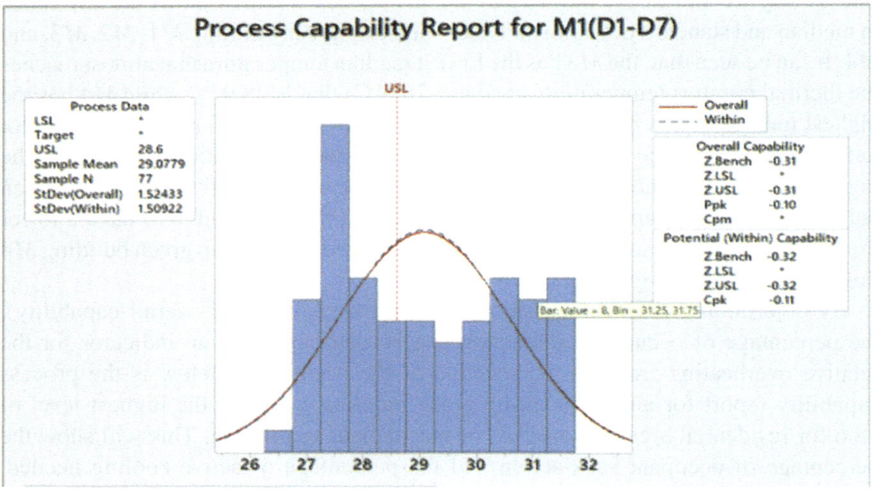

Fig. 35.4 Process capability report of $M1$

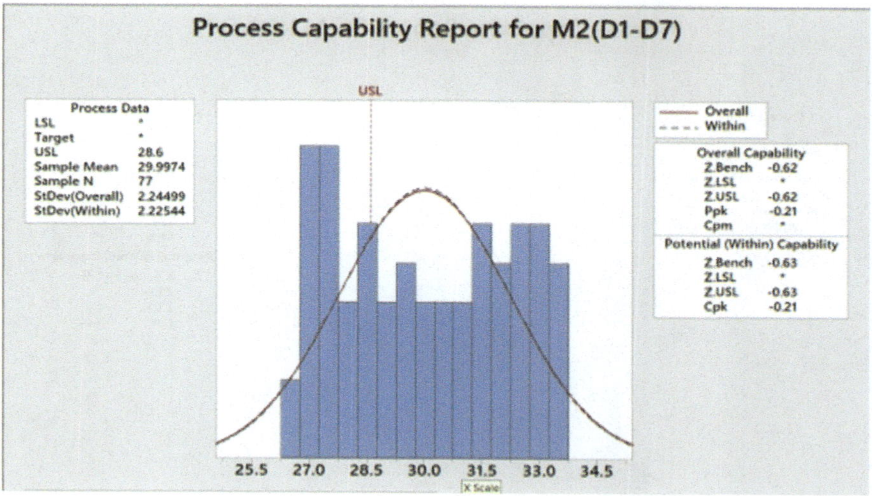

Fig. 35.5 Process capability report of $M2$

below in Fig. 35.6, Z bench value is -0.27 means that only in 39% the occupants can be satisfied which makes almost 61% more active cooling is necessary. From the result of process capability of $M4$ below in Fig. 35.7, Z bench value is -1.37 means that only in 8.53% the occupants can be satisfied which makes almost 91.47% more active cooling is necessary (Table 35.6).

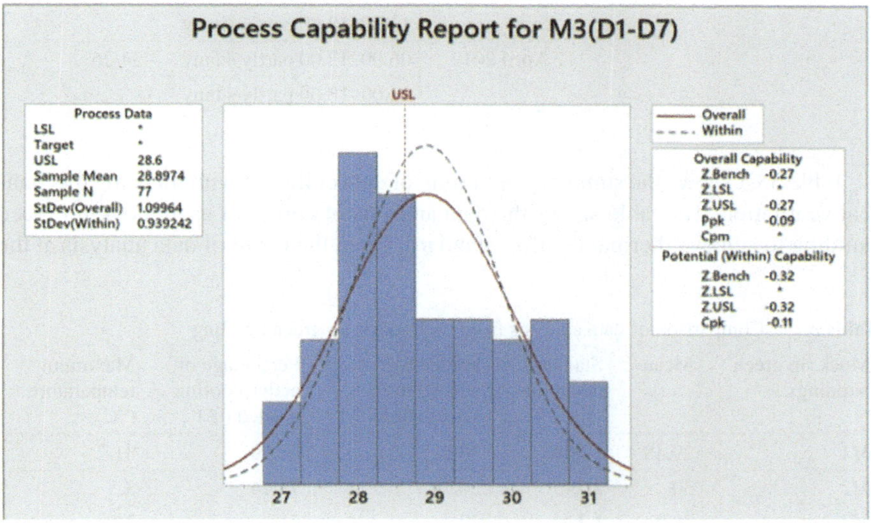

Fig. 35.6 Process capability report of $M3$

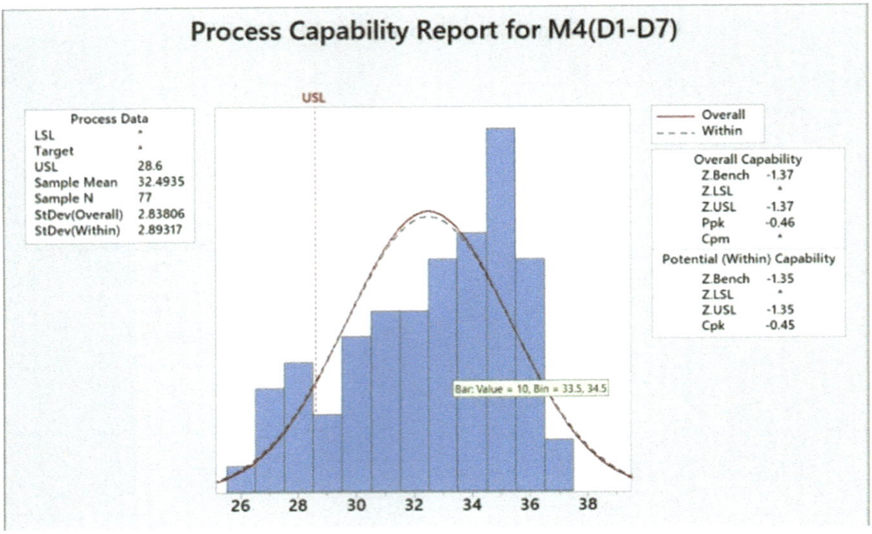

Fig. 35.7 Process capability report of $M4$

Table 35.6 Comparison of data analysis (temperature) from each mock-up green building

Date	Weather condition	Temperature Hi/Low (°C)
15 April 2019	06.00–12.00 partly sunny	34/24
	12.00–18.00 partly sunny	
16 April 2019	06.00–12.00 partly sunny	34/26
	12.00–18.00 partly sunny	
17 April 2019	06.00–12.00 partly sunny	34/26
	12.00–18.00 partly sunny	

Table 35.7 shows the summary of finding of data collected without using mechanical ventilation. The table shows the data analysis of ability of each mock-up green building to achieve thermal comfort standard. From the result of data analysis at the

Table 35.7 Comparison of data analysis from each mock-up green building

Mock-up green buildings	Mean	Standard deviation	Percentage of occupant satisfaction (%)	Percentage of active cooling needed (%)	Maximum temperature (°C)
$M1$	29.48	1.524	38.21	61.79	31.7
$M2$	30	2.245	26.76	73.24	3.5
$M3$	28.9	1.1	39	61	31
$M4$	32.09	2.838	8.53	91.47	36.7

Table 35.8 Total energy consumption and difference of energy consumption between *M*4 and *M*3

Dates	Energy consumption needed to provide a 100% satisfaction to occupants for *M*3 (kWh)	Difference of energy consumption between *M*4 and *M*3 (kWh)
April 15, 2019	$[(40/100) \times 7.6] + 0.1892 =$ 3.23 kWh	7.6 KWh–3.23 kWh = 4.37 kWh
April 16, 2019	$[(40/100) \times 7.7] + 0.1818 =$ 3.26 kWh	7.7 kwh–3.26 kWh = 4.44 kWh
April 17, 2019	$[(40/100) \times 7.6] + 0.1855 =$ 3.22 kWh	7.6 kwh–3.22 kWh = 4.38 kWh
Total	9.71 kWh	13.19 kWh

Table 35.8, it can be seen that mock-up green building that is built from green material can provide better temperature environment than the mock-up green building that is built from non-green material.

From the capability process report of *M*3 for 3 observed days, the energy consumption needed is reduced as green building *M*3 with exhaust fans provide 60% of requests temperature, so only 40% of energy consumption needed from the air conditioner or fans in order to achieve thermal comfort temperature standard. The total energy consumption of green building *M*4 is used as an indicator in order to estimate the energy needed by green building *M*3. 40% of energy consumption from *M*4 of each observed day has been calculated and will be added to the energy consumption of green building *M*3 in order to estimate the total energy consumption needed by green building *M*3 to provide a 100% satisfaction of thermal comfort temperature to occupants.

Table 35.8 shows the total energy consumption needed by green building *M*3 to provide a 100% satisfaction to occupants and the difference in energy consumption between green building *M*4 and *M*3 that indicate the total energy that can be saved by using green building *M*3. From the Table 35.8, it can be seen that the total energy consumption of green building *M*3 that provides 100% satisfaction to occupants is 9.71 kWh, less than the total energy consumption of green building *M*4 that is 22.9 kWh from Table 35.8. It also can be seen that the difference in energy consumption between green building *M*4 and *M*3 indicates the total energy that can be saved from green building *M*3 that is 13.19 kWh. This proves that the mock-up green building *M*3 that has been built from green material can reduce energy consumption two times better than mock-up green building *M*4 that has been built from non-green material.

35.4 Conclusion

Based on the research studies, it can be concluded that mock-up green buildings that have been built from the composite green materials (*M*1, *M*2, and *M*3) can provide a better temperature environment inside the building compared to the non-composite $M4$. This is because of specification of the construction material that gives effect toward the heat penetration of each wall of mock-up green building. From all the mock-up green building, $M3$ is the most ideal in terms of providing thermal comfort temperature to surrounding. In terms of energy consumption, it can be seen that green buildings that have been built from green material can reduce energy consumption. The research has proven that mock-up green buildings that are built from green material need less active cooling than mock-up green building that is built from non-green material.

Specification of construction material is very important to be considered because it can be seen that there is huge difference between mock-up green buildings $M1$, $M2$, $M3$, and $M4$ in terms of providing a better temperature environment. Mock-up green building $M3$ that built from fired clay brick is the ideal because of its specification that have low thermal conductivity and high specific heat capacity. So, build a building with a good specification is also one of many methods to reduce energy consumption in hot and humid climate countries such as Malaysia for the long term.

References

1. Adam EA, Agib ARA (2001) Compressed stabilised earth block manufacture in Sudan. United Nations Educational, Scientific and Cultural Organization (UNESCO), Paris, pp 11–22
2. Al-Tamimi NAM, Syed FSF (2011) Thermal performance analysis for ventilated and unventilated glazed rooms in (comparing simulated and field data). Indoor Built Environ 20(5):534–542
3. Balaras CA (2003) Energy conservation potential, HVAC installation and operational issues in residential buildings. Energy Build 35(3):1105–1120
4. Bayer C, Crow S, Fischer J (2000) Causes of indoor air quality problems in school. Build Res Inf. 55:133–147
5. Dudzińska A, Kotowicz A (2015) Features of materials versus thermal comfort in a passive building. Procedia Eng 108:108–115
6. Elovitz KM (1999) Understanding what humidity does and why. ASHRAE J 41(4):84

The manufacturer's authorised representative in the EU is Springer Nature Customer Service Centre GmbH, Europaplatz 3, 69115 Heidelberg, Germany. If you have any concerns regarding our products, please contact ProductSafety@springernature.com

Printed and bound by CPI Group (UK) Ltd, Croydon, CR0 4YY
30/09/2026
02225250-0012